This series aims to report new developments in mathematical research and teaching – quickly, informally and at a high level. The type of material considered for publication includes:

1. Preliminary drafts of original papers and monographs

2. Lectures on a new field, or presenting a new angle on a classical field

3. Seminar work-outs

4. Reports of meetings

Texts which are out of print but still in demand may also be considered if they fall within these categories.

The timeliness of a manuscript is more important than its form, which may be unfinished or tentative. Thus, in some instances, proofs may be merely outlined and results presented which have been or will later be published elsewhere.

Publication of *Lecture Notes* is intended as a service to the international mathematical community, in that a commercial publisher, Springer-Verlag, can offer a wider distribution to documents which would otherwise have a restricted readership. Once published and copyrighted, they can be documented in the scientific literature.

Manuscripts

Manuscripts are reproduced by a photographic process; they must therefore be typed with extreme care. Symbols not on the typewriter should be inserted by hand in indelible black ink. Corrections to the typescript should be made by sticking the amended text over the old one, or by obliterating errors with white correcting fluid. Should the text, or any part of it, have to be retyped, the author will be reimbursed upon publication of the volume. Authors receive 75 free copies.

The typescript is reduced slightly in size during reproduction; best results will not be obtained unless the text on any one page is kept within the overall limit of 18 x 26.5 cm (7 x 10 ½ inches). The publishers will be pleased to supply on request special stationery with the typing area outlined.

Manuscripts in English, German or French should be sent to Prof. Dr. A. Dold, Mathematisches Institut der Universität Heidelberg, Tiergartenstraße or Prof. Dr. B. Eckmann, Eidgenössische Technische Hochschule, Zürich.

Die „*Lecture Notes*" sollen rasch und informell, aber auf hohem Niveau, über neue Entwicklungen der mathematischen Forschung und Lehre berichten. Zur Veröffentlichung kommen:

1. Vorläufige Fassungen von Originalarbeiten und Monographien.

2. Spezielle Vorlesungen über ein neues Gebiet oder ein klassisches Gebiet in neuer Betrachtungsweise.

3. Seminarausarbeitungen.

4. Vorträge von Tagungen.

Ferner kommen auch ältere vergriffene spezielle Vorlesungen, Seminare und Berichte in Frage, wenn nach ihnen eine anhaltende Nachfrage besteht.

Die Beiträge dürfen im Interesse einer größeren Aktualität durchaus den Charakter des Unfertigen und Vorläufigen haben. Sie brauchen Beweise unter Umständen nur zu skizzieren und dürfen auch Ergebnisse enthalten, die in ähnlicher Form schon erschienen sind oder später erscheinen sollen.

Die Herausgabe der „*Lecture Notes*" Serie durch den Springer-Verlag stellt eine Dienstleistung an die mathematischen Institute dar, indem der Springer-Verlag für ausreichende Lagerhaltung sorgt und einen großen internationalen Kreis von Interessenten erfassen kann. Durch Anzeigen in Fachzeitschriften, Aufnahme in Kataloge und durch Anmeldung zum Copyright sowie durch die Versendung von Besprechungsexemplaren wird eine lückenlose Dokumentation in den wissenschaftlichen Bibliotheken ermöglicht.

Lecture Notes in Mathematics

A collection of informal reports and seminars
Edited by A. Dold, Heidelberg and B. Eckmann, Zürich

112

Colloquium on Methods of Optimization

Held in Novosibirsk/USSR, June 1968

Edited by N. N. Moiseev, Akad. Nauk SSSR, Moscow/USSR

Springer-Verlag
Berlin · Heidelberg · New York 1970

Contents

ON A NEW COMPUTING TECHNIQUE IN OPTIMAL CONTROL AND ITS APPLICATION TO MINIMAL TIME FLIGHT PROFILE OPTIMIZATION

A. V. Balakrishnan

1. Introduction

In this paper we present the first results in applying the new computing technique for optimal control problems studied in [1] to a minimal time flight trajectory optimization problem. We also indicate how the technique in [1] may be extended to the isoperimetric problem of Lagrange [see [2]] with equality constraints. In particular the method yields a constructive way of obtaining the Lagrange multipliers whose existence alone is usually proved. We actually obtain an approximate (or ' epsilon' in the terminology below) maximum principle in which the optimal solution is shown to exist for example for bounded controls and with the usual growth restriction on the dynamics. In this way we obtain (and motivate) the approximate Hamiltonian and the integral form of the maximum principle which yields the familiar maximum principle in the limit. The computing method is applied to a "time-optimal" problem with nonlinear state and control dynamics.

The basic method can be stated quite simply. Suppose it is required to minimize the functional

$$\int_0^T g\big(t, x(t),\ u(t)\big)\, dt$$

where

$$\dot{x}(t) = f\big(t, x(t),\ u(t)\big) \ ; \ \ x(o),\ x(T) \text{ given.}$$

is the state equation, the control $u(.)$ being also constrained to a class Δ. The virtue of the method is that it avoids having to solve dynamic equations of any kind, including adjoint systems etc. Thus we formulate a non-dynamic problem for fixed $\epsilon > 0$. We seek to minimize:

$$\frac{1}{2\epsilon} \int_0^T \| \dot{x}(t) - f\big(t, x(t),\ u(t)\big) \|^2\, dt + \int_0^T g\big(t, x(t),\ u(t)\big)\, dt$$

in the class of absolutely continuous state functions x(.) satisfying the stipulated end conditions in addition to the controls u(.) in Δ. As shown in [1] this yields as close an approximation to the optimal solution of the original problem for sufficiently small ϵ. In fact, we show in the particular example studied that the solution may be relatively insensitive to how small ϵ has to be.

In section 2 we extend this method to the isoperimetric problem of Lagrange with equality constraints, and show that under some conditions usually postulated, the epsilon problem has a solution even though the Lagrange problem need not necessarily have an 'ordinary' (as opposed to 'relaxed') solution. We obtain necessary conditions for the epsilon solution, and actually indicate how the Hamiltonian enters naturally. The behavior as epsilon goes to zero is briefly discussed in section 4, since the main aim here is to present computational results. The computational aspects are studied in section 5 for time optimal problems. Two methods are suggested and these are then studied for a particular example of an aerodynamic problem. In both methods a Rayleigh-Ritz procedure is used to reduce dimensionality. Advantage is also taken in the particular problem for reducing the number of functions to be optimized. While no exhaustive comparison with other methods is presented, the relative ease of mechanization and the increase in speed obtainable by avoiding having to solve dynamic equations make the present method an attractive one.

2. General Theory

In this section we shall outline the general theory with reference to the isoperimetric problem of Lagrange with equality constraints (as formulated in Hestenes [2]). Thus let the dynamic system be governed by the equation:

$$\dot{x}(t) = f\big(t;x(t);\, u(t)\big) \; ; \; x(0) = x_o \text{ given} \tag{2.1}$$

where f(.) is continuous in all the variables (locally) and of class C^1 in the state variable x. The optimal control problem is to minimize the scalar functional:

$$\int_0^T g\Big(t, x(t),\ u(t)\Big)\,dt \tag{2.2}$$

with the end condition:

$$x(T) = x_1 \tag{2.3}$$

and the equality constraints:

$$I_L = \int_0^T L\Big(t; x(t)\ ; u(t)\Big)\,dt + g = 0 \tag{2.4}$$

$$\phi\Big(t; x(t)\ ; u(t)\Big) = 0 \qquad 0 < t < T \tag{2.5}$$

where we assume that the functions $g(\ldots)$, $L(\ldots)$, $\phi(\ldots)$ are locally continuous in all the variables and of class C^1 in the state variable x. In addition there may be constraints on the control function $u(.)$. We shall denote the set of admissible control functions by Δ. Every function in Δ will be assumed to be Lebesgue measurable.

To obtain a constructive method of obtaining the optimal solution without having to solve any dynamic equations, we proceed by first defining the following non-dynamic 'epsilon' problem. We seek to minimize, for each ϵ greater than zero:

$$\begin{aligned} h\Big(\epsilon; T; x(.); u(.)\Big) = \frac{1}{2\epsilon}\Bigg[&\int_0^T \|\dot{x}(t) - f\Big(t; x(t)\ ; u(t)\Big)\|^2\,dt \\ &+ \Big\|\int_0^T L\Big(t; x(t); u(t)\Big)\,dt + g\Big\|^2 \\ &+ \int_0^T \Big[\phi\Big(t; x(t)\ ; u(t)\Big),\ \phi\Big(t; x(t)\ ; u(t)\Big)\Big]\,dt\Bigg] \\ &+ \int_0^T g\Big(t; x(t); u(t)\Big)\,dt \end{aligned} \tag{2.6}$$

in the class of functions $x(t)$ absolutely continuous and satisfying the end conditions:

$$x(0) = x_0 \ ;\ x(T) = x_1$$

and $u(.)$ in the class of admissible control functions.

Existence of solutions for the epsilon problem

Among the many favorable features of this formulation is that the approximate non-dynamic problem has a solution (that is, the infimum of (2.6) is actually attained) even when the original optimal control problem does not have a solution. For instance, the conditions:

$$\text{(bounded control)}\quad u(t) \in U \quad \text{for each } t \tag{2.7}$$

where U is compact, and (growth condition)

$$\left|\left[x,\ f(t,x,u)\right]\right| \le c\left(1 + \|x\|^2\right),\quad u \in U \tag{2.8}$$

in each finite interval, the constant c depending on the interval, are known to be [3] in general insufficient to ensure existence of an optimal control for (2.2). However we shall show here that these conditions are sufficient for existence of solutions of (2.6), by extending a similar result given by A. Chaudhury.†

Theorem 2.1 Suppose that the admissible terminal times in (2.6) are bounded, and suppose that conditions (2.7) and (2.8) hold. Then the infimum of (2.6) is attained.

Proof Let $u_n(.)$, $x_n(.)$ be a minimizing sequence for (2.6), with corresponding terminal times T_n. We may clearly take T_n converging to T_ϵ say. Let us denote the infimum of (2.6) by $h(\epsilon)$. Then it is clear that

$$h(\epsilon) = \operatorname{Inf}\ h\left(\epsilon;\ T_\epsilon;\ x(.)\ ;\ u(.)\right)$$

where the terminal time T_ϵ is fixed. Let $x_n(.)$, $u_n(.)$ again denote a minimizing sequence for fixed terminal time T_ϵ. Then because of (2.7), (2.8) it follows that the sequence $x_n(.)$ is equicontinuous, and hence we may take it to be converging uniformly to a function $x_o(t)$, also absolutely continuous, and further,

$$\lim_n h\left(\epsilon;\ T_\epsilon;\ x_n(.)\ ;\ u_n(.)\right) \ge \lim_n\ h\left(\epsilon;\ T_\epsilon;\ x_o(.)\ ;\ u_n(.)\right) \tag{2.9}$$

† A. K. Chaudhury: Forthcoming Ph. D. thesis, Department of Engineering, UCLA.

Next let us use the notation:

$$\phi(t; u) = \frac{1}{2\epsilon}\left(\left\|\dot{x}_o(t) - f\left(t, x_o(t); u\right)\right\|^2 + \left\|\phi\left(t; x_o(t)\ ; u\right)\right\|^2\right) + g\left(t; x_o(t); u\right) \tag{2.10}$$

and let $\psi(t, u)$ denote the mapping (into the appropriate finite dimensional product space)

$$\psi(t, u) = \frac{1}{2\epsilon} L\left(t\ ; x_o(t)\ ; u\right), \phi(t, u) \tag{2.11}$$

Then $\psi(t, u)$ is obviously continuous in both variables.

Let $q(t)$ be any measurable function such that

$$q(t) \epsilon \{\psi(t, u)\ ,\ u \epsilon U\}$$

Since U is compact we can apply the Blackwell theorem as in Neustadt [4] and follow it with an application of the Fillipov Lemma [3] also as in Neustadt [4], and thus obtain that there exists a measurable function $u_o(.)$, such that $u_o(t) \epsilon U$ for each t (or a.e.) and

$$\lim_n \int_o^{T_\epsilon} \psi\left(t, u_n(t)\right) dt = \int_o^{T_\epsilon} \psi\left(t, u_o(t)\right) dt \tag{2.12}$$

It readily follows from this that:

$$\lim h\left(\epsilon\ ; T_\epsilon; x_o(.)\ ; u_n(.)\right) = h\left(\epsilon\ ; T_\epsilon; x_o(.)\ ; u_o(.)\right)$$
$$= h(\epsilon)$$

or, the existence proof is complete.

3. Necessary conditions for optimality for the epsilon problem

In this section we shall assume that the problem (2.6) has a solution, and denote this solution by $x_o(\epsilon, t)$, and let the corresponding terminal time to be T_ϵ. We shall obtain the necessary conditions to be satisfied by the solution, and obtain an 'epsilon' maximum principle extending the previous results in [1]. In this way we shall obtain the Hamiltonian in a straightforward manner also. For this, let us note that we can rewrite $h\left(\epsilon\ ; T_\epsilon; x_o(\epsilon, .); u(t)\right)$ as:

$$\epsilon\, h\Big(\epsilon ; T_\epsilon ; x_o(\epsilon, .) ; u(t)\Big)$$

$$= \frac{1}{2}\int_o^{T_\epsilon} \| \dot{x}_o(\epsilon, t) \|^2\, dt - \frac{1}{2}\int_o^{T_\epsilon} \Big\| f\Big(t,\ x_o(\epsilon, t); u_o(\epsilon, t)\Big) \Big\|^2\, dt$$

$$+ \frac{1}{2}\int_o^{T_\epsilon} \Big\| f\Big(t, x_o(\epsilon, t); u(t)\Big) - f\Big(t; x_o(\epsilon, t); u_o(\epsilon, t)\Big) \Big\|^2\, dt$$

$$+ \frac{1}{2}\, \| I_L - I_{L_0}(\epsilon) \|^2$$

$$+ \frac{1}{2}\int_o^{T_\epsilon} \Big\| \phi\Big(t, x_o(\epsilon, t) ; u_o(\epsilon, t)\Big) - \phi\Big(t, x_o(\epsilon, t) ; u(t)\Big) \Big\|^2\, dt$$

$$- \int_o^{T_\epsilon} H\Big(\epsilon; z_o(\epsilon, t) ; x_o(\epsilon, t) ; u(t) ; t\Big)\, dt \qquad (3.1)$$

where

$$I_L = \int_o^{T_\epsilon} L\Big(t,\ x_o(\epsilon, t); u(t)\Big)\, dt \qquad (3.2)$$

$$I_{L_o}(\epsilon) = \int_o^{T_\epsilon} L\Big(t, x_o(\epsilon, t) ; u_o(\epsilon, t)\Big)\, dt \qquad (3.3)$$

$$z_o(\epsilon, t) = \dot{x}_o(\epsilon, t) - f\Big(t, x_o(\epsilon, t) ; u_o(\epsilon, t)\Big)$$

$$H(\epsilon, z, x, u, t) = [z, f(t, x, u)] - \epsilon\, g(t, x, u)$$

$$+ \Big[I_{L_o}(\epsilon),\ L(t, x, u)\Big]$$

$$+ \Big[\phi\Big(t, x_o(\epsilon, t) ; u_o(\epsilon, t)\Big),\ \phi(t, x, u)\Big] \qquad (3.4)$$

It readily follows by inspection of (3.1) that (the epsilon maximum principle):

$$\max_{u(.)\in\Delta} \int_o^{T_\epsilon} H\Big(\epsilon, z_o(\epsilon, t) ; x_o(\epsilon, t) ; u(t) ; t\Big)\, dt$$

$$= \int_0^{T_\epsilon} H\left(\epsilon, z_o(\epsilon, t) \; ; \; x_o(\epsilon, t) \; ; \; u_o(\epsilon, t)\right) dt \tag{3.5}$$

As in [1], we can go from an integral maximum principle to a point-wise maximum principle upon appropriate point-wise constraints characterizing the class of admissible controls.

We now proceed to obtain the first order necessary conditions that the solution must satisfy. Let h(.) be any infinitely smooth (state variable) function vanishing outside compact subsets of $(0, T_\epsilon)$. Then exactly as in [1], it follows that:

$$\int_0^{T_\epsilon} \left[z_o(\epsilon, t), \; \dot{h}(t) - f_1\left(t \; ; \; x_o(\epsilon, t) \; ; \; u_o(\epsilon, t)\right) h(t)\right] dt$$

$$+ \int_0^{T_\epsilon} \left[I_{L_o}(\epsilon) \, , \; L_1\left(t, x_o(\epsilon, t) \; ; \; u_o(\epsilon, t)\right) h(t)\right] dt$$

$$+ \int_0^{T_\epsilon} \left[\phi\left(t, x_o(\phi, t) \; ; \; u_o(\phi, t)\right), \; \phi_1\left(t ; x_o(\epsilon, t) \; ; \; u_o(\epsilon, t)\right) h(t)\right] dt$$

$$+ \int_0^{T_\epsilon} \left[g_1\left(t, \; x_o(\epsilon, t) \; ; \; u_o(\epsilon, t)\right), \; h(t)\right] dt = 0 \tag{3.6}$$

where

$$f_1 = \nabla_x f$$
$$L_1 = \nabla_x L$$
$$\phi_1 = \nabla_x \phi$$
$$g_1 = \nabla_x g(t, x, u)$$

∇_x denoting gradient (or Frechet derivative) with respect to x. Next by appropriate integration by parts as in [1] we can obtain:

$$- \dot{z}_o(\epsilon, t) - f_1\left(t ; x_o(\epsilon, t); \; u_o(\epsilon, t)\right)^* \; z_o(\epsilon, t)$$
$$+ L_1\left(t \; ; \; x_o(\epsilon, t) \; ; \; u_o(\epsilon, t)\right)^* \; I_{L_o}(\epsilon)$$

$$+ \phi_1\left(t, x_o(\epsilon, t) ; u_o(\epsilon, t)\right)^* \phi\left(t, x_o(\epsilon, t); u_o(\epsilon, t)\right)$$

$$+ \epsilon\, g_1\left(t, x_o(\epsilon, t) ; u_o(\epsilon, t)\right) = 0 , \qquad 0 < t < T_\epsilon \qquad (3.7)$$

where * denotes adjoint (or transpose).

4. Behavior as epsilon goes to zero

We shall now consider the limiting behavior as epsilon goes to zero. For this we shall assume that the epsilon problem has a solution and use the same notation as in section 3. We have

$$h(\epsilon) = \frac{1}{2\epsilon} \int_0^{T_\epsilon} \left[\| z_o(\epsilon, t) \|^2 + \| \phi\left(t, x_o(\epsilon, t) ; u_o(\epsilon, t)\right) \|^2 \right] dt$$

$$+ \frac{1}{2\epsilon} \left\| \int_0^{T_\epsilon} L\left(t, x_o(\epsilon, t) ; u_o(\epsilon, t)\right) dt + g \right\|^2$$

$$+ \int_0^{T_\epsilon} g\left(t, x_o(\epsilon, t); u_o(\epsilon, t)\right) dt$$

which we shall rewrite as:

$$h(\epsilon) = \delta(\epsilon) + \int_0^{T_\epsilon} g\left(t, x_o(\epsilon, t) ; u_o(\epsilon, t)\right) dt \qquad (4.1)$$

If we denote the infimum of (2.2) by h(0), we have of course that

$$h(\epsilon) \leqq h(0)$$

Just as in [1] we can show that $\delta(\epsilon)$ converges monotonically to zero as epsilon goes to zero, while the second term in (4.1) increases monotonically to h(0). If we assume conditions (2.7) and (2.8) and also that the sequence T_ϵ is bounded, then almost exactly as in McShane [5], we can show that there exists an optimal limiting generalized or 'relaxed' control. More precisely, if $x_o(\epsilon_n, t)$, $u_o(\epsilon_n, t)$ is a minimizing sequence with corresponding terminal times T_{ϵ_n} converging to T_o, then $x_o(\epsilon_n, t)$ will again be an equicontinuous sequence and we may assume that we are working with a convergent subsequence converging uniformly on

$[0, T_o]$ to $x_o(t)$ say. We can show that there exists a generalized control $u_o(t,\omega)$ such that

$$\dot{x}_o(t) = E\Big(f(t, x_o(t); u_o(t,\omega))\Big)$$

where E(.) denotes mean value or expected value, and

$$\int_0^{T_o} E\Big(L(t, x_o(t) ; u_o(t,\omega))\Big)\, dt + g = 0$$

$$E\Big(\phi(t, x_o(t) ; u_o(t,\omega))\Big) = 0$$

We omit the details; these and more general results may be found in [6]. As we have remarked before, our main emphasis in this paper is on the computational aspects of the problem.

It may be also of interest to note that under the conditions above it is possible to show that

$$z_o(\epsilon, t)/\epsilon ; \ \phi\Big(t, x_o(\epsilon, t) ; u_o(\epsilon, t)\Big)/\epsilon$$

converge, and so does

$$I_{L_o}(\epsilon)/\epsilon$$

as epsilon goes to zero, and denoting these limits by $\psi(t)$, $\phi^1(t)$ and $I^1_L(0)$, we can obtain the limiting Hamiltonian by dividing through by epsilon in (3.4) and then passing to the limit:

$$\tilde{H}(\psi, x, u, t) = \Big[\psi, \ f(t, x, u)\Big] - g(t, x, u)$$

$$+ \Big[I^1_{L_o}(0) \ , \ L(t, x, u)\Big]$$

$$+ \Big[\phi^1(t) \ , \ \phi(t, x, u)\Big] \qquad (4.2)$$

Also, the function $\psi(.)$ will satisfy the equation one obtains by dividing through by epsilon in (3.7), and reinterpreting (3.7) with the aid of the above functions and the 'relaxed' control. Rigorous statements are given in [6]. We can thus arrive at the Pontrjagin Maximum principle in this way; but what is more important, the "Lagrange parameters" (whose existence is usually proved, as in [2]), can thus actually be obtained in a constructive fashion.

5. Computational Aspects: Time Optimal Problems

We shall now turn to computational aspects of the epsilon method. We shall do this with reference to a specific class of problems: the time optimal control problems. Let the dynamics be described by:

$$\dot{x}(t) = f\left(t, x(t), u_{*}(t)\right)$$

We are given the initial state $x(0) = x_1$, and we are required to find the control that will 'steer' the system to a given terminal state x_2 in minimal time, the control function being constrained to be chosen from a class Δ. In other words, in the notation of the previous sections, the function $g(\ldots)$ is identically equal to one, and we also set $L(\ldots)$ and $\phi(\ldots)$ to be zero. The known algorithms for such problems (see [7] for a recent method for the linear case as well as for earlier references) involve the solution of dynamic equations. We shall discuss not only the non-dynamic epsilon method studied in the previous sections but also a variant suggested by it for the time-optimal problem. We shall first outline these methods and then describe an application of both methods to a minimal time flight trajectory problem in which the sufficient conditions (2.8) and (2.7) do not hold.

Method 1: Suppose that a rough estimate of the minimal time is available. Then we take a terminal time T which is definitely less than the minimal time T_m. For this fixed T, we minimize:

$$q(t, x(.)) = \int_0^T \| \dot{x}(t) - f(t, x(t), u(t)) \|^2 \, dt \tag{5.1}$$

in the class of absolutely continuous functions $x(.)$ with specified initial and terminal values, and $u(.)$ in the given class Δ. Let us denote the minimum of (5.1) by $q_{min}(T)$. Then, in theory at any rate, $q_{min}(T)$ will be a monotonic decreasing function of T, and the first time it becomes zero will correspond to T_m. We note that the need for performing the differentiation operation in (5. 1) can be avoided by carrying out the minimization over the class of functions y(t), and setting

$$x(t) = x_o + \int_o^t y(s)\, ds$$

with y(.) restricted so that

$$x_1 = x_o + \int_o^t y(s)\, ds$$

We minimize q(.) for fixed T, and step forward in T, until q(.) is as small as is acceptable.

The particular quadratic form of (5. 1) can be exploited in the gradient or Newton-Raphson techniques. Thus let

$$x(t) = \hat{x}_n(t) + \lambda\, x(t)$$

$$\hat{u}(t) = u_n(t) + \lambda\, u(t)$$

with x(.) vanishing outside compact subsets of the open interval (0, T), and u(.) in an appropriate subset so that $\hat{u}(.)$ is admissible for sufficiently small λ. Then we note that

$$\begin{aligned} q\left(T; x(.) ; u(.)\right) &= q\left(T; x_n(.) ; u_n(.)\right) \\ &+ 2\lambda\left[z_n(.) ,\ \mathcal{L}\left(x(.),\ u(.)\right)\right] \\ &+ \lambda^2 \left\| \mathcal{L}\left(x(.),\ u(.)\right) \right\|^2 + \lambda^2\left[z_n(.),\ Q_n\left(x(.),\ u(.)\right)\right] \\ &+ \text{higher order tems in } \lambda \end{aligned} \qquad (5.\ 2)$$

where [,] denote the inner product in the appropriate $L_2[0, T]$ space,

$$z_n(t) = \dot{x}_n(t) - f\left(t, x_n(t),\ u_n(t)\right)$$

and $\mathcal{L}(.)$ is a linear transformation defined by

$$\mathcal{L}\left(x(.),\ u(.)\right) = \dot{x}\,(t) - f_1\left(t, x_n(t),\ u_n(t)\right) x(t)$$

$$- f_2\left(t,\ x_n(t),\ u_n(t)\right) u(t)\ , \qquad 0 < t < T$$

where

$$f_1(t, x, u) = \nabla_x\, f(t, x, u)$$

$$f_2(t, x, u) = \nabla_u\, f(t, x, u)$$

and

$$Q_n\left(x(.),\ u(.)\right) \doteqdot -\frac{d^2}{d\lambda^2}\ f\left(t,\ x_n(.) + \lambda\, x(t),\ u_n(t) + \lambda\, u(t)\right)\Big|_{\lambda=0},\ 0<t<T$$

The main thing to be noted in (5. 2) is the fact that of the two final terms of the second degree in λ, the first one is always nonnegative, while the second may be of either sign. Moreover, for calculating the latter, second derivatives are required. Thus if the gradient method is too slow, we may determine the perturbing functions $x(.)$, $u(.)$ by minimizing the quadratic form:

$$2\left[z_n(.)\ ,\ \mathcal{L}\left(x(.),\ u(.)\right)\right] + \|\ \mathcal{L}\left(x(.),\ u(.)\right)\|^2 \tag{5. 3}$$

in the stipulated linear class. Now, since the functions $x(.)$ are required to vanish at zero and at T, and we may assume $z_n(t)$ to be of class C^1, we can rewrite (5. 3) as:

$$\|\ \mathcal{L}\left(x(.),\ u(.)\right)\|^2 - 2\int_0^T \left[\dot{z}_n(t) + f_1(t, x_n(t),\ u_n(t))^*\ z_n(t),\ x(t)\right] dt$$

$$- 2\int_0^T \left[f_2\left(t, x_n(t),\ u_n(t)\right)^*\ z_n(t),\ u(t)\right] dt \tag{5. 4}$$

by an integration by parts. The minimization of (5. 4) is of course further simplified if the functions $x(.)$, and $u(.)$ are restricted to lie in finite dimensional subspaces. In particular, the gradient in the latter case becomes:

$$P_x\ \left(-\ \dot{z}_n(.) - f_1\left(.\,, x_n(.),\ u_n(.)\right)^*\ z_n(.)\right)$$

$$P_u\ \left(-\ f_2\left(.\,, x_n(.),\ u_n(.)\right)^*\ z_n(.)\right)$$

where P_x, P_u denote the projections on to the appropriate state and control subspaces. We shall return to some details in the case of an example below. In any event we note that (5. 3) or (5. 4) can be made the basis for various Newton-Raphson techniques.

Method 2: This is the nondynamic epsilon method proper. We set

$$h\Big(\epsilon, T, x(.), u(.)\Big) = T + \frac{1}{2\epsilon}\int_0^T \left\| \dot{x}(t) - f\Big(t, x(t), u(t)\Big) \right\|^2 dt$$

and minimize h(.) varying T, x(.) and u(.), with x(.) subject to the necessary terminal conditions and u(.), constrained to the class Δ. As before, a gradient or Newton-Raphson method or the usual variations thereof may be used. However, since the parameter T is involved and derivatives with respect to T are not well defined, it is usually preferable to use a Rayleigh-Ritz procedure by expressing the function x(.) in terms of a finite number of known functions in which the parameter T enters in a sufficiently smooth manner. We shall give more details on this below.

We shall now proceed to a more detailed study for a specific nonlinear problem. This is the problem of rocket flight in a resisting medium [8], and we take a simplified version of motion in the plane. The equations of motion assuming no mass flow, and zero angle between the thrust and velocity vectors, (Cf [8]) can be written:

$$\begin{aligned} &\dot{h}(t) - v(t) \sin \gamma(t) = 0 \\ &\dot{v}(t) + g \sin \gamma(t) - f_1\Big(h(t), v(t)\Big) + f_2\Big(h(t), v(t)\Big) L(t)^2 = 0 \\ &v(t)\, \dot{\gamma}(t) + g \cos \gamma(t) + L(t) = 0 \end{aligned} \tag{5. 5}$$

where h(.) is a vertical coordinate, v(.) the magnitude of the velocity vector, $\gamma(t)$ the flight angle (the inclination of the flight path with respect to the horizon), L(t) the lift, and g the acceleration due to gravity. The lift program L(t) is taken as the control. The functions $f_1(.)$, $f_2(.)$ are continuously differentiable, and what is more important:

$$f_1(h, v) < m < \infty \tag{5.6}$$

$$f_2(h, v) \geq D > 0 \tag{5.7}$$

in the regions of interest. The initial values for $h(0)$, $\gamma(0)$, $v(0)$ are given and the terminal values are such that

$$h_1 = h(T) > h(0) \tag{5.8}$$

$$v_1 = v(T) > v(0) > 0 \tag{5.9}$$

Under these conditions it is known that the system is 'controllable' in the sense that it is possible to find a lift program for a finite T, with velocity never zero. The problem is to find the smallest such T. We note that in this example the sufficient conditions of Theorem 2.1 do not necessarily hold.

Method 1: In applying this method it is convenient to take advantage of the possibility of reducing the number of functions to be optimized by simplifying the integrand in (5.1) to be minimized as:

$$q(T) = \int_0^T \Bigg[\Big(\dot{h}(t) - v(t) \sin \gamma(t)\Big)^2 + \Bigg\{ \Big(\dot{v}(t) + g \sin \gamma(t) - f_1\big(h(t), v(t)\big) + f_2\big(h(t), v(t)\big)\Big(v(t)\,\dot{\gamma}(t) + g \cos \gamma(t)\Big)^2 \Bigg\}^2 \Bigg] dt \tag{5.10}$$

by substituting (5.5) into (5.4). Here of course we take T to be less than the minimal time estimated for the problem.

In the calculations the following representation for the various functions was used:

$$\begin{aligned} h(t) &= h(0) + \big(h_1 - h(0)\big)\, t/T + \sum_1^N a_k \sin(k\pi t/T) \\ v(t) &= v(0) + \big(v_1 - v(0)\big)\, t/T + \sum_1^N b_k \sin(k\pi t/T) \\ \gamma(t) &= \gamma(0) + \big(\gamma_1 - \gamma(0)\big)\, t/T + \sum_1^N c_k \sin(k\pi t/T) \end{aligned} \tag{5.11}$$

Values for the functions $f_1(.)$, $f_2(.)$ were obtained from tables and interpolated as necessary. A Newton-Raphson algorithm based on (5.3) (minimizing (5.3) in the indicated finite dimensional subspaces) was used. The performance of the algorithm is shown in Figure 1 where q(. .) is plotted as a function of the number of iterations, with the number of functions, N, in (5. 11) set equal to 8. The dependence on the number of functions is studied in Figure 2 which shows the minimum q(.) value attained as a function of N. It is seen that about 8 to 10 functions suffice. Now for fixed N, it would be reasonable to expect (and as actually demonstrable for linear dynamics by sample calculations), that the behavior of the minimum as a function of the interval T used would be roughly parabolic; whereas in the ideal case (or N = infinity) the function $q_{min}(T)$ would vanish for T larger than the T_{min} for the terminal conditions stipulated, for fixed number of approximating functions, the corresponding $q_{min}(T)$ decreases at first and then increases. This behavior is plotted in Figure 3 for two values of N. From this figure a good approximation for the minimal time (about 145 sec) may be obtained. The time taken for computations (on SDS 9-300, floating point arithmetic) is shown in Figure 4, based on 30 subdivisions of the time intervals in calculating the integrals. The total time typically was about 270 secs.

Method 2: We shall now discuss the results obtained by applying the epsilon method to the same problem (5. 5). As in method 1 we may reduce the number of functions to be optimized by formulating the epsilon problem (5.) as that of minimizing

$$h(\epsilon; T) = q(T) + T \qquad (5.12)$$

for each fixed epsilon, with q(T) defined by (5. 10). We note that T is also to be varied, in addition to the functions h(.), v(.), (.) subject to the appropriate end conditions. For this purpose we again use the representation (5. 11), and observe that for fixed N the functional (5. 12) is differentiable in the variable T. We now seek the minimum over the variables a_k, b_k, c_k and T (positive), and we again use a Newton-Raphson method based on (5. 3), where we now also have to add the term corresponding to perturbing T. The results for $\epsilon = 1$ are shown in Figure 5, and the corresponding values for the variable T for the same iterations are plotted in Figure 6, with N = 8 and beginning with a starting value for T = 250 secs. The minimal value was reached in 7 iterations, and the actual minimal value is again within about one percent of the value obtained in method 1. The corresponding value of T (denoted T_{final}) obtained for various values of epsilon are shown in Table 1 from which it is seen that it is relatively insensitive to the actual value of epsilon used in a wide range. The behavior of the iterations (not shown) for the various values of epsilon were also very similar.

In conclusion it was felt that the feasibility of the method and the relative ease and speed of mechanization was such as to merit further experimentation.

Acknowledgment

All the calculations presented in this paper were carried out under the direction of L. W. Taylor, NASA Flight Research Center, Edwards, California, who also contributed many essential ideas throughout this work.

REFERENCES

1. A. V. Balakrishnan: On a new computing technique in Optimal control, SIAM Journal on Control, May 1968.

2. M. R. Hestenes: Calculus of Variations and Optimal Control Theory, John Wiley & Sons, 1966.

3. A. F. Fillipov: On certain questions in the theory of optimal control, SIAM Journal on Control, Vol. 1, No. 1, 1962.

4. L. W. Neustadt: The existence of optimal controls in the absence of convexity conditions, Journal of Mathematical Analysis and Applications, August 1963.

5. E. J. McShane, Relaxed controls and variational problems, SIAM Journal on Control, August 1967.

6. A. V. Balkrishnan, To appear in Proceedings of the Workshop on Calculus of Variations and Control Theory, UCLA July 1968, Academic Press.

7. T. Fujisawa, and Y. Yasuda, An iterative procedure for solving the time-optimal regulator problem, SIAM Journal on Control, Nov. 1967.

8. A Miele, The Calculus of Variations in Applied Aerodynamics and Flight Mechanics, in 'Optimization Techniques' edited by A. Leitman, Academic Press, 1962.

TABLE 1

Value of epsilon	T_{Final} Obtained
100	142.00
10	141.9
1	141.78
0.1	138.97

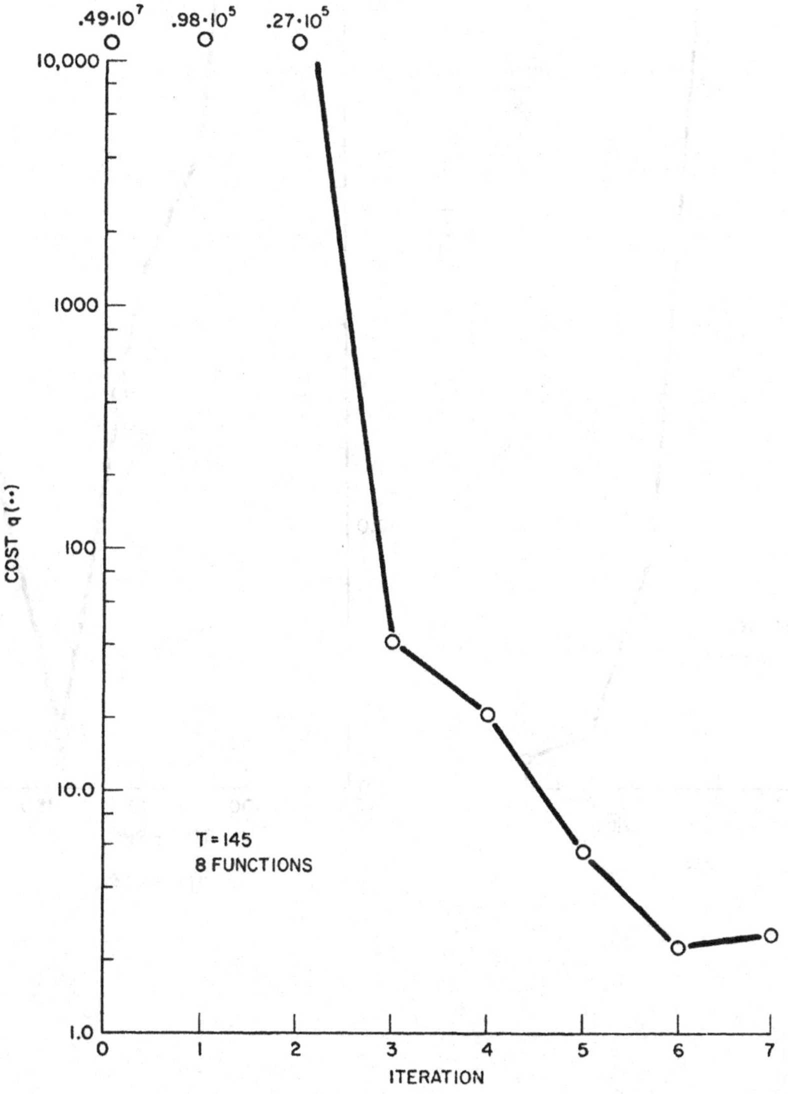

Fig. 1

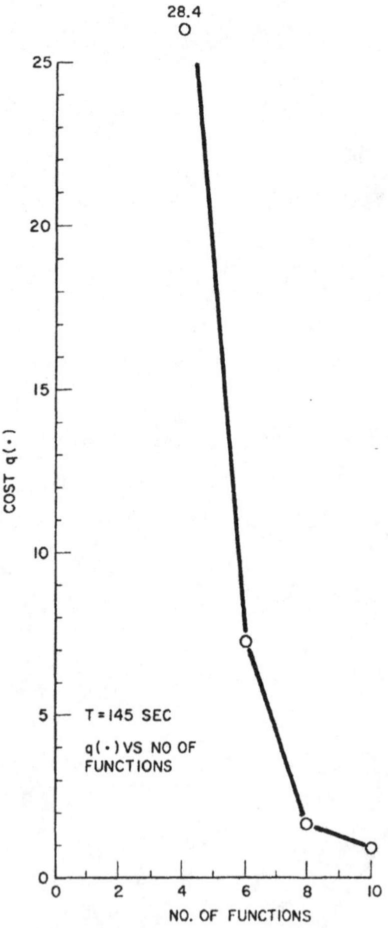
28.4
25
20
15
10
5
0
COST q(·)
T = 145 SEC
q(·) VS NO OF FUNCTIONS
0
2
4
6
8
10
NO. OF FUNCTIONS

Fig. 2

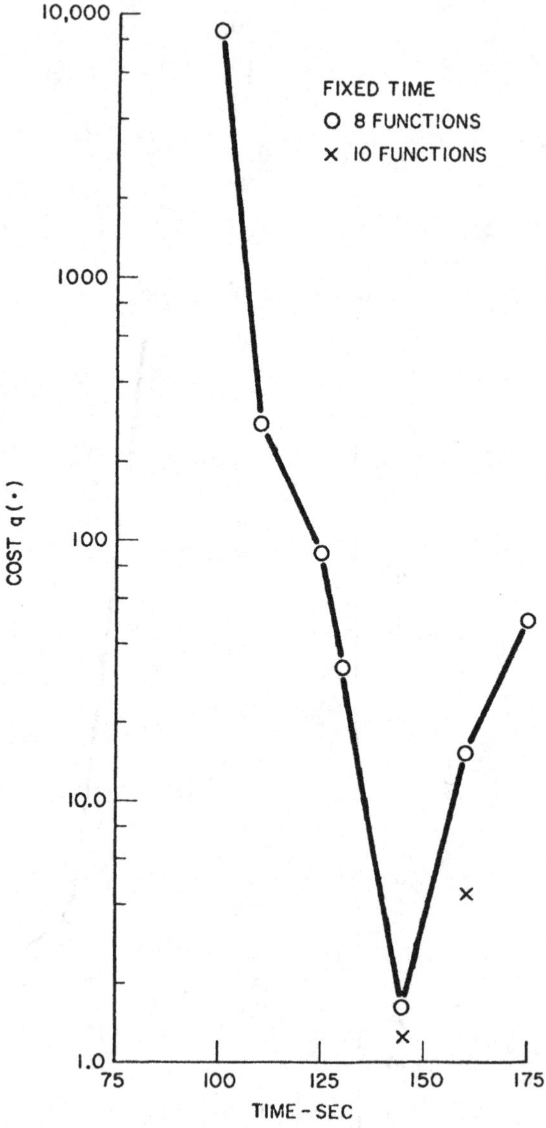
10,000
1000
100
10.0
1.0
COST q(·)
FIXED TIME
O 8 FUNCTIONS
X 10 FUNCTIONS
75
100
125
150
175
TIME - SEC

Fig. 3

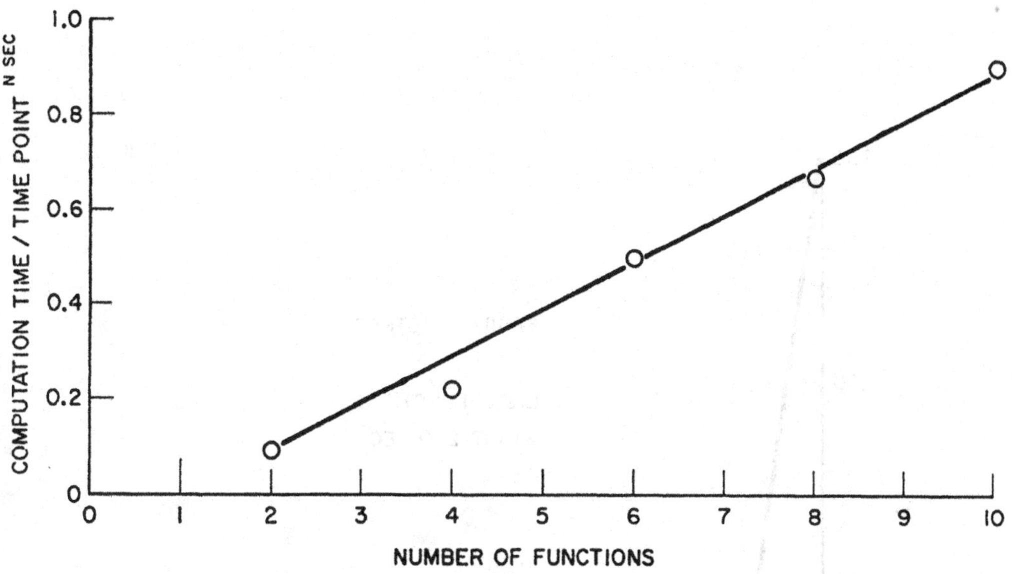
COMPUTATION TIME / TIME POINT N SEC
1.0
0.8
0.6
0.4
0.2
0
0 1 2 3 4 5 6 7 8 9 10
NUMBER OF FUNCTIONS

Fig. 4

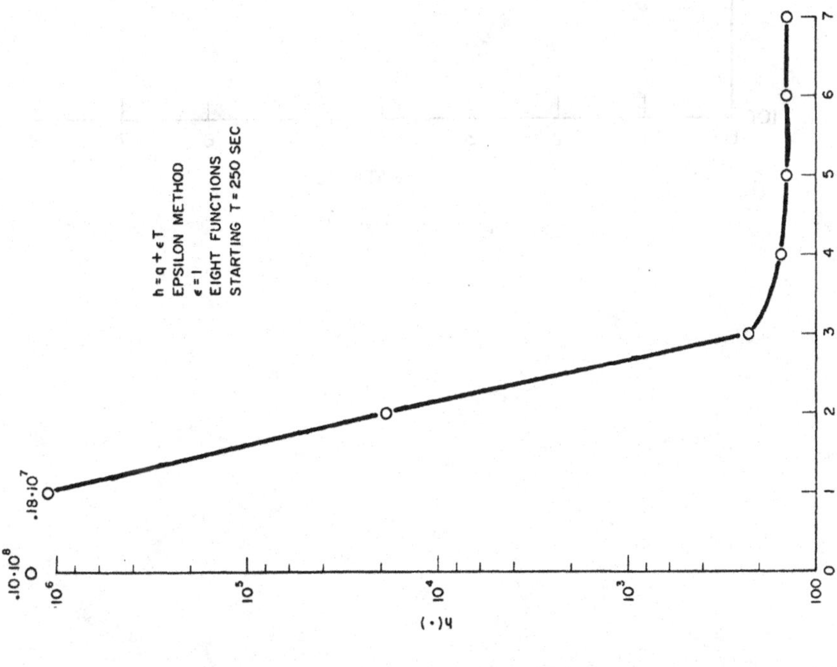
h=q+εT
EPSILON METHOD
ε=1
EIGHT FUNCTIONS
STARTING T=250 SEC
.18·10^7
.10·10^8
10^6
10^5
10^4
10^3
100
h(·)
ITERATIONS
0 1 2 3 4 5 6 7

Fig. 5

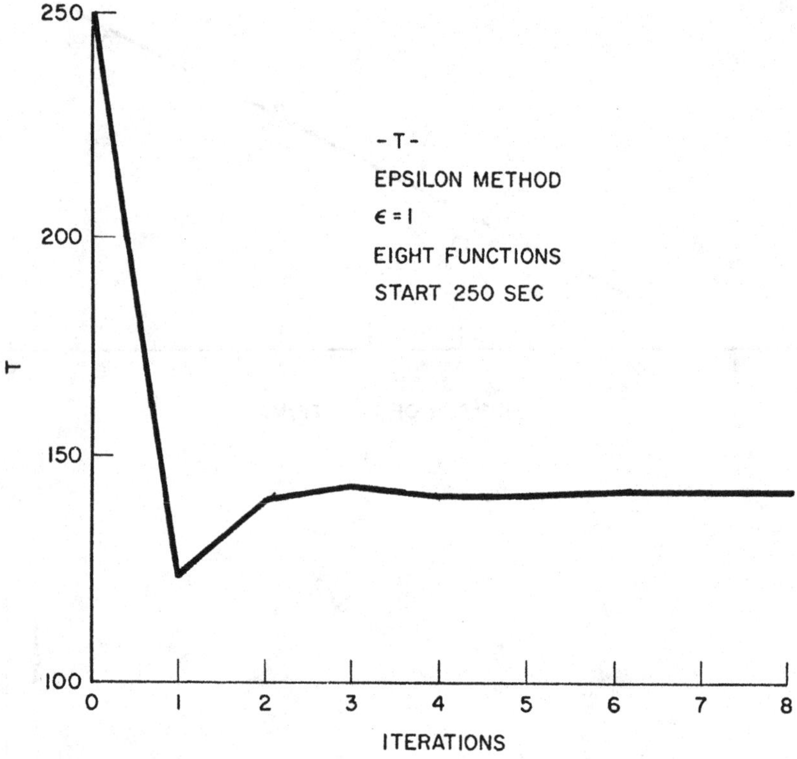

250
200
150
100
T
-T-
EPSILON METHOD
$\epsilon = 1$
EIGHT FUNCTIONS
START 250 SEC
0
1
2
3
4
5
6
7
8
ITERATIONS

Fig. 6

REMARQUES SUR LA METHODE DE PENALISATION ET APPLICATIONS

J.L. LIONS

INTRODUCTION.

On indique (Section 1) comment on peut, par une simple variante de la méthode de pénalisation [8] , introduire les pénalisations sur les équations d'état - ce qui admet des applications numériques aux problèmes de contrôle optimal, d'identification de systèmes etc... et cela pour des systèmes très variés, gouvernés par des équations aux dérivées partielles, linéaires ou non.

On indique ensuite (Section 2) comment une idée tout à fait analogue conduit à des procédés de décomposition de type nouveau en Analyse Numérique.

On indique enfin brièvement (Section 3) une autre application de l'idée de pénalisation à des équations aux dérivées partielles non linéaires et les liens de la pénalisation et de la régularisation.

Le plan de la Conférence est le suivant :

1. PENALISATION SUR L'EQUATION D'ETAT.

1.1. Un exemple.

Soit Ω un ouvert de $\mathbb{R}^n$, de frontière Γ régulière et soit un système dont l'état $y = y(x, t) = y(x, t; v)$ est donné par

(1.1) $\frac{\partial y}{\partial t} - \Delta y = f(x, t), \; x \in \Omega, \; t \in]o, T[$ [1],

avec la condition initiale

(1.2) $y(x, o) = y_o(x), \; y_o$ donné dans $L^2(\Omega)$,

(1.3) $\frac{\partial y}{\partial n}(x, t) = v(x, t), \; x \in \Gamma, \; t \in]o, T[$,

(où $\frac{\partial}{\partial n}$ = dérivée normale à Γ) ; dans (1.3), v désigne le contrôle ; on suppose que

(1.4) $v \in L^2(\Sigma), \; \Sigma = \Gamma \times]o, T[$,

et, plus précisément, que

(1.5) $v \in \mathcal{U}_{ad}, \; \mathcal{U}_{ad}$ = ensemble convexe fermé de $L^2(\Sigma)$.

Il est bien connu que (1.1) (1.2) (1.3) définissent y de façon unique ; y dépend (d'ailleurs de façon affine) de v ; on utilise l'une des notations :

$$y = y(x, t) = y(x, t; v) = y(v).$$

Considérons, par exemple, le problème de contrôle suivant :

---

(1) $\Delta y = \sum_{j=1}^{n} \frac{\partial^2 y}{\partial^2 x_j}$; f est donné dans $\Omega \times]o, T[$.

(1.6) minimiser, pour v dans $\mathcal{U}_{ad}$, la fonction coût J(v) donnée par

$$J(v) = \int_\Omega (y(x , T ; v) - z_d(x))^2 \, d x + \nu \int_\Sigma v^2 \, d \Sigma ,$$

où $\nu > o$ et z_d est donné dans $L^2(\Omega)$.

On vérifie sans peine (cf. [11] pour des situations beaucoup plus générales) qu'il existe $u \in \mathcal{U}_{ad}$ unique tel que

(1.7) $J(u) \leq J(v) \quad \forall \, v \in \mathcal{U}_{ad}$.

On considère maintenant :

(i) y , y_o et v comme des "variables indépendantes" ;

(ii) les relations (1.1) (1.2) (1.3) comme des contraintes devant être satisfaites par le triplet $\{ y , y_o , v \}$.

Soit

(1.8) $\varepsilon = \{ \varepsilon_1 , \varepsilon_2 , \varepsilon_3 \}$, $\varepsilon_i > o$;

nous considérons la nouvelle fonctionnelle (où $Q = \Omega \times]o , T[$) :

(1.9)
$$\begin{aligned} J_\varepsilon(y,y_o,v) = & \int_\Omega (y(x , T) - z_d(x))^2 \, dx + \nu \int_\Sigma v^2 \, d \Sigma + \\ & + \frac{1}{\varepsilon_1} \int_Q \left(\frac{\partial y}{\partial t} - \Delta y - f\right)^2 dx \, dt + \\ & + \frac{1}{\varepsilon_2} \int_\Omega (y(x , o) - y_o(x))^2 \, dx + \\ & + \frac{1}{\varepsilon_3} \int_\Sigma \left(\frac{\partial y}{\partial n}(x , t) - v(x)\right)^2 d \Sigma \quad . \end{aligned}$$

Dans (1.9) nous supposons que

(1.10) $y \in Y$, $y_o \in L^2(\Omega)$, $v \in \mathcal{U}_{ad} \; (\subset L^2(\Sigma))$,

où

$$Y = \{ \psi \mid \psi , \frac{\partial \psi}{\partial x_i} , \frac{\partial \psi}{\partial t} - \Delta \psi \text{ sont dans } L^2(Q) \} ,$$

l'espace Y étant muni de la norme (hilbertienne)

$$\left(\| \psi \|^2_{L^2(Q)} + \sum_{i=1}^{n} \| \frac{\partial \psi}{\partial x_i} \|^2_{L^2(Q)} + \| \frac{\partial \psi}{\partial t} - \Delta \psi \|^2_{L^2(Q)} \right)^{1/2} .$$

Les termes en $\frac{1}{\epsilon_i}$ sont les termes de pénalisation.

On vérifie sans peine le (cf. résultats plus généraux dans [11]).

Théorème 1.1. Il existe une solution $\{ y_\epsilon , y_{o\epsilon} , v_\epsilon \}$ unique minimisant, sur $Y \times L^2(\Omega) \times \mathcal{U}_{ad}$, la fonctionnelle $J_\epsilon(y , y_o , v)$.

Lorsque $\epsilon \longrightarrow o$, on a :

(1.11) $$\text{Inf. } J_\epsilon (y , y_o , v) \longrightarrow \underset{v \in \mathcal{U}_{ad}}{\text{Inf. }} J(v)$$

et

(1.12) $v_\epsilon \longrightarrow u$ dans $L^2(\Sigma)$ (u solution de (1.7)) .

1.2. Application numérique.

On discrétise les équations et les intégrales dans (1.9) et on est alors ramené à un problème de programmation mathématique, que l'on traite par exemple par une méthode de gradient conjugué.

D'excellents résultats numériques ont été obtenus par ces méthodes dans plusieurs exemples par Yvon [25] .

Remarque 1.1

On peut également introduire des pénalisations sur les contraintes "$v \in \mathcal{U}_{ad}$" - Cf. [9] [10] pour plusieurs applications numériques.

1.3. Extensions et variantes.

1) La méthode indiquée au 1.1 peut s'étendre, avec des adaptations convenables, à des systèmes gouvernés par des équations non linéaires, par ex. des équations aux dérivées partielles non linéaires ; cf. [6] .

2) La méthode s'étend aussi - avec une "double-pénalisation" - au cas de systèmes gouvernés par des <u>inéquations</u> (du type des problèmes d'élasto-plasticité) cf. [6].

3) La méthode s'adapte aux problèmes <u>d'identification</u>. Cf. [4].

4) La méthode peut s'étendre aussi à certains jeux (jeux différentiels, jeux "aux équations à dérivées partielles") , cf. [22] .

2. PENALISATION ET TECHNIQUES DE DECOMPOSITION.

2.1. Décomposition d'un problème elliptique.

Les notations étant analogues à celles du N°1 , soit u la solution du problème classique de Dirichlet :

(2.1) $\quad -\Delta u = f \text{ dans } \Omega , \quad u = o \text{ sur } \Gamma .$

De façon générale, $\mathcal{O}$ étant un ouvert de $\mathbb{R}^n$, on pose

$$a_{\mathcal{O}}(u , v) = \int_{\mathcal{O}} \text{grad } u \, . \, \text{grad } v \, dx , \quad u , v \in H_o^1(\mathcal{O}) ,$$

$$H_o^1(\mathcal{O}) = \{ v \mid v , \frac{\partial v}{\partial x_i} \in L^2(\Omega) , \quad v = o \text{ sur } \Gamma \} .$$

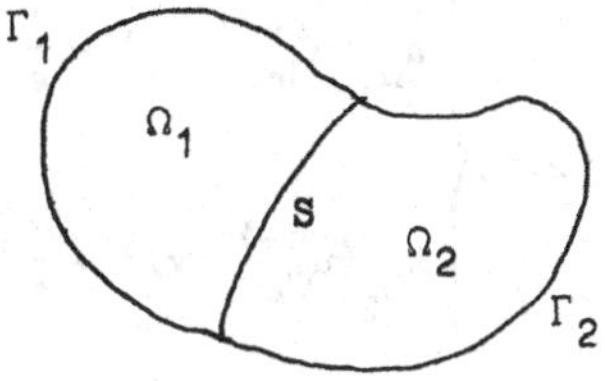

fig. 1

Alors (2.1) <u>équivaut</u> à

(2.2) $$\begin{cases} a_\Omega(u , v) = \int_\Omega f \, v \, dx \quad \forall v \in H_o^1(\Omega) , \\ u \in H_o^1(\Omega) . \end{cases}$$

Divisons maintenant Ω par <u>"l'interface artificielle"</u> S (comme indiqué fig. 1) ([1]).

On va considérer (2.2) comme consistant de <u>deux</u> problèmes, de solutions respectives u_1 et u_2 dans Ω_1 et Ω_2 , avec la <u>contrainte</u>

$$u_1 = u_2 \text{ sur } S .$$

On arrive de la sorte au problème suivant : on définit d'abord

---

([1]) Tout ce qui suit s'étend à une <u>décomposition finie quelconque</u> de Ω .

$V_i = \{ \psi \mid \psi , \frac{\partial \psi}{\partial x_i} \in L^2(\Omega_i) , \psi = o$ sur Γ_i (cf. fig. 1) $\}$;

on cherche $\{ u_{1\epsilon}, u_{2\epsilon} \} \in V_1 \times V_2$ solution de

$$(2.3) \begin{cases} a_{\Omega_1}(u_{1\epsilon}, v_1) + a_{\Omega_2}(u_{2\epsilon}, v_2) + \frac{1}{\epsilon} \int_S (u_{1\epsilon} - u_{2\epsilon})(v_1 - v_2)\, dS = \\ = \int_{\Omega_1} f v_1\, dx + \int_{\Omega_2} f v_2\, dx \quad \forall \{v_1, v_2\} \in V_1 \times V_2 . \end{cases}$$

On montre sans peine le

Théorème 2.1. Pour tout $\epsilon > o$, (2.3) admet une solution unique.

Lorsque $\epsilon \longrightarrow o$, on a :

$$(2.4) \qquad u_{i\epsilon} \longrightarrow u , \quad \frac{\partial u_{i\epsilon}}{\partial x_j} \longrightarrow \frac{\partial u}{\partial x_j} \quad \text{dans } L^2(\Omega_i) , \quad i = 1,2 .$$

On peut maintenant introduire une méthode itérative associée à (2.3) :

on définit $u_{i\epsilon}^{n+1}$ à partir de $u_{i\epsilon}^n$ par

$$(2.5) \begin{cases} a_{\Omega_1}(u_{1\epsilon}^{n+1}, v_1) + \frac{1}{\epsilon} \int_S u_{1\epsilon}^{n+1} v_1\, dS = \frac{1}{\epsilon} \int_S u_{2\epsilon}^n v_1\, d\,S + \int_{\Omega_1} f v_1\, dx , \\ a_{\Omega_2}(u_{2\epsilon}^{n+1}, v_2) + \frac{1}{\epsilon} \int_S u_{2\epsilon}^{n+1} v_2\, dS = \frac{1}{\epsilon} \int_S u_{1\epsilon}^n v_2\, d\,S + \int_{\Omega_2} f v_2\, dx . \end{cases}$$

On vérifie le

Théorème 2.2. Lorsque $n \longrightarrow \infty$, on a

$$(2.6) \qquad u_{i\epsilon}^n \longrightarrow u_{i\epsilon} \quad \text{dans } V_i , \quad i = 1,2 .$$

Remarque 2.1.

Le schéma itératif (2.5) conduit à une "décomposition" du problème initial - les calculs des $u_{i\epsilon}^{n+1}$, $i = 1,2$, pouvant être conduits "en parallèle" .

Remarque 2.2.

Comme on a déjà indiqué, les résultats précédents s'étendent au cas d'une décomposition arbitraire du domaine Ω . Ils s'étendent également à "tous" les problèmes aux limites variationnels linéaires . Le Théorème 2.1 s'étend aux problèmes non linéaires "monotones" , l'extension du procédé itératif (2.5) nécessitant quelques modifications.

2.2. Décomposition d'un problème parabolique.

Soit $u = u(x, t)$, $x \in \Omega$, $t > o$, la solution de

$$(2.7) \quad \begin{cases} \frac{\partial u}{\partial t} - \Delta u = f \text{ dans } Q = \Omega \times]o,T[, \\ u = o \text{ sur } \Sigma = \Gamma \times]o,T[, \\ u(x, o) = u_o(x) , \quad x \in \Omega . \end{cases}$$

Uniquement pour simplifier l'exposé, supposons la dimension d'espace égale à 2 . On va donner une décomposition du problème (2.7) qui a la particularité d'être globale en t (par opposition avec les méthodes usuelles de Directions Alternées ou de Pas Fractionnaires - cf. Yanenko [24] et la Bibliographie de ce travail - qui sont locales en t) .

Soient θ_1 , θ_2 deux constantes vérifiant

$$(2.8) \quad o < \theta_i < 1 , \quad \theta_1 + \theta_2 = 1 .$$

Si l'on décompose $f = f_1 + f_2$, on considère le système

$$\begin{cases} \theta_1 \frac{\partial u_1}{\partial t} - \frac{\partial^2 u_1}{\partial x_1^2} + \frac{1}{\varepsilon}(u_1 - u_2) = f_1 \text{ dans } Q , \\ \theta_2 \frac{\partial u_2}{\partial t} - \frac{\partial^2 u_2}{\partial x_2^2} + \frac{1}{\varepsilon}(u_2 - u_1) = f_2 \text{ dans } Q , \end{cases}$$

avec

$$(2.10) \quad u_1(x, o) = u_2(x, o) = u_o(x) , \; x \in \Omega$$

(2.11) $\begin{cases} u_i = o \text{ sur les parties de } \Gamma \text{ non parallèles à l'axe des} \\ x_i \ , \ i = 1,2 \ . \end{cases}$

Ce système <u>admet une solution</u> $\{ u_{i\epsilon} \ , \ i = 1,2 \}$ <u>unique</u> , ($\epsilon > o$) .

On montre le

<u>Théorème</u> 2.3. <u>Lorsque</u> $\epsilon \longrightarrow o$, <u>on a</u>

$$(2.12) \quad \begin{cases} u_{1\epsilon} \longrightarrow u \ , \ \dfrac{\partial u_{1\epsilon}}{\partial x_1} \longrightarrow \dfrac{\partial u}{\partial x_1} \quad \text{dans } L^2(Q) \\ u_{2\epsilon} \longrightarrow u \ , \ \dfrac{\partial u_{2\epsilon}}{\partial x_2} \longrightarrow \dfrac{\partial u}{\partial x_2} \quad \text{dans } L^2(Q) \end{cases}$$

<u>Remarque</u> 2.3.

Le procédé (2.9) s'étend (avec un résultat analogue à (2.12)):

- à des équations paraboliques non linéaires;
- aux équations hyperboliques etc...

<u>Par exemple</u>, si u est la solution de l'équation des ondes :

$$(2.13) \quad \begin{cases} \dfrac{\partial^2 u}{\partial t^2} - \Delta u = f \quad \text{dans } Q \ , \\ u = o \quad \text{sur } \Sigma \ , \\ u(x \ , o) = u_o(x) \ , \ \dfrac{\partial u}{\partial t}(x \ , o) = u_1(x) \ , \ x \in \Omega \ , \end{cases}$$

on introduira le système

$$(2.14) \quad \begin{cases} \theta_1 \dfrac{\partial^2 u_1}{\partial t^2} - \dfrac{\partial^2 u_1}{\partial x_1^2} + \dfrac{1}{\epsilon}\left(\dfrac{\partial u_1}{\partial t} - \dfrac{\partial u_2}{\partial t} \right) = f_1 \quad \text{dans } Q \ , \\ \theta_2 \dfrac{\partial^2 u_2}{\partial t^2} - \dfrac{\partial^2 u_2}{\partial x_2^2} + \dfrac{1}{\epsilon}\left(\dfrac{\partial u_2}{\partial t} - \dfrac{\partial u_1}{\partial t} \right) = f_2 \quad \text{dans } Q \ , \end{cases}$$

avec les conditions initiales

$$(2.15)\quad u_1(x\ ,\ o) = u_2(x\ ,\ o) = u_o(x)\ ,\ \frac{\partial u_1}{\partial\ t}(x\ ,\ o) = \frac{\partial u_2}{\partial\ t}(x,o) = u_1(x)\ ,x \in \Omega,$$

et les conditions aux limites (2.11).

On a alors un résultat analogue à (2.12) , avec en outre

$$\frac{\partial u_{i\varepsilon}}{\partial\ t} \longrightarrow \frac{\partial\ u}{\partial\ t} \quad \text{dans} \quad L^2(Q)\quad ,\quad i = 1,2\ .$$

3. PENALISATION ET APPROXIMATION.

3.1. Un problème aux limites non linéaire.

Soit Q un ouvert non cylindrique de $\mathbb{R}^n_x \times \mathbb{R}_t$ (cf. schéma sur fig. 2); avec les notations de la figure, on cherche une fonction u solution de

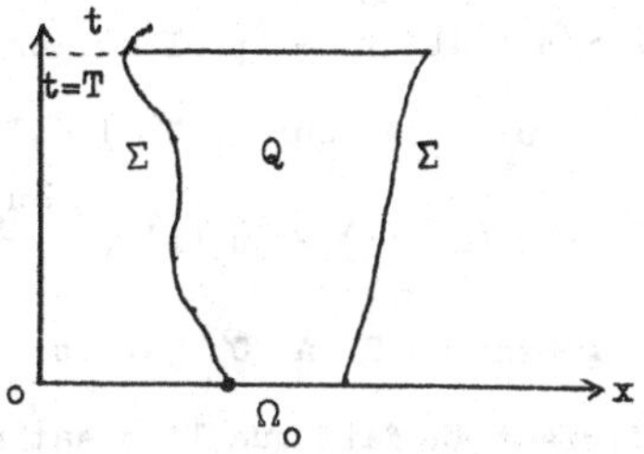

fig. 2

$$(3.1)\quad \frac{\partial^2 u}{\partial t^2} - \Delta u + |\ u\ |^\rho\ u = f \ \text{dans}\ Q\ ,\ (\ \rho > o)$$

$$(3.2)\quad u = o \ \text{sur}\ \Sigma\ ,$$

$$(3.3)\quad u(x\ ,\ o) = u_o(x)\ ,\ \frac{\partial\ u}{\partial\ t}(x\ ,\ o) = u_o^*(x)\quad ,\quad x \in \Omega_o\ .$$

On peut montrer l'existence d'une solution de ce problème (non linéaire) sous l'hypothèse

$$(3.4)\quad \begin{cases} \text{si}\ \ x_o\ ,\ t_o \in \complement\, Q = \text{complémentaire de } Q \text{ dans la bande} \\ o < t < T\ ,\ \text{alors}\ \{x_o\ ,\ \theta\ t_o\} \in \complement\, Q \ \text{pour}\ \ o < \theta < 1\ , \end{cases}$$

la méthode de démonstration utilisant une pénalisation .

On introduit $\mathcal{O}$ = ouvert de $\mathbb{R}^n$ tel que

(3.5) $\quad Q \subset \mathcal{O} \;\;]o,T[$

et

$$M = 0 \text{ dans } Q \,, \quad 1 \text{ dans } \mathcal{O} \;\;]o,T[\, - Q \,.$$

Soit alors u_ε une solution (pour $\varepsilon > o$) du problème non linéaire

(3.6) $$\frac{\partial^2 u_\varepsilon}{\partial t^2} - \Delta u_\varepsilon + |u_\varepsilon|^\rho u_\varepsilon + \frac{1}{\varepsilon} M \frac{\partial u_\varepsilon}{\partial t} = \tilde{f} \text{ dans } \mathcal{O} \;\;]o,T[\,,$$

où $\varepsilon > o$ et $\tilde{f}$ = prolongement de f par 0 hors de Q , avec

(3.7) $\quad u_\varepsilon = o$ sur $\partial \mathcal{O} \;\;]o,T[$,

(3.8) $\quad u_\varepsilon(x\,,\,o) = \tilde{u}_o(x)$, $\frac{\partial u_\varepsilon}{\partial t}(x\,,\,o) = \tilde{u}^*_o(x)$ (où $\sim$ désigne le prolongement de Ω à $\mathcal{O}$ par o hors de Ω).

Bénéficiant du fait que l'on est maintenant dans le cas d'un domaine cylindrique (le terme en $\frac{1}{\varepsilon}$ étant la pénalisation), on montre l'existence d'une solution u_ε de (3.6). Si la dimension d'espace est ≤ 3 et si $\rho \leq 2$, il y a unicité de u_ε .

On montre [12]

<u>Théorème</u> 3.1. <u>Sous l'hypothèse</u> (3.4) <u>le problème</u> (3.1)(3.2)(3.3) <u>admet une solution qui peut être obtenue comme limite de solutions</u> u_ε <u>de</u> (3.6)(3.7)(3.8)

Cf. aussi [13] [18] [20] et le travail [7] .

3.2. <u>Pénalisation et régularisation</u>.

Les méthodes de pénalisation et de régularisation sont (essentiellement) équivalentes , cf. [5] .

Pour des applications à l'Analyse Numérique, cf. [1] [2] . Cela peut être étendu aux inéquations variationnelles, cf. [17] .

Au même ordre d'idées se rattache l'approximation par des systèmes du type de Cauchy Kowaleska [23] [14] .

BIBLIOGRAPHIE.

[1] J.P. AUBIN. Approximation des espaces de distributions et des opérateurs différentiels. Bulletin S.M.F. , Mémoire 12 , 1967 .

[2] J.P. AUBIN - L. LIONS . Remarques sur l'approximation régularisée de problèmes aux limites elliptiques. C.I.M.E. , Juillet 1967.

[3] A.V. BALAKRISHNAN . On a new computing technique in optimal control theory and and the maximum principle. Proc. Nat. Acad. Sciences U.S.A. (cf. aussi ce volume).

[4] A. BENSOUSSAN . Thèse, Paris, 1969 .

[5] A. BENSOUSSAN - P. KENNETH . Sur l'analogie entre les méthodes de régularisation et de pénalisation. A paraître.

[6] A. BENSOUSSAN - J.L. LIONS. A paraître.

[7] D. CABY . A paraître.

[8] R. COURANT. Variational methods for the solution of problems of equilibrium and vibrations. Bull. A.M.S. 49 (1943), p. 1 - 23 -

[9] P. KENNETH - M. SIBONY . A paraître .

[10] P. KENNETH - J.P. YVON . A paraître.

[11] J.L. LIONS . Contrôle optimal de systèmes gouvernés par des équations aux dérivées partielles. Paris, Dunod, 1968 .

[12] J.L. LIONS . Une remarque sur les problèmes d'évolution non linéaires dans les domaines non cylindriques. Revue Roumaine M. Pures et Appl. IX(1964)p.11-18

[13] J.L. LIONS . Sur l'approximation des solutions de certains problèmes aux limites. Rend. Sem. Mat. Padova, XXXII (1962), p. 3 - 54 .

[14] J.L. LIONS . On the numerical approximation of some equations arising in hydrodynamics. Amer. Math. Soc. Symposium, Durham, Avril 1968.

[15] J.L. LIONS . Problèmes aux limites non homogènes à données irrégulières; une méthode d'approximation - C.I.M.E. , Juillet 1967 .

[16] J.L. LIONS - E. MAGENES . Problèmes aux limites non homogènes et applications. Vol. 1 et 2 . Paris, Dunod, 1968 .

[17] J.L. LIONS - G. STAMPACCHIA . Variational Inequalities. Comm. Pure Applied Math., 20(1967), p. 493 - 519 .

[18] A. MIGNOT . Thèse, Paris, 1967 .

[19] N.N. MOISEIV .

[20] V.K. SAULEV . Méthode du domaine auxiliaire.

[21] S. SOBOLEV . Applications de l'Analyse Fonctionnelle à la Physique Mathématique. Leningrad 1950 .

[22] L. TARTAR . A paraître .

[23] R. TEMAM . Sur l'intégration numérique des équations de Narier-Stokes . A paraître.

[24] N.N. YANENKO . La méthode des pas fractionnaires. Leningrad, 1967 . Armand Colin, Paris, 1968 .

[25] J.P. YVON . A paraître .

-:-:-:-

RECHERCHE NUMERIQUE
D'UN OPTIMUM DANS UN ESPACE PRODUIT

Jean CEA

INTRODUCTION

Il s'agit de donner dans ce travail une méthode d'approximation numérique de la solution u d'un problème de minimisation d'une fonctionnelle J(v), la variable v étant soumise ou non à des contraintes. La méthode proposée est une généralisation des méthodes de relaxation, plus particulièrement de la méthode de Gauss-Siedel.

Des applications numériques à des problèmes en relation avec la théorie des équations aux dérivées partielles sont données.

1. - Position du problème.

Tous les éléments introduits sont réels.

On donne m espaces de Hilbert V_i, $i = 1,\dots,m$, le produit scalaire dans V_i est noté par $(u_i,v_i)_i$ et la norme par $|u_i|_i$.

On désigne par $\mathcal{V}$ l'espace produit $\prod_{i=1}^{m} V_i$, et on pose :

$$((u,v)) = \sum_{i=1}^{m} (u_i,v_i)_i$$

$$\| u \| = ((u,u))^{1/2}$$

où

$$u = (u_1,\dots,u_m), \quad v = (v_1,\dots,v_m).$$

On désigne par K_i un sous-ensemble convexe fermé donné dans V_i, et ceci pour $i = 1,\dots,m$:

$$K_i \subset V_i.$$

On pose $K = \prod_{i=1}^{m} K_i$; K est un sous-ensemble convexe fermé donné dans $\mathcal{V}$.

On donne maintenant une forme bilinéaire $u,v \longrightarrow a(u,v)$ définie sur V x V, continue, symétrique et coercive, c'est à dire que

$$a(v,v) \geq \|v\|^2 \qquad \forall\, v \in \mathcal{V}$$

où $\alpha > 0$, α indépendante de v.

On donne une forme linéaire $v \longrightarrow L(v)$ définie et continue sur V.

On pose :

$$J(v) = a(v,v) - 2\, L(v).$$

Problème 1.1

Déterminer $u \in K$ tel que :

(1.1) $\quad J(u) \leq J(v) \qquad \forall\, v \in K.$

sous les données et hypothèses précédentes, on peut démontrer le :

Théorème 1.1

i) Le problème 1.1 *a une solution et une seule.*

ii) Cette solution est caractérisée par :

$$(1.2)\qquad \begin{cases} a(u,v-u) - L(v-u) \geq 0 & \forall\, v \in V \\ u \in K. \end{cases}$$

(Autrement dit (1.1) et (1.2) sont des relations équivalentes).

Nous allons déduire du théorème 1.1 deux relations utiles pour la suite ; nous avons :

$$a(v-u,v-u) + (a(u,v-u) - L\,(v-u)) = a(v,v-u) - L(v-u)$$

ce qui, avec (1.2), entraîne

$$(1.3)\qquad a(v-u,v-u) \leq a(v,v-u) - L(v-u) \qquad \forall\, v \in K.$$

On vérifierait, en explicitant, la relation suivante :

$$(1.4)\qquad 2\{a(v,v-u) - L(v-u)\} = J(v) - J(u) + a(v-u,v-u).$$

A partir de (1.3) et de (1.4) il vient trivialement :

$$(1.5)\qquad a(v-u,v-u) \leq J(v) - J(u) \qquad \forall\, v \in K$$

et compte tenu de la coercivité de $a(u,v)$, on a finalement :

$$(1.6)\qquad \alpha\, \|v-u\|^2 \leq J(v) - J(u) \qquad \forall\, v \in K.$$

Nous nous proposons maintenant d'approcher la solution u du problème 1.1.

2. - Une méthode itérative.

Si $w \in K$, on pose :

$$K_j(w) = \{v \mid v \in K,\ v_i = w_i \quad \forall\, i \neq j\}$$

ou encore

$$K_j(w) = \{v \mid v_i = w_i \quad \forall\, i \neq j,\ v_j \in K_j\}.$$

La méthode :

On donne $u^0 \in K$, on construit une suite d'itérés $u^{k+j/m}$, $j = 1,\ldots,m$, $k = 1, 2, \ldots$; le passage de $u^{k+j/m}$ à $u^{k+(j+1)/m}$ se fait par :

$u^{k+(j+1)/m}$ est solution du problème suivant :

Déterminer $u^{k+(j+1)/m} \in K_{j+1}(u^{k+j/m})$ tel que :

$$(2.1) \quad \begin{cases} J(u^{k+(j+1)/m}) \leq J(v) \\ \forall\ v \in K_{j+1}(u^{k+j/m}) \end{cases}$$

on a le :

Théorème 2.1

Sous les données et hypothèses précédentes, la suite $u^{k+j/m}$ *converge vers* u *dans* V *faible lorsque* $k \to +\infty$.

Démonstration.

1er point. La suite $J(u^{k+j/m})$ est décroissante.

En effet, $u^{k+j/m} \in K_{j+1}(u^{k+j/m})$ et donc d'après (2.1) :

$$J(u^{k+(j+1)/m}) \leq J(u^{k+j/m})$$

par suite :

$$(2.2) \quad J(u) \leq \ldots \leq J(u^{k+(j+1)/m}) \leq J(u^{k+j/m}) \leq \ldots \leq J(u^{0}).$$

2ème point. La suite $u^{k+(j+1)/m} - u^{k+j/m}$ converge vers 0.

En appliquant la relation (1.6) au problème (2.1) avec $v = u^{k+j/m}$, $u = u^{k+(j+1)/m}$, il vient :

$$(2.3) \quad \alpha \|u^{k+(j+1)/m} - u^{k+j/m}\|^{2} \leq J(u^{k+j/m}) - J(u^{k+(j+1)/m})$$

or d'après (2.2),

$$\lim_{k\to+\infty} J(u^{k+j/m}) - J(u^{k+(j+1)/m}) = 0$$

par suite :

$$\lim_{k\to+\infty} \|u^{k+(j+1)/m} - u^{k+j/m}\| = 0$$

et donc :

(2.4) $$\lim_{k\to+\infty} \|u^{k+j/m} - u^{k+i/m}\| = 0 \ ; \quad j = 0,\dots,m \ ; \quad i = 0,\dots,m.$$

3ième point. Extraction de suites faiblement convergentes.

Compte tenu de la coercivité, la relation (2.2) entraîne

(2.5) $$\|u^{k+j/m}\| < \text{constante} < +\infty \quad \forall\, k,\ \forall\, j$$

on peut donc extraire une sous-suite $u^{k'+j'/m}$ qui converge faiblement vers un élément $w \in V$; mais $u^{k'+j'/m} \in K$ qui est un ensemble convexe et fermé, par suite $w \in K$.

De plus à l'aide de (2.4), il vient :

(2.6) $$\lim_{k'\to+\infty} u^{k'+i(k')/m} = w \quad \text{dans } V \text{ faible}$$

où $i(k')$ est quelconque dans $\{0,1,2,\dots,m\}$.

4ième point. Convergence faible.

Par définition de $u^{k+(j+1)/m}$ on a :

(2.7) $$\begin{cases} a(u^{k+(j+1)/m}, z - u^{k+(j+1)/m}) - L(z - u^{k+(j+1)/m}) \geq 0 \\ \forall\, z \in K_{j+1}(u^{k+j/m}). \end{cases}$$

Soit v un élément quelconque fixé dans K ; dans l'inégalité précédente, on choisit :

$$z \in K_{j+1}(u^{k+j/m}), \quad z_{j+1} = v_{j+1}$$

c'est à dire :

$$z = (u_1^{k+j/m}, \dots, u_j^{k+j/m}, v_{j+1}, u_{j+2}^{k+j/m}, \dots, u_m^{k+j/m}).$$

Nous emploierons la notation suivante :

$P_j\, y = (0,\dots,0, y_j, 0, \dots, 0)$ où $y = (y_1, \dots, y_m)$.

Dans (2.7) il vient avec le choix de z :

$$a(u^{k+(j+1)/m}, P_{j+1}(v - u^{k+(j+1)/m})) - L(P_{j+1}(v - u^{k+(j+1)/m})) \geq 0, \ \forall\, v \in K.$$

d'où en remplaçant j+1 par j, et en disposant autrement les termes,

$$(2.8)\qquad a(u^{k+j/m}, P_j v) - L(P_j(v - u^{k+j/m})) \geq a(u^{k+j/m}, P_j u^{k+j/m}) \qquad \forall\, v \in K.$$

La convergence faible de la suite $u^{k'+i/m}$ nous permet de passer à la limite dans le 1er membre de (2.8), mais pas dans le second. Pour remédier à cette difficulté nous allons faire la somme des inégalités (2.8) pour j = 1, ..., m et dans le second membre nous allons faire intervenir les différences $u^{k+1} - u^{k+j/m}$. Nous avons :

$$\begin{cases} a(u^{k+j/m}, P_j u^{k+j/m}) = a(u^{k+1}, P_j u^{k+1}) + a(u^{k+j/m} - u^{k+1}, P_j\, u^{k+j/m}) + \\ \qquad + a(u^{k+1}, P_j\, u^{k+j/m} - P_j\, u^{k+1}) \end{cases}$$

d'où

$$(2.9)\quad \begin{cases} \sum_{j=1}^{m} a(u^{k+j/m}, P_j u^{k+j/m}) = a(u^{k+1}, u^{k+1}) + \sum_{j=1}^{m} a(u^{k+j/m} - u^{k+1}, P_j\, u^{k+j/m}) + \\ \qquad + \sum_{j=1}^{m} a(u^{k+1}, P_j u^{k+j/m} - P_j\, u^{k+1}). \end{cases}$$

Lorsque $k \to +\infty$, compte tenu de (2.4) et de (2.5) les 2 derniers termes de (2.9) tendent vers 0, par suite :

$$(2.10)\qquad \liminf_{k\to+\infty} \sum_{j=1}^{m} a(u^{k+j/m}, P_j\, u^{k+j/m}) \geq \liminf_{k\to+\infty} a(u^{k+1}, u^{k+1}).$$

En particulier pour k = k' ; mais alors la relation (2.6), la symétrie et la positivité de a(u,v), entraînent :

$$(2.11)\qquad \liminf_{k'\to+\infty} a(u^{k'+1}, u^{k'+1}) \geq a(w,w).$$

Par sommation des inégalités (2.8), il vient :

$$\sum_{j=1}^{m} a(u^{k+j/m}, P_j v) - \sum_{j=1}^{m} L(P_j(v - u^{k+j/m})) \geq \sum_{j=1}^{m} a(u^{k+j/m}, P_j\, u^{k+j/m}).$$

D'où, par passage à la limite, $k' \to +\infty$, compte tenu de (2.6), (2.10), (2.11)

$$\sum_{j=1}^{m} a(w,P_j v) - \sum_{j=1}^{m} L(P_j(v - w)) \geq a(w,w)$$

d'où

$$a(w,v) - L(v - w) \geq a(w,w)$$

ce qui montre que w est une solution du problème 1.1. Comme la solution u est unique, on a $w = u$.

On vérifierait avec un argument classique qu'il y a convergence non seulement de la sous-suite $u^{k'+j'/m}$ mais encore de la suite originelle $u^{k+j/m}$.

3. - Applications.

Dans toute la suite A désignera une matrice symétrique et coercive, f un élément donné.

3.1. Cas d'un espace et d'un convexe particulier.

On prend :

$$V = \mathbb{R}^m, \quad J(v) = (Av,v)_{\mathbb{R}^m} - 2\,(f,v)_{\mathbb{R}^m} \quad .$$

On définit K par la donnée d'une fonction g définie dans $\mathbb{R}^m$ à valeurs dans $\mathbb{R}$. On suppose que g est concave, différentiable, son gradient est noté par ∇g et on suppose que :

$$|g(v)| + \|\nabla g(v)\|_{\mathbb{R}^m} \neq 0 \qquad \forall\, v \in \mathbb{R}^m.$$

Par définition :

$$v \in K \Longleftrightarrow g(v) \geq 0$$

K est donc un sous ensemble convexe fermé dans $\mathbb{R}^m$.

On sait que :

(3.1) $\{u \in K \text{ est solution de } J(u) \leq J(v) \quad \forall\, v \in K\} \quad \Longleftrightarrow$

(3.2) $\{\exists\, \lambda\ \mathbb{R} \text{ tel que } \nabla J(u) = \lambda \,.\, \nabla g(u),\ \lambda \geq 0,\ g(u) \geq 0,\ \lambda \,.\, g(u) = 0\}$

Notons que $\nabla J(u) = 2\,(Au - f)$.

3.2. Cas d'un produit d'espaces.

On prend :

$$V = \mathbb{R}^{P_1} \times \dots \times \mathbb{R}^{P_m} \qquad (\text{autrement dit } V_j = \mathbb{R}^{P_j},\ j = 1, \dots, m)$$

$$v \in V \Longleftrightarrow v = (v_1, \dots, v_m)\ ,\ v_j \in \mathbb{R}^{P_j}\ ,\ \ v_j = (v_{j,1}, \dots, v_{j,P_j})$$

On définit K_j par la donnée d'une fonction g_j définie dans $\mathbb{R}^{P_j}$ à valeurs dans $\mathbb{R}$ et ayant les propriétés de la fonction g précédente. Par définition :

$$v_j \in K_j \subset R^{P_j} \iff g_j(v_j) \geq 0$$

$$v \in K \iff v_j \in K_j, \quad j = 1, \ldots, m.$$

Dans le cas présent la définition de l'itéré partiel $u^{k+i/m}$ est la suivante :

$$(3.3) \quad \begin{cases} \text{Déterminer } \tilde{u}_i \in K_i \text{ tel que} \\ J(\ldots, u_{i-1}^{k+i/m}, \tilde{u}_i, u_{i+1}^{k+(i-1)/m}, \ldots) \leq J(\ldots, u_{i-1}^{k+i/m}, v_i, u_{i+1}^{k+(i-1)/m}, \ldots) \\ \forall\, v_i \in K_i \end{cases}$$

alors

$$u_i^{k+i/m} = \tilde{u}_i.$$

Le problème partiel est donc de même nature que celui du 3.1.

Si nous désignons par $\nabla_i\, J(u)$ le gradient de la fonction $u_i \longrightarrow J(u)$, alors $\tilde{u}_i$ solution de (3.3) est aussi solution de

$$(3.4) \quad \begin{cases} \nabla_i\, J(\ldots, u_{i-1}^{k+i/m}, \tilde{u}_i, u_{i+1}^{k+(i-1)/m}, \ldots) = \lambda^{k+i/m}.\ \nabla g_i(\tilde{u}_i) \\ \lambda^{k+i/m} \in \mathbb{R}, \quad \lambda^{k+i/m} \geq 0, \quad g_i(\tilde{u}_i) \geq 0, \quad \lambda^{k+i/m}.\ g_i(u_i) = 0. \end{cases}$$

Afin de simplifier les notations, tous les itérés partiels seront désignés par $\tilde{u}$, étant entendu que dans la ième partielle seule la ième composante intervient, les autres composantes conservant les valeurs de l'étape précédente. Ainsi (3.4) devient :

$$(3.5) \quad \begin{cases} \tilde{u}_i \text{ est définie par } \quad \nabla_i\, J(\tilde{u}) = \tilde{\lambda}\, \nabla g_i(\tilde{u}_i) \\ \tilde{\lambda} \in \mathbb{R}, \ \tilde{\lambda} \geq 0, \quad g_i(\tilde{u}_i) \geq 0, \quad \tilde{\lambda}\ .\ g_i(\tilde{u}_i) = 0. \end{cases}$$

Lorsque certaines composantes v_i sont libres $[K_i = v_i]$, les formules (3.5) deviennent evidemment :

$$(3.6) \quad \begin{cases} \tilde{u}_i \text{ est définie par} \\ \nabla_i\, J(\tilde{u}) = 0. \end{cases}$$

3.3. La méthode de Gauss-Siedel par points :

Il s'agit de résoudre

$Au = f$

où la matrice A est symétrique définie positive, où $f \in \mathbb{R}^m$ et u est cherché dans $\mathbb{R}^m$, on sait que u est aussi solution du problème de minimisation

$$J(u) \leq J(v) \quad \forall\, v \in \mathbb{R}^m$$

avec

$$J(v) = (Av,v)_{\mathbb{R}^m} - 2(f,v)_{\mathbb{R}^m}.$$

L'application de (3.6) conduit dans ce cas à :

$$\sum_{j=1}^{m} a_{i,j}\,\tilde{u}_j - f_i = 0$$

d'où

$$\tilde{u}_i = \frac{1}{a_{ii}}\left[-\sum_{j=1}^{i-1} a_{i,j}\,\tilde{u}_j - \sum_{j=i+1}^{m} a_{i,j}\,\tilde{u}_j + f_i\right] \tag{3.7}$$

on reconnait alors la méthode de Gauss-Siedel par points.

On retrouverait de la même façon la méthode de Gauss-Siedel par blocs. Le théorème du N° 2 établit donc la convergence de ces méthodes.

3.4. Le cas des contraintes ponctuelles linéaires.

On prend $V = \mathbb{R}^m$; on sépare les indices 1,...,m en 2 ensembles complémentaires disjoints I_l et I_c.

L'ensemble K est défini par la donnée de nombres $a_i \in \mathbb{R}$:

$$v \in K \Longleftrightarrow v_i \geq a_i \quad \forall\, i \in I_c.$$

Dans ce cas on a donc :

$$g_i(v_i) = v_i - a_i.$$

Si bien que les formules (3.5) et (3.6) deviennent :

$$\left\{\begin{array}{l} \tilde{u}_i \text{ est définie par} \\ \sum_{j=1}^{m} a_{i,j}\,\tilde{u}_j - f_i = \tilde{\lambda}\,.\,1 \\ \tilde{\lambda} \in \mathbb{R},\ \tilde{\lambda} \geq 0,\ \tilde{u}_i - a_i \geq 0,\ \tilde{\lambda}\,(\tilde{\tilde{u}}_i - a_i) = 0 \end{array}\right. \tag{3.8}$$

lorsque $i \in I_c$ et (3.7) lorsque $i \in I_l$.

La résolution de (3.8) est immédiate : désignons par $\hat{u}_i$ la solution de

(3.9) $$\sum_{j=1}^{i-1} a_{i,j}\,\tilde{u}_j + a_{ii}\,\hat{u}_i + \sum_{j=i+1}^{m} a_{i,j}\,\tilde{u}_j = f_i.$$

(Cas de la méthode de Gauss-Siedel sans contrainte). Alors, ou bien $\hat{u}_i \geq a_i$ et alors $\tilde{u}_i = \hat{u}_i$, ou bien $\hat{u}_i < a_i$ et alors $\tilde{u}_i = a_i$.

Par suite, dans tous les cas, que $i \in I_1$ ou $i \in I_c$, on a :

(3.10) $$\begin{cases} \tilde{u}_i = P_{[a_i,+\infty)}\,\hat{u}_i \\ \hat{u}_i \text{ définie par (3.9)} \end{cases}$$

où $$P_{[a_i,+\infty)}\,\hat{u}_i = \begin{cases} \hat{u}_i & \text{si } \hat{u}_i \in [a_i,\,+\infty) \\ a_i & \text{si } \hat{u}_i < a_i \end{cases}$$

(P est une projection sur un ensemble convexe fermé).

Un autre cas : Supposons que

$$v \in K \Longleftrightarrow a_i \leq v_i \leq f_i, \quad \forall\, i \in I_c$$

on trouverait que :

(3.10)' $$\begin{cases} \tilde{u}_i = P_{[a_i,f_i)}\,\hat{u}_i \\ \hat{u}_i \text{ définie par (3.9).} \end{cases}$$

3.5. Le cas de la programmation quadratique.

Il s'agit du problème suivant : Minimisation de $\frac{1}{2}\,(Av,v)_{\mathbb{R}^m} - (f,v)_{\mathbb{R}^m}$ sous la contrainte $Bv - g \geq 0$, $Bv - g \in \mathbb{R}^m$, B désignant une matrice donnée. On vérifierait que le problème dual du précédent est le suivant :

Minimisation de $\frac{1}{2}\,(B^* v + f,\, A^{-1}\,(B^* v + f))_{\mathbb{R}^m} - (g,v)_{\mathbb{R}^m}$ sous la contrainte $v \geq 0$, $v \in \mathbb{R}^m$. On peut donc appliquer au problème dual de la programmation quadratique la méthode du 3.4 ; on retrouve ici un résultat de HAUGAZEAU.

3.6. Un exemple de contraintes non linéaires.

Soit à minimiser :

$$J(v) = (Av,v)_{\mathbb{R}^{2p}} - 2(f,v)_{\mathbb{R}^{2p}}$$

où f est donné dans $\mathbb{R}^{2p}$ et où les coefficients de la matrice A sont donnés par :

$$(3.11)\quad \begin{cases} a_{i,i} = 2/h^2 + 1 & i = 1, \ldots, 2p \\ a_{i,i+1} = a_{i+1,i} = -1/h^2 & i = 1, \ldots, 2p-1 \\ a_{i,j} = 0 \text{ dans tous les autres cas.} \end{cases}$$

(h représente un paramètre donné).

Les contraintes sont données à l'aide des fonctions g_i

$$(3.12)\quad \begin{cases} g_1(u_1) = 1 - (u_1/h)^2 \\ g_i(u_{2i-2}, u_{2i-1}) = 1 - \left(\dfrac{u_{2i-1} - u_{2i-2}}{h}\right)^2, & i = 2, \ldots, p \\ g_{p+1}(u_{2p}) = 1 - (u_{2p}/h)^2 \end{cases}$$

l'ensemble convexe $\dot{K}$ est défini par :

$$u \in K \Longleftrightarrow u \in \mathbb{R}^{2p}, \quad g_i(u) \geq 0 \qquad \forall\, i = 1, \ldots, p+1.$$

L'examen des contraintes (3.12) montre qu'on peut appliquer la méthode itérative en groupant les variables v_i de la manière suivante :

$$\{v_i\} \; ; \ldots \; ; \{v_{2i-2}, v_{2i-1}\} \; ; \{v_{2p}\}.$$

On posera donc $V_1 = \mathbb{R}$; $V_2 = \ldots = V_p = \mathbb{R}^2$; $V_{p+1} = \mathbb{R}$

et on pourra appliquer la méthode itérative.

Nous allons expliciter les relations analogues aux relations (3.5) :

Cas i = 1.

$\tilde{u}_1$ est définie par :

$$((2/h^2) + 1)\,\tilde{u}_1 - (1/h^2)\,\tilde{u}_2 = -\tilde{\lambda}_1\,(\tilde{u}_1/h^2)$$

$$\tilde{\lambda}_1 \geq 0, \quad 1 - ((\tilde{u}_1/h)^2 \geq 0\,,\; \lambda_1 \,.\, (1 - (\tilde{u}_1/h)^2) = 0$$

système que l'on peut résoudre explicitement.

Cas i = 2, ..., p :

$\tilde{u}_{2i-2}$, $\tilde{u}_{2i-1}$ sont définies par :

$$- (1/h^2)\, \tilde{u}_{2i-3} + (2/h^2 + 1)\, \tilde{u}_{2i-2} - (1/h^2)\, \tilde{u}_{2i-1} = + \lambda_i .\, (\tilde{u}_{2i-1} - \tilde{u}_{2i-2})/h^2$$

$$- (1/h^2)\, \tilde{u}_{2i-2} + (2/h^2+1)\, \tilde{u}_{2i-1} - (1/h^2)\, \tilde{u}_{2i} = - \lambda_i \,.\, (\tilde{u}_{2i-1} - \tilde{u}_{2i-2}) \,/\, h^2$$

$$\lambda_i \geq 0, \quad 1 - ((u_{2i-1} - u_{2i-2})/h)^2 \geq 0 \,, \quad \lambda_i(1 - ((u_{2i-1} - u_{2i-2})/h)^2) = 0$$

On vérifierait qu'on peut résoudre explicitement ce système.

Cas $i = p+1$:

Ce cas est semblable au cas $i = 1$.

4. - Application numérique I :

4.1. Position du problème.

Il s'agit de trouver $u \in H^1(\Omega)$ définie sur un domaine $\Omega \subset \mathbb{R}^2$ de frontière Γ, assez régulière, telle que :

$$(P) \quad \begin{cases} - \Delta u + u = f & \text{dans } \Omega \\ u \geq 0 & \text{sur } \Gamma \\ \dfrac{\partial u}{\partial n} \geq 0 & \text{sur } \Gamma \\ u \,.\, \dfrac{\partial u}{\partial n} = 0 & \text{sur } \Gamma \end{cases}$$

(où $\frac{\partial}{\partial n}$ désigne la dérivée normale ...).

Ce problème est équivalent au problème suivant : (cf. cours J.L. LIONS, Analyse Numérique, Paris 1966-1967) déterminer u vérifiant :

$$\begin{cases} J(u) \leq J(v) \quad \forall\, v \in H^1(\Omega), \quad v(x) \geq 0 \quad \text{p.p. sur } \Gamma. \\ u \in H^1(\Omega) \quad , \quad u(x) \geq 0 \quad \text{p.p. sur } \Gamma \end{cases}$$

où

$$J(v) = \int_\Omega |\text{grad } v(x)|^2 \, dx - 2 \int_\Omega f(x).v(x) \, dx.$$

Dans les essais numériques, on choisira d'abord Ω puis une fonction $u \in H^1(\Omega)$

vérifiant $u \geq 0$, $\frac{\partial u}{\partial n} \geq 0$, $u.\frac{\partial u}{\partial n} = 0$ sur Γ et on calculera f par $f = -\Delta u + u$. Il s'agira ensuite de résoudre le problème (sous forme variationnelle) associé à la fonction f ainsi trouvée. La connaissance de la solution exacte u permettra le calcul de l'erreur.

4.2. La discrétisation.

On emploit la méthode de discrétisation variationnelle comme dans CEA [1]. Il s'introduira un vecteur u représentant les valeurs aux noeuds du réseau de la solution u. Les contraintes $u(x) \geq 0$ p.p. sur Γ se traduiront ici par $u_i \geq 0$ pour les indices i qui repèrent les points frontaliers. Le problème discrétisé sera donc analogue a celui étudié au N° 3.4. Il sera donc résolu par utilisation des formules (3.9) et (3.10).

4.3. Un premier exemple.

Ω est alors le rectangle $0 \leq x \leq 3\pi/2$; $\pi \leq y \leq 3\pi$. La fonction u vaut : $u(x,y) = (1 + \cos y) \cos x$.

Les pas employés dans la discrétisation sont : $h_x = h_y = \pi/10$ si bien qu'il y aura 336 points dont 70 points frontaliers.

Résultats :

Numéro d'itération	$\frac{\|u^k - u\|_{l^2}}{\|u\|_{l^2}}$
20	0,13
40	0,021
60	0,007
80	0,005
100	0,0047

Calculs effectués sur C.D.C. 3600 en 17 secondes.

Dans un autre essai, la valeur $\hat{u}_i$ qui figure dans (3.10) a été calculée en modifiant (3.9) par l'introduction (classique) d'un paramètre de relaxation w. Les résultats sont les suivants :

numéro d'itération	$\dfrac{\lVert u^k - u \rVert_{l^2}}{\lVert u \rVert_{l^2}}$
5	0,20
10	0,08
15	0,03
20	0,007
25	0,0047
27	0,0046

le temps est cette fois de 22 secondes (à cause de la recherche de w optimal).

4.4. Un deuxième exemple :

Ω est le disque $(x - 3R/2)^2 + (y - 3R/2)^2 \leq R^2$ avec $R = 2$.

La fonction u est définie en coordonnées polaires par :

$$u(\rho,\theta) = \begin{cases} \rho^2(3R - \rho)\sin^4 \dfrac{\theta}{m} & \text{pour} \quad 0 \leq \theta \leq m\pi \\ \rho^2(\rho - R)\sin^4 \dfrac{\theta}{2-m} & \text{pour} \quad (m-2)\pi < \theta < 0 \end{cases}$$

on prendra $h_x = h_y = 2/9$ et on fera 3 essais en choisissant $m = 1$, $m = 1/2$, $m = 3/2$.

Résultats :

	Nombre d'itérations	$\dfrac{\lVert u^k - u \rVert_{l^2}}{\lVert u \rVert_{l^2}}$
m = 1	32	0,039
m = 1/2	25	0,178
m = 3/2	21	0,029

temps total : 2 mn 10 s.

5. - Application numérique II.

5.1. Position du problème.

Il s'agit de déterminer $u \in H_0^1(\Omega)$ et vérifiant

$u \in K$

$J(u) \leq J(v) \quad \forall\, v \in K$

où

$$v \in K \Longleftrightarrow v \in H_0^1(\Omega), \qquad |\text{grad } v(x)| \leq 1 \quad \text{p.p. sur } \Omega$$

$$J(v) = \int_\Omega |\text{grad } v(x)|^2\, dx + \int_\Omega |v(x)|^2\, dx - 2\int_\Omega f(x)\, v(x)\, dx.$$

Dans un essai numérique, nous avons choisi $\Omega =]0,3[\subset \mathbb{R}$ si bien que le problème est le suivant : déterminer u $H_0^1(\Omega)$ vérifiant :

$$(5.1) \quad \begin{cases} |u'(x)| \leq 1 \quad \text{p.p. sur } (0,3) \\ J(u) \leq J(v) \\ \forall\, v \in H_0^1(\Omega) \quad \text{vérifiant} \quad |v'(x)| \leq 1 \quad \text{p.p. sur } (0,3) \end{cases}$$

où

$$J(v) = \int_0^3 |v'(x)|^2\, dx + \int_0^3 |v(x)|^2\, dx - 2\int_0^3 f(x)\, v(x)\, dx.$$

5.2. La discrétisation :

On introduit un réseau de maille $h = 3 / (2p + 1)$, p = nombre entier positif. La variable sera un vecteur $v = (v_1, \ldots, v_{2p}) \in \mathbb{R}^{2p}$, on introduira aussi $v_0 = v_{2p+1} = 0$. La fonction économique discrétisée sera désignée par $J_h(v)$

$$J_h(v) = \sum_{i=0}^{2p} ((v_{i+1} - v_i)/h)^2 \,.\, h + \sum_{i=1}^{2p} v_i^2\, h - 2 \sum_{i=1}^{2p} f_i \,.\, v_i h.$$

La discrétisation de la contrainte $|v'(x)| \leq 1$ p.p. $x \in (0,3)$ ou encore $1 - v'^2(x) \geq 0$ p.p. $x \in (0,3)$ se fera de la façon suivante :

$$\begin{cases} 1 - (v_1/h)^2 \geq 0 \\ 1 - ((v_{2i-1} - v_{2i-2})/h)^2 \geq 0 \qquad i = 2, \ldots, p. \\ 1 - (v_{2p}/h)^2 \geq 0. \end{cases}$$

On remarquera que la contrainte a été seulement introduite sur les segments $(2i.h, (2i+1).h)$, $i = 0, \ldots, 2p$; autrement dit un pas sur deux. Ce procédé est général pour les contraintes de type local (par exemple $|\text{grad } u(x)| \leq 1$, $-\Delta u(x) + u(x) - f(x) = 0$)

une étude de cette méthode paraîtra dans un article ultérieur.

La résolution du problème discrétisé se fait comme dans le cadre du N° 3.6.

5.3. Un exemple numérique.

On vérifierait ceci :

si

$$f(x) = \begin{cases} 2 + x(1 - x) & x \in [0,1] \\ -2 + (1 - x)(2 - x) & x \in]1,2] \\ 2 + (x - 2)(3 - x) & x \in]2,3] \end{cases}$$

alors la solution optimale u est la suivante :

$$u(x) = \begin{cases} x(1 - x) & x \in [0,1] \\ (1 - x)(2 - x) & x \in]1,2] \\ (x - 2)(3 - x) & x \in]2,3] \end{cases}$$

le pas est de 1/25 et il y a 76 points.

Résultats :

itérations	$J(u^k) - J(u)$	$\|u^k - u\|_{l^2}$
60	$1{,}8.10^{-2}$	4.10^{-2}
100	$1{,}2.\ 10^{-3}$	$1.\ 10^{-2}$
140	$1{,}2.\ 10^{-4}$	$4.\ 10^{-3}$
180	$2.\ 10^{-5}$	$1{,}8.\ 10^{-3}$
200	$1.\ 10^{-5}$	$1{,}2.\ 10^{-3}$

Remarquons ceci : Le problème précédent a été résolu par la méthode de Gauss-Siedel, généralisée au cas où il y a des contraintes. La résolution par la méthode de Gauss-Siedel classique du problème de minimisation avec la même fonction économique mais sans contrainte nous a conduit à des résultats très voisins des précédents.

Autres résultats : Dans la résolution numérique du problème approché, nous avons introduit un paramètre de relaxation. Les équations qui fournissent $\tilde{u}_{2i-2}$ et $\tilde{u}_{2i-1}$ dans

3.6 sont modifiées en conséquence, lorsque $\lambda_i = 0$, on retrouve alors les formules qui permettent la résolution d'un système linéaire par la méthode de relaxation par blocs de 2 variables avec paramètre optimal. Les résultats sont les suivants :

itérations	$J(u^k) - J(u)$	$\|\|u^k - u\|\|_{l^2}$
50	$1.\ 10^{-4}$	$2.\ 10^{-3}$
60	$3.\ 10^{-5}$	$1.\ 10^{-3}$
70	$1.\ 10^{-5}$	$4.\ 10^{-4}$
80	$1.\ 10^{-6}$	$2.\ 10^{-4}$
100	$1.\ 10^{-7}$	$1.\ 10^{-5}$

BIBLIOGRAPHIE

J.P. AUBIN : Approximation des espaces de distribution et des opérateurs différentiels S. MF 1967.

J. CEA : Approximation variationnelle des problèmes aux limites. Ann. Inst. Fourier 14, 2 (1964).

Y. HAUGAZEAU : Sur la minimisation des formes quadratiques avec contraintes C.R. Acad. Sci. Paris, 13 février 1967.

J.L. LIONS : Cours d'Analyse Numérique. Faculté des Sciences. Paris 1966-67.

R.S. VARGA : Matrix iterative analysis. Prentice Mall - 1963.

MULTIPLE-STAGE QUANTITATIVE GAMES

A. Blaquière · G. Leitmann

This paper contains a discussion of a class of two-player, zero-sum games. The state of the game is determined by a given set of difference equations. A play takes place during a given number of stages and terminates when a state in given set of states is reached at the last stage.

One player seeks to minimize while the other player seeks to maximize a payoff which is additive over the stages of the play. Optimality is expressed by a saddle-point condition. Optimal strategy pairs are those satisfying the saddle-point condition; the corresponding value of the payoff is called the value of the game.

Necessary conditions are derived under certain assumptions on the value of the game and under the assumption of locally directional convexity of the stage-to-stage reachable set of states. These conditions are applied to a simple example.

1. PROBLEM STATEMENT

We shall consider a multiple-stage quantitative game; that is, a quantitative game whose state is characterized by n-1 real numbers $x_1, x_2, \ldots, x_{n-1}$, and by the stage k of the process, where $k \in \{0,1,\ldots,K\}$.

It is convenient to think of the state of the game as a point

$$x \triangleq (x_1, x_2, \ldots, x_n) \in E^n$$

where $x_n \in \{0, 1, \ldots, K\}$.

For $x = x^k \triangleq (x_1, x_2, \ldots, x_{n-1}, k)$ we say that the state is at the k-th stage.

The state changes, i.e., x takes on different values in E^n , as x_n varies from 0 to K . The evolution of the state is governed by a set of difference equations; in particular, we shall suppose that the change of x^k to x^{k+1} depends on x^k and some parameters, the control variables $u_1, u_2, \ldots, u_r$ and $v_1, v_2, \ldots, v_s$, which the players J_P and J_E , respectively, choose at each stage of the play. Let $u \triangleq (u_1, u_2, \ldots, u_r)$ and $v \triangleq (v_1, v_2, \ldots, v_s)$. We shall assume that $u \in U$ and $v \in V$ where U and V are fixed domains of E^r and E^s , respectively. With each change or transfer of the state, we shall associate a real number or cost.

It is our purpose to discuss some geometric aspects of optimal play, and we shall illustrate some consequences of geometric properties by presenting a derivation of necessary conditions for a restricted class of problems. We shall assume, without loss of generality, that the multiple-stage game that we are discussing now is the discrete version of a differential game which has been quantized, and that the stage k is the time variable. Of course, all the results will hold if k is not the time.

Since x_n takes on integer values on $[0, K]$, we shall define

$$E_k \triangleq \{x: x_n = k\}$$

Let D be a fixed domain of E^n, such that $G_k \triangleq D \cap E_k \neq \emptyset$ for $k = 0, 1, \ldots, K$.

In what follows we shall assume that state x belongs to the set $G \triangleq \bigcup_k G_k$, $k \in \{0, 1, \ldots, K\}$.

2. STATE EQUATIONS, STRATEGIES AND TARGET σ

We shall consider the state of the game as a function of stage k, namely $\tilde{x} : k \to x^k = \tilde{x}(k)$, $k \in \{i,i+1,\ldots,j\}$, $0 \leq i \leq j \leq K$, satisfying difference equation

$$\tilde{x}(k+1) - \tilde{x}(k) = f^k(\tilde{x}(k),\tilde{u}(k),\tilde{v}(k)) \qquad (1)$$

where $\tilde{u}$ and $\tilde{v}$ are functions of k, namely $\tilde{u} : k \to u = \tilde{u}(k)$, $\tilde{v} : k \to v = \tilde{v}(k)$, $k \in \{i,i+1,\ldots,j\}$, $0 \leq i \leq j \leq K-1$; and $f^k(x,u,v) \triangleq (f_1^k(x,u,v), f_2^k(x,u,v),\ldots,f_n^k(x,u,v))$

Of course, since $x_n^k = k$, we have $f_n^k(x,u,v) \equiv 1$.

We shall assume that the functions f_ν^k , $\nu = 1,2,\ldots,n-1$ are of class C^1 on $D \times U \times V$, for $k = 0,1,\ldots,K-1$.

Let there be a fixed domain $R \subseteq D$ such that $X_k \triangleq R \cap E_k \neq \emptyset$ for $k = 0,1,\ldots,K$. Let $X \triangleq \bigcup_k X_k$, $k \in \{0,1,\ldots,K-1\}$.

Strategies will be functions of x, $p : x \to p(x)$ and $e : x \to e(x)$, $x \in X$, belonging to prescribed classes, such that $p(x) \in K_u(x) \subseteq U$, $e(x) \in K_v(x) \subseteq V$, where $K_u(x)$ and $K_v(x)$ are given subsets of U and V respectively, which may depend on x .

We shall be interested in transferring the state of the game from an initial state $\tilde{x}(i) = x^i$ to any one in prescribed set of states $\sigma \triangleq X_K$.

3. PATHS IN STATE SPACE E^n

For given strategy pair (p,e) and an initial state $\tilde{x}(i) = x^i$ in G , equation

$$\tilde{x}(k+1) - \tilde{x}(k) = f^k(\tilde{x}(k),p(\tilde{x}(k)),e(\tilde{x}(k))) \qquad (2)$$

defines a unique sequence of states $\Gamma = \{\tilde{x}(i),\tilde{x}(i+1),\ldots,\tilde{x}(i+d)\}$, where $i+d$ is the largest value of stage k for which $\tilde{x}(i+d)$ is defined. Of course, $i+d \leq K$.

If $\tilde{x}(i) = x^i$ does not belong to X, or if $i = K$, or both, $\tilde{x}(i+1)$ is not defined; that is, $d = 0$, and Γ reduces to the single point x^i.

Any subsequence π^{ij} of Γ, namely

$$\pi^{ij} = \{\tilde{x}(i), \tilde{x}(i+1), \ldots, \tilde{x}(j)\} \quad \text{where} \quad j = i+l, \; l \in \{0,1,\ldots,d\}$$

is a <u>path</u>.

Note that

$$\tilde{x}(i+m) \notin \sigma \quad \forall m \in \{0,1,\ldots,l-1\}$$

since $j \leq K$ and $\sigma \subset E_K$.

A point $\tilde{x}(k)$, $k \in \{0,1,\ldots,K\}$, is a null path π^o.

4. COST OF TRANSFER

Next let us consider the transfer of the state from a point $\tilde{x}(i) = x^i$ to a point $\tilde{x}(j) = x^j$, along a path π^{ij} generated by a strategy pair (p,e). With each such transfer we shall associate a <u>cost</u> $V(x^i,x^j;p,e,\pi^{ij})$.

We shall assume that

(i) the cost of transfer from x^k to x^{k+1} along path π^{ij} is given by

$$f_o^k(\tilde{x}(k),\tilde{u}(k),\tilde{v}(k)) \tag{3}$$

where

$$\tilde{u}(k) = p(\tilde{x}(k))$$

$$\tilde{v}(k) = e(\tilde{x}(k))$$

and the functions f_o^k are of class C^1 on $D \times U \times V$ for $k = 0,1,\ldots,K-1$.

(ii) the cost of transfer from x^i to x^j, $j = i + \ell$, along π^{ij} is given by

$$V(x^i,x^j,p,e,\pi^{ij}) \triangleq \sum_{k=i}^{i+\ell-1} f_o^k(\tilde{x}(k),\tilde{u}(k),\tilde{v}(k)) \tag{4}$$

(iii) the cost of transfer associated with a null path is zero; that is,

$$V(x^k,x^k;p,e,\pi^o) = 0 \tag{5}$$

for all $k \in \{0,1,\dots,K\}$ and for all strategy pairs (p,e) .

5. PLAYABLE STRATEGY PAIR

A play is the evolution of state x along a path.

A terminating play is a play whose end state belongs to θ . The corresponding path is called a terminating path.

We shall say that (p,e) is a playable strategy pair at point x^i if it generates a terminating play from initial state x^i . We shall denote by $I(x^i)$ the set of all playable strategy pairs at point x^i .

From the properties of difference equation (2) it follows that each playable strategy pair at point x^i generates a unique terminating play from initial state x^i .

We shall denote by $V(x^i,\theta;p,e)$ the cost of transfer from x^i to θ, along the terminating path π^{if} generated by $(p,e) \in I(x^i)$; namely,

$$V(x^i,\theta;p,e) \triangleq V(x^i,x^f;p,e,\pi^{if}) \tag{6}$$

6. OPTIMAL STRATEGY PAIR, VALUE OF THE GAME

Let there be a fixed set $S \subseteq G$; we shall say that strategy pair (p^*, e^*) is optimal on S if

(i) $(p^*, e^*) \in I(x^1)$ $\forall\, x^1 \in S$

(ii) the saddle-point condition

$$V(x^1, \theta; p^*, e) \leq V(x^1, \theta; p^*, e^*) \leq V(x^1, \theta; p, e^*) \tag{7}$$

holds for all $x^1 \in S$ and for all playable strategy pairs $(p^*, e) \in I(x^1)$ and $(p, e^*) \in I(x^1)$.

ASSUMPTION 1 : There exists a strategy pair (p^*, e^*) that is optimal on X .

By (5), (p^*, e^*) is optimal on $X^* \triangleq X \cup \theta = \bigcup_k X_k$, $k = \{0, 1, \ldots, K\}$.

Since $V(x^1, \theta; p^*, e^*)$ is defined for all $x^1 \in X^*$, $V^*: x^1 \to V^*(x^1)$, where

$$V^*(x^1) \triangleq V(x^1, \theta; p^*, e^*) \tag{8}$$

is a (single valued) function on X^* . We call $V^*(x^1)$ the value of the game for play starting at x^1 .

7. GAME SURFACES

Let us introduce another variable x_o , and consider an (n+1)-dimensional Euclidean space E^{n+1} of points $\mathfrak{x}$, where $\mathfrak{x} = (x_o, x) = (x_o, x_1, \ldots, x_n)$ is the vector which defines a point relative to a rectangular coordinate system in E^{n+1} , the augmented state space.

Since $V^* : x^1 \to V^*(x^1)$ is defined on X^* , we can define a game surface $\Sigma(C)$ in $\mathfrak{X}^* \triangleq X^* \times \{x_o\}$ by

$$\Sigma(C) \triangleq \{\mathfrak{x} : x_o + V^*(x) = C\} \tag{9}$$

where C is a constant parameter.

As the value of parameter C is varied, equation (9) defines a one-parameter family of surfaces, namely $\{\Sigma(C)\}$.

For each given C , Eq. (9) defines a single-sheeted surface in $\mathfrak{X}^*$; that is, $\Sigma(C)$ is a set of points which are in one-to-one correspondence with the points of X^* .

Consider now two game surfaces $\Sigma(C_1)$ and $\Sigma(C_2)$, corresponding to two different parameter values C_1 and C_2 , respectively. Let x_{o1} and x_{o2} denote the values of x_o on $\Sigma(C_1)$ and $\Sigma(C_2)$, respectively, for the same state. Then it follows from (9) that

$$x_{o1} - x_{o2} = C_1 - C_2 \qquad \forall\, x \in X^*$$

Consequently the members of the one-parameter family $\{\Sigma(C)\}$ may be deduced from one another by translation parallel to the x_o-axis. Furthermore these surfaces are ordered along the x_o-axis in the same way as the parameter value C ; that is, the "higher" a surface in the family the greater is the parameter value C . Thus one and only one surface $\Sigma(C)$ passes through a given point $\mathfrak{x}$ in $\mathfrak{X}^*$.

A given game surface $\Sigma(C)$ separates the set $\mathfrak{X}^*$ into two disjoint sets; that is, sets which have no point in common. We shall denote theses sets by $A/\Sigma(C)$ ("above" $\Sigma(C)$) and $B/\Sigma(C)$ ("below" $\Sigma(C)$), respectively. For a game surface $\Sigma(C)$ corresponding to parameter value C , we have

$$A/\Sigma(C) \triangleq \{\mathfrak{x} : x_o > C - V^*(x)\} \qquad (10)$$

$$B/\Sigma(C) \triangleq \{\mathfrak{x} : x_o < C - V^*(x)\} \qquad (11)$$

A point $\mathfrak{x} \in A/\Sigma(C)$ will be called an A-point relative to $\Sigma(C)$, and a point $\mathfrak{x} \in B/\Sigma(C)$ a B-point relative to $\Sigma(C)$.

8. A DECOMPOSITION OF GAME SURFACES

Further, we shall define a surface $\Sigma_k(C)$, namely

$$\Sigma_k(C) \triangleq \Sigma(C) \cap \tilde{E}_k , \quad \tilde{E}_k \triangleq E_k \times \{x_o\} \tag{12}$$

for $k = 0,1,\ldots,K$.

$$\{\Sigma_o(C), \Sigma_1(C),\ldots, \Sigma_K(C)\}$$

constitutes a decomposition of game surface $\Sigma(C)$.

A surface $\Sigma_k(C)$, $k \in \{0,1,\ldots,K\}$, is a single sheeted surface in $\mathfrak{X}_k \triangleq X_k \times \{x_o\}$ that is a set of points which are in one-to-one correspondence with the points of X_k.

A surface $\Sigma_k(C)$, $k \in \{0,1,\ldots,K\}$, separates $\mathfrak{X}_k$ into two disjoint sets $A/\Sigma_k(C)$ and $B/\Sigma_k(C)$, where

$$A/\Sigma_k(C) \triangleq (A/\Sigma(C)) \cap \tilde{E}_k$$
$$B/\Sigma_k(C) \triangleq (B/\Sigma(C)) \cap \tilde{E}_k$$

The equation of $\Sigma_k(C)$ is deduced from (9). It is

$$\Phi(\mathfrak{x}) \triangleq x_o + V^*(x) = C , \qquad x \in X_k \tag{13}$$

where $V^*(x)$ is the value of the game for play starting at point x and, accordingly

$$\Sigma_k(C) = \{\mathfrak{x} = (x_o,x) : x \in X_k , \quad \Phi(\mathfrak{x}) = C\} \tag{14}$$

$$A/\Sigma_k(C) = \{\mathfrak{x} = (x_o,x) : x \in X_k , \quad \Phi(\mathfrak{x}) > C\} \tag{15}$$

$$B/\Sigma_k(C) = \{\mathfrak{x} = (x_o,x) : x \in X_k , \quad \Phi(\mathfrak{x}) < C\} \tag{16}$$

As the value of parameter C is varied, Eq. (13) defines a one-parameter family of surfaces, namely $\{\Sigma_k(C)\}$.

For a given C, we shall say that $\mathfrak{x}^k \in \Sigma_k(C)$ is an interior point of $\Sigma_k(C)$ if there exists an n-dimensional ball $\Delta(\mathfrak{x}^k)$ with center $\mathfrak{x}^k$ such that $\Delta(\mathfrak{x}^k) \subset \mathfrak{X}^k$.

Since R is open in E^n, $X_k \triangleq R \cap E_k$ is open in E_k and $\mathfrak{X}_k \triangleq X_k \times \{x_o\}$ is open in $E_k \times \{x_o\}$. Hence, every point of $\Sigma_k(C)$ is an interior point of $\Sigma_k(C)$, $k = 0,1,\ldots,K$.

9. PATHS IN AUGMENTED STATE SPACE E^{n+1}

Let us define paths $\Pi^{is}(C)$ in $\mathfrak{G} \triangleq G \times \{x_o\}$, namely

$$\Pi^{is}(C) \triangleq \{\mathfrak{x}^k : x_o^k + V(x^k, x^s; p, e, \pi^{ks}) = C\,, \quad \pi^{ks} \subset \pi^{is}\} \tag{17}$$

where π^{is} is a path in G from x^i to x^s generated by strategy pair (p,e), and C is a constant parameter.

Thus path π^{is} is the projection on G of paths $\Pi^{is}(C)$ in $\mathfrak{G}$.

By varying the value of parameter C in (17) one generates a one-parameter family of paths $\{\Pi^{is}(C)\}$. This family belongs to an x_o-cylindrical surface whose intersection with G is path π^{is}.

Let us also define three families of paths emanating from points in $\mathfrak{X}^*$; namely, $\{\Pi_P^{ij}(C)\}$, $\{\Pi_E^{lm}(C)\}$, and $\{\Pi_{PE}^{nq}(C)\}$. Let

$$\Pi_P^{ij}(C) \triangleq \{\mathfrak{x}^k : x_o^k + V(x^k, x^j, p^*, e, \pi_P^{kj}) = C\,, \quad \pi_P^{kj} \subset \pi_P^{ij}\} \tag{18}$$

$$\Pi_E^{lm}(C) \triangleq \{\mathfrak{x}^k : x_o^k + V(x^k, x^m; p, e^*, \pi_E^{km}) = C\,, \quad \pi_E^{km} \subset \pi_E^{lm}\} \tag{19}$$

$$\Pi_{PE}^{nq}(C) \triangleq \{\mathfrak{x}^k : x_o^k + V(x^k, x^q; p^*, e^*, \pi_{PE}^{kq}) = C\,, \quad \pi_{PE}^{kq} \subset \pi_{PE}^{nq}\} \tag{20}$$

Paths π_P^{ij}, π_E^{lm} and π_{PE}^{nq} are paths in G, emanating from points x^i, x^l and x^n of X^*, generated by the strategy pairs (p^*,e), (p,e^*) and (p^*,e^*) and ending at points x^j, x^m and x^q, respectively.

Since (p^*,e^*) is playable at all points $x^n \in X^*$, there exists a path $\Pi_{PE}^{nf}(C)$ which reaches $\Theta \triangleq \mathcal{O} \times \{x_o\}$ at point $\mathfrak{x}^f$. It is called an <u>optimal path</u>.

Hereafter we shall denote an optimal path by $\Pi^*(C)$.

When (p^*,e) and (p,e^*) are playable strategy pairs at points x^i and x^ℓ , respectively, there also exist paths $\Pi_P^{if}(C)$ and $\Pi_E^{lf}(C)$ which reach θ . They are called P-optimal and E-optimal paths, respectively.

Hereafter we shall write simply $\Pi_P(C)$ and $\Pi_E(C)$ to indicate theses paths.

Now let $\tilde{x}_o : k \to x_o = \tilde{x}_o(k)$, $k \in \{i,i+1,\dots,j\}$, $0 \leq i \leq j \leq K$, be a function satisfying difference equation

$$\tilde{x}_o(k) = \tilde{x}_o(k-1) + f_o^{k-1}(\tilde{x}(k-1),\tilde{u}(k-1),\tilde{v}(k-1)) \tag{21}$$

Equations (1) and (21) constitute a set of n+1 scalar difference equations which we shall write in vector form

$$\tilde{r}(k+1) - \tilde{r}(k) = F^k(\tilde{r}(k),\tilde{u}(k),\tilde{v}(k)) \tag{22}$$

where $\tilde{r} : k \to r^k = \tilde{r}(k)$, $r^k \triangleq (x_o,x_1,\dots,x_{n-1},k)$, $\tilde{r} \triangleq (\tilde{x}_o,\tilde{x}_1,\dots,\tilde{x}_{n-1},\tilde{x}_n)$

$F^k(r,u,v) \triangleq (f_o^k(x,u,v),\dots,f_{n-1}^k(x,u,v),1)$.

From the definition of paths in $\mathfrak{C}$ it follows that the motion of state r along a path in $\mathfrak{C}$ is governed by Eq. (22), and

(i) if points $\tilde{r}(k)$ and $\tilde{r}(k+1)$ belong to a path $\Pi_P^{ij}(C)$ we have

$$\tilde{r}(k+1) - \tilde{r}(k) = F^k(\tilde{r}(k),\tilde{u}^*(k),\tilde{v}(k)) \tag{23}$$

(ii) if points $\tilde{r}(k)$ and $\tilde{r}(k+1)$ belong to a path $\Pi_E^{lm}(C)$ we have

$$\tilde{r}(k+1) - \tilde{r}(k) = F^k(\tilde{r}(k),\tilde{u}(k),\tilde{v}^*(k)) \tag{24}$$

(iii) if points $\tilde{r}(k)$ and $\tilde{r}(k+1)$ belong to a path $\Pi_{PE}^{nq}(C)$ we have

$$\tilde{r}(k+1) - \tilde{r}(k) = F^k(\tilde{r}(k),\tilde{u}^*(k),\tilde{v}^*(k)) \tag{25}$$

where

$$\tilde{u}(k) = p(\tilde{x}(k)) \qquad \tilde{u}^*(k) = p^*(\tilde{x}(k))$$
$$\tilde{v}(k) = e(\tilde{x}(k)) \qquad \tilde{v}^*(k) = e^*(\tilde{x}(k))$$

10. SOME PROPERTIES OF GAME SURFACES AND PATHS

From the additivity property of cost one can prove

LEMMA 1 : No point of a path $\Pi_P^{is}(C')$ which emanates from $\mathfrak{r}^i$ is an A-point relative to the game surface through $\mathfrak{r}^i$; and

LEMMA 2 : No point of a path $\Pi_E^{is}(C'')$ which emanates from $\mathfrak{r}^i$ is a B-point relative to the game surface through $\mathfrak{r}^i$.

Lemmas 1 and 2 have

COROLLARY 1 : All points in $\mathfrak{X}^*$ of a path $\Pi_{PE}^{is}(C''')$ which emanates from $\mathfrak{r}^i$ belong to the game surface through $\mathfrak{r}^i$.

This corollary is a direct consequence of Lemmas 1 and 2 together with the definitions of A and B-points relative to a game surface.

COROLLARY 2 : A path $\Pi_P^{is}(C')$ whose initial point is a B-point relative to $\Sigma(C)$ has no A-point relative to $\Sigma(C)$ and, of course, no point in $\Sigma(C)$.

COROLLARY 3 : A path $\Pi_E^{is}(C'')$ whose initial point is an A-point relative to $\Sigma(C)$ has no B-point relative to $\Sigma(C)$ and, of course, no point in $\Sigma(C)$.

These corollaries are direct consequences of Lemmas 1 and 2, respectively, together with the translation property of game surfaces.

11. SETS Ω_P AND Ω_E

Now consider optimal path $\Pi^*(C)$ which starts at point $\mathfrak{r}^i = (x_o, x^i) \in \mathfrak{X}^*$, $\mathfrak{X}^* \triangleq X^* \times \{x_o\}$, is generated by (p^*, e^*), and is given by $\tilde{\mathfrak{r}}^* : k \longrightarrow \mathfrak{r}^k = \tilde{\mathfrak{r}}^*(k)$, $k \in \{i, i+1, \ldots, K\}$, $\tilde{\mathfrak{r}}^* = (\tilde{x}_o^*, \tilde{x}^*) = (\tilde{x}_o^*, \tilde{x}_1^*, \ldots, \tilde{x}_n^*)$.

One can prove easily that

$$\tilde{\mathfrak{r}}^*(j) \in \mathfrak{X}^* \quad \forall\, j \in \{i, i+1, \ldots, K\} \tag{26}$$

It follows that

$$\Pi^*(C) \subset \mathfrak{X}^* . \tag{27}$$

Finally, from (27) and from Corollary 1 it follows that $\Pi^*(C)$ belongs to the Game Surface through x^i. Now let us consider non-null paths $\Pi_E^{jl}(C')$ and $\Pi_P^{jm}(C'')$ which emanate from $\tilde{\mathfrak{r}}^*(j)$, $j < K$ and which are generated by strategy pairs (p, e^*) and (p^*, e) respectively. Let us represent $\Pi_E^{jl}(C')$ and $\Pi_P^{jm}(C'')$ by $\tilde{\mathfrak{r}}_E : k \longrightarrow \mathfrak{r}^k = \tilde{\mathfrak{r}}_E(k)$, $k \in \{j, j+1, \ldots, l\}$ and $\tilde{\mathfrak{r}}_P : k \longrightarrow \mathfrak{r}^k = \tilde{\mathfrak{r}}_P(k)$, $k \in \{j, j+1, \ldots, m\}$, respectively.

From Lemmas 1 and 2 it follows that $\Pi_P^{jm}(C'')$ has no A-point and $\Pi_E^{jl}(C')$ has no B-point, relative to $\Sigma(C)$.

Since $\tilde{\mathfrak{r}}^*(j)$ belongs to $\mathfrak{X}^*$ and $j < K$, points $\tilde{\mathfrak{r}}_P(j+1)$ and $\tilde{\mathfrak{r}}_E(j+1)$ of path $\Pi_P^{jm}(C'')$ and $\Pi_E^{jl}(C')$, respectively, are defined. From the properties of these paths, that is from Lemmas 1 and 2, we deduce that

$$\tilde{\mathfrak{r}}_P(j+1) \notin A/\Sigma_{j+1}(C) \tag{28}$$

$$\tilde{\mathfrak{r}}_E(j+1) \notin B/\Sigma_{j+1}(C) \tag{29}$$

Next let us define the sets $\Omega_P(j+1)$ and $\Omega_E(j+1)$, $j = 0, 1, \ldots, K-1$; namely,

$$\Omega_P(j+1) \triangleq \{\tilde{\mathfrak{r}}_P(j+1) \quad \text{for all strategy pairs } (p^*, e)\} \tag{30}$$

$$\Omega_E(j+1) \triangleq \{\tilde{\mathfrak{r}}_E(j+1) \quad \text{for all strategy pairs } (p, e^*)\} \tag{31}$$

From (23) and (24) we have

$$\Omega_P(j+1) = \{\tilde{r}(j+1) : \tilde{r}(j+1) = \tilde{r}^*(j) + F^j(\tilde{r}^*(j),\tilde{u}^*(j),\tilde{v}(j)) \quad \forall\ \tilde{v}(j) \in K_v(\tilde{x}^*(j))\} \tag{32}$$

$$\Omega_E(j+1) = \{\tilde{r}(j+1) : \tilde{r}(j+1) = \tilde{r}^*(j) + F^j(\tilde{r}^*(j),\tilde{u}(j),\tilde{v}^*(j)) \quad \forall\ \tilde{u}(j) \in K_u(\tilde{x}^*(j))\} \tag{33}$$

and from (28) and (29), since j is arbitrary, it follows that

$$\Omega_P(j+1) \cap (A/\Sigma_{j+1}(C)) = \emptyset \tag{34}$$

and

$$\Omega_E(j+1) \cap (B/\Sigma_{j+1}(C)) = \emptyset \tag{35}$$

for all $j \in \{0,1,\dots,K-1\}$. Clearly

$$\tilde{r}^*(j+1) \in \Omega_P(j+1) \cap \Omega_E(j+1) .$$

ASSUMPTION 2 : Set $\Omega_E(j+1)$ is x_0^+-directionally convex, and set $\Omega_P(j+1)$ is x_0^--directionally convex, for $j = i,i+1,\dots,K-1$.

12. LEMMA 3

Let us represent by $\tilde{x}^* : k \longrightarrow x^k = \tilde{x}^*(k)$, $k \in \{i,i+1,\dots,K\}$, the projection π^* of $\Pi^*(C)$ on X^* .

ASSUMPTION 3 : At every point $\tilde{x}^*(k)$, $k = i,i+1,\dots,K-1$, there exists a neighborhood $\Delta(\tilde{x}^*(k))$ in X_k of $\tilde{x}^*(k)$ on which p^* and e^* are of class C^1 .

Since strategy pair (p^*,e^*) is optimal on X , it is optimal on $\Delta(\tilde{x}^*(k))$, $k = i,i+1,\dots,K-1$.

From (4) and the definition of the value of the game, we have

$$V^*(x^k) = \sum_{\nu=k}^{K-1} f_0^\nu(\tilde{x}(\nu),p^*(\tilde{x}(\nu)),e^*(\tilde{x}(\nu))) \tag{36}$$

$$\forall\ x^k \in \Delta(\tilde{x}^*(k))$$

where $\tilde{x} : \nu \longrightarrow x^\nu = \tilde{x}(\nu)$ is the solution of difference equation

$$\tilde{x}(\nu+1) - \tilde{x}(\nu) = f^\nu(\tilde{x}(\nu), p^*(\tilde{x}(\nu)), e^*(\tilde{x}(\nu)))$$

with initial condition $\tilde{x}(k) = x^k$, defined on $\{k,k+1,\ldots,K\}$.

Then we have

LEMMA 3 : At all points $\tilde{\mathfrak{r}}^*(k)$, $k = i,i+1,\ldots,K$, of optimal path $\Pi^*(C)$ on game surface $\Sigma(C)$, grad $\Phi(\mathfrak{r}^k)$ is defined, where

$$\text{grad } \Phi(\mathfrak{r}^k) \triangleq (1, \frac{\partial V^*}{\partial x_1}, \ldots, \frac{\partial V^*}{\partial x_{n-1}}, 0)$$

For $k = i,i+1,\ldots,K-1$, Lemma 3 is a direct consequence of Assumption 3, together with (36) and with the fact that the functions f_o^ν , $\nu = k,k+1,\ldots,K-1$, are of class C^1 on $D \times U \times V$.

For $k = K$, it follows from (5) that $V^*(x^K) = 0$ for all $x^K \in \mathcal{O}$, and hence

$$\text{grad } \Phi(\mathfrak{r}^k) = (1,0,\ldots,0) .$$

13. VARIATIONAL DIFFERENCE EQUATION

ASSUMPTION 4 : Matrix M^k given by

$$M^k \triangleq I + \left.\frac{dF^k(\mathfrak{r}^k, p^*(x^k), e^*(x^k))}{d\mathfrak{r}}\right|_{\mathfrak{r}^k = \tilde{\mathfrak{r}}^*(k)}$$

where $\frac{dF^k}{d\mathfrak{r}} = \frac{\partial F^k}{\partial \mathfrak{r}} + \frac{\partial F^k}{\partial u}\frac{\partial p^*}{\partial \mathfrak{r}} + \frac{\partial F^k}{\partial v}\frac{\partial e^*}{\partial \mathfrak{r}}$ and I is the $n + 1$ unit matrix, is defined and nonsingular for $k = i,i+1,\ldots,K-1$.

$$\frac{\partial F^k}{\partial \mathfrak{r}} \triangleq \left[\frac{\partial f_\nu^k(\mathfrak{r},u,v)}{\partial x_\alpha}\right] \qquad \nu,\alpha = 0,1,\ldots,n$$

$$\frac{\partial p^*}{\partial \mathfrak{r}} \triangleq \left[\frac{\partial p_\nu^*(x)}{\partial x_\alpha}\right] \qquad \begin{matrix}\nu = 1,2,\ldots,r\\ \alpha = 0,1,\ldots,n\end{matrix}$$

$$\frac{\partial e^*}{\partial \mathfrak{r}} \triangleq \left[\frac{\partial e_\nu^*(x)}{\partial x_\alpha}\right] \qquad \begin{matrix}\nu = 1,2,\ldots,s\\ \alpha = 0,1,\ldots,n\end{matrix}$$

evaluated at $\mathfrak{r}^k = \tilde{\mathfrak{r}}^*(k)$, $u = p^*(\tilde{x}^*(k))$, $v = e^*(\tilde{x}^*(k))$, $k = i,i+1,\ldots,K-1$.

Now consider the <u>variational difference equation</u>

$$\tilde{\eta}(k+1) - \tilde{\eta}(k) = \left[\left.\frac{dF^k(\mathfrak{r}^k,p^*(x^k),e^*(x^k))}{d\mathfrak{r}}\right|_{\mathfrak{r}^k=\tilde{\mathfrak{r}}^*(k)} \right] \tilde{\eta}(k) \tag{37}$$

for $k = i,i+1,\ldots,K-1$.

For given $\tilde{\eta}(i) = \eta^i$, we have

$$\tilde{\eta}(k+1) = M^k\tilde{\eta}(k) = M^k(M^{k-1}\tilde{\eta}(k-1)) = M^kM^{k-1}\ldots M^i\tilde{\eta}(i) \ .$$

Let

$$A^k \triangleq M^kM^{k-1} \ldots M^i$$

for $k = i,i+1,\ldots,K-1$.

Since matrices M^k , $k = i,i+1,\ldots,K-1$, are assumed nonsingular, matrices A^k , $k = i,i+1,\ldots,K-1$, are nonsingular. Hence the linear transformation

$$\tilde{\eta}(k+1) = A^k\tilde{\eta}(i) \ , \quad k = i,i+1,\ldots,K-1 \tag{38}$$

is nonsingular. Then one can prove the following.

LEMMA 4 : If **Assumptions** 3 and 4 are satisfied, then the transform due to linear transformation A^{k-1} of the tangent plane $P_{\Sigma_i}(\tilde{r}^*(i))$ of $\Sigma_i(C)$ at $\tilde{r}^*(i)$ is the tangent plane $P_{\Sigma_k}(\tilde{r}^*(k))$ of $\Sigma_k(C)$ at $\tilde{r}^*(k)$; that is,

$$P_{\Sigma_k}(\tilde{r}^*(k)) = A^{k-1} P_{\Sigma_i}(\tilde{r}^*(i)) \qquad \text{for } k = i,i+1,\dots,K .$$

14. ADJOINT EQUATIONS

Next let us introduce $\tilde{\lambda}(k) \in E^{n+1}$, $k = i,i+1,\dots,K$, satisfying the adjoint equations

$$\tilde{\lambda}(k+1) - \tilde{\lambda}(k) = -\left[\left.\frac{dF^k(r^k,p^*(x^k),e^*(x^k))}{dr}\right|_{r^k=\tilde{r}^*(k)}\right]^T \tilde{\lambda}(k+1) \qquad (39)$$

for $k = i,i+1,\dots,K-1$.

These equations can be rewritten as

$$\tilde{\lambda}(k) = \left[I + \left.\frac{dF^k(r^k,p^*(x^k),e^*(x^k))}{dr}\right|_{r^k=\tilde{r}^*(k)} \right]^T \tilde{\lambda}(k+1)$$

for $k = i,i+1,\dots,K-1$.

Since matrices $I + \left.\frac{dF^k(r^k,p^*(x^k),e^*(x^k)}{dr}\right|_{r^k=\tilde{r}^*(k)}$, $k = i,i+1,\dots,K-1$ are assumed to be nonsigular , so are matrices $\left[I + \left.\frac{dF^k(r^k,p^*(x^k),e^*(x^k))}{dr}\right|_{r^k=\tilde{r}^*(k)}\right]^T$. Thus their inverses exists; that is,

$$\tilde{\lambda}(k+1) = \left\{ \left[I + \left.\frac{dF^k(r^k,p^*(x^k),e^*(x^k))}{dr}\right|_{r^k=\tilde{r}^*(k)} \right]^T \right\}^{-1} \tilde{\lambda}(k) \qquad (40)$$

for $k = i,i+1,\dots,K-1$.

Hence, vectors $\tilde{\lambda}(k)$, $k = i,i+1,\ldots,K$, are defined for any given $\tilde{\lambda}(i)$ or $\tilde{\lambda}(k)$.

From Eq. (40) and variational equation (37) it follows that

$$\tilde{\lambda}(i)\cdot\tilde{\eta}(i) = \tilde{\lambda}(i+1)\cdot\tilde{\eta}(i+1) = \ldots = \tilde{\lambda}(K)\cdot\tilde{\eta}(K) . \tag{41}$$

Since x_o and x_n do not occur in the argument of $F^k(\mathfrak{r}^k, p^*(x^k), e^*(x^k))$, $k = 0,1,\ldots,K-1$, we deduce from (39) that

$$\tilde{\lambda}_o(k+1) - \tilde{\lambda}_o(k) = 0 \; ; \quad \text{and}$$

$$\tilde{\lambda}_n(k+1) - \tilde{\lambda}_n(k) = 0$$

for $k = i,i+1,\ldots,K-1$.

It follows that

$$\tilde{\lambda}_o(i) = \tilde{\lambda}_o(i+1) = \ldots = \tilde{\lambda}_o(K) \tag{42}$$

$$\tilde{\lambda}_n(i) = \tilde{\lambda}_n(i+1) = \ldots = \tilde{\lambda}_n(K) \tag{43}$$

$$i = 0,1,\ldots,K .$$

From (41) we can draw the conclusion that, if initial vectors $\lambda^i = \tilde{\lambda}(i)$ and $\eta^i = \tilde{\eta}(i)$ are chosen in such a way that

$\lambda^i\cdot\eta^i = 0$ with $\|\lambda^i\|\|\eta^i\| \neq 0$

that is, if λ^i is perpendicular to η^i , then

$$\tilde{\lambda}(k)\cdot\tilde{\eta}(k) = 0 \qquad k = i,i+1,\ldots,K . \tag{44}$$

Consequently, $\tilde{\lambda}(k)$ is perpendicular to $\tilde{\eta}(k)$ at all points of $\Pi^*(C)$.

Also if $P(\tilde{\mathfrak{r}}^*(i))$ is an $(n-1)$-dimensional plane containing point $\tilde{\mathfrak{r}}^*(i)$ of $\Pi^*(C)$, and if $\lambda^i = \tilde{\lambda}(i)$ is chosen perpendicular to that plane, then $\tilde{\lambda}(k)$ is perpendicular to the transform

$$P(\tilde{\mathfrak{r}}^*(k)) = A^{k-1}P(\tilde{\mathfrak{r}}^*(i)) \tag{45}$$

of $P(\tilde{\mathfrak{r}}^*(i))$ due to linear transformation A^{k-1} , $k = i,i+1,\ldots,K$.

Finally, from (44) and Lemma 4, it follows that if $\lambda^i = \tilde{\lambda}(i)$ is chosen perpendicular to the tangent plane $P_{\Sigma_i}(\tilde{\mathfrak{r}}^*(i))$ of $\Sigma_i(C)$ at $\tilde{\mathfrak{r}}^*(i)$, then $\tilde{\lambda}(k)$ is perpendicular to the tangent plane $P_{\Sigma_k}(\tilde{\mathfrak{r}}^*(k))$ of $\Sigma_k(C)$ at $k = i, i+1, \ldots, K$.

REMARK: If the control constraints are inequalities independent of state x, the adjoint equation reduces to

$$\tilde{\lambda}(k+1) - \tilde{\lambda}(k) = -\left(\frac{\partial F^k}{\partial \mathfrak{r}}\right)^T \tilde{\lambda}(k+1) . \tag{46}$$

15. GRADIENT AND ADJOINT VECTOR

The x_0-component and the x_n-component of $\tilde{\lambda}(k)$ are constant. Thus if we choose $\tilde{\lambda}_0(i) = 1$ and $\tilde{\lambda}_n(i) = 0$, we have $\tilde{\lambda}_0(k) = 1$ and $\tilde{\lambda}_n(k) = 0$ for $k = i, i+1, \ldots, K$. Furthermore, if we choose $\tilde{\lambda}(i)$ perpendicular to $P_{\Sigma_i}(\tilde{\mathfrak{r}}^*(i))$, then, as pointed out above, $\tilde{\lambda}(k)$ is perpendicular to $P_{\Sigma_k}(\tilde{\mathfrak{r}}^*(k))$ for $k = i, i+1, \ldots, K$.

Clearly, this choice of $\tilde{\lambda}(i)$ implies that $\tilde{\lambda}(i) = \text{grad}\ \Phi(\mathfrak{r}^i)$ at point $\mathfrak{r}^i = \tilde{\mathfrak{r}}^*(i)$.

Since $\text{grad}\ \Phi(\mathfrak{r}^k)$ at point $\mathfrak{r}^k = \tilde{\mathfrak{r}}^*(k)$ and $\tilde{\lambda}(k)$

(i) belong to $E_k \times \{x_0\}$,

(ii) have zeroth component equal to 1,

(iii) are perpendicular to $P_{\Sigma_k}(\tilde{\mathfrak{r}}^*(k))$

for $k = i, i+1, \ldots, K$

it follows that

$$\tilde{\lambda}(k) = \text{grad}\ \Phi(\mathfrak{r}^k) \text{ at point } \mathfrak{r}^k = \tilde{\mathfrak{r}}^*(k) \tag{47}$$

for $k = i, i+1, \ldots, K$.

16. TRANSVERSALITY CONDITIONS

At the terminal point $\tilde{\mathfrak{r}}^*(K)$ of optimal path $\Pi^*(C)$ we have

$$\tilde{\lambda}(K) \cdot \tilde{\eta}(K) = 0 \qquad \forall\ \tilde{\eta}(K) \in P_{\Sigma_K}(\tilde{\mathfrak{r}}^*(K)) .$$

The intersection of game surface $\Sigma(C)$ with Θ is

$$\Theta \cap \Sigma(C) = \{\mathfrak{r} : \Phi(\mathfrak{r}) , \quad x \in \sigma\} .$$

Because of (5) the condition $x \in \sigma$ implies that

$$\Phi(\mathfrak{r}) \triangleq x_o + V^*(x) = x_o .$$

Hence we have

$$\Theta \cap \Sigma(C) = \{\mathfrak{r}: x_o = C , \quad x \in \sigma\} .$$

Consequently, $\Theta \cap \Sigma(C)$ may be deduced from σ by translation parallel to the x_o-axis, and since $\sigma \triangleq X_K \triangleq R \cap E_K$, $P_{\Sigma_K}(\tilde{\mathfrak{r}}^*(K))$ is an (n-1)-dimensional plane perpendicular to the x_o-axis.

It follows that

$$\tilde{\lambda}_1(K) = \ldots = \tilde{\lambda}_{n-1}(K) = \tilde{\lambda}_n(K) = 0 \tag{48}$$

at the terminal point of $\Pi^*(C)$.

These are the transversality conditions at the point where optimal path $\Pi^*(C)$ reaches Θ .

17. A MIN-MAX PRINCIPLE

Upon introducing the Hamiltonian

$$\mathfrak{H}^{k+1}(\tilde{\lambda}(k+1),\tilde{\mathfrak{r}}(k),\tilde{u}(k),\tilde{v}(k)) \triangleq \tilde{\lambda}(k+1)\cdot F^k(\tilde{\mathfrak{r}}(k),\tilde{u}(k),\tilde{v}(k))$$

for $k = i,i+1,\ldots,K-1$, one can deduce from the geometric arguments of the preceding sections the following theorem.

THEOREM 1 : Let $\Pi^*(C)$ be an optimal path, generated by strategies p^*,e^*, represented by $\tilde{\mathfrak{r}}^* : k \longrightarrow \mathfrak{r}^k = \tilde{\mathfrak{r}}^*(k)$, and let π^* be its projection on X^* represented by $\tilde{x}^* : k \longrightarrow x^k = \tilde{x}^*(k)$, $k \in \{i,i+1,\ldots,K\}$. If Assumptions 1-4 are satisfied, then there exist nonzero vectors $\tilde{\lambda}(k)$, $k = i,i+1,\ldots,K$ which satisfy adjoint equation (39), such that

(a) $$\min_{\tilde{u}(k) \in K_u(\tilde{x}^*(k))} \mathfrak{H}^{k+1}(\tilde{\lambda}(k+1), \tilde{r}^*(k), \tilde{u}(k), e^*(\tilde{x}^*(k)))$$

$$= \max_{\tilde{v}(k) \in K_v(\tilde{x}^*(k))} \mathfrak{H}^{k+1}(\tilde{\lambda}(k+1), \tilde{r}^*(k), p^*(\tilde{x}^*(k)), \tilde{v}(k))$$

$$= \mathfrak{H}^{k+1}(\tilde{\lambda}(k+1), \tilde{r}^*(k), p^*(\tilde{x}^*(k)), e^*(\tilde{x}^*(k)))$$

for $k = i, i+1, \ldots, K-1$

(b) $\tilde{\lambda}_o(i) = \tilde{\lambda}_o(i+1) = \ldots = \tilde{\lambda}_o(K) > 0$

(c) $\tilde{\lambda}_n(i) = \tilde{\lambda}_n(i+1) = \ldots = \tilde{\lambda}_n(K) = 0$

(d) $\tilde{\lambda}_1(K) = \ldots = \tilde{\lambda}_{n-1}(K) = \tilde{\lambda}_n(K) = 0$.

18. EXAMPLE

Let us consider a multiple-stage game governed by

$$\begin{aligned} \tilde{x}_1(k+1) - \tilde{x}_1(k) &= \tilde{x}_2(k) \\ \tilde{x}_2(k+1) - \tilde{x}_2(k) &= \tilde{u}(k) + \tilde{v}(k) \\ \tilde{x}_3(k+1) - \tilde{x}_3(k) &= 1 \end{aligned} \tag{49}$$

Let

$$\tilde{x}_o(k+1) - \tilde{x}_o(k) = \tilde{x}_1(k)\tilde{x}_2(k) + \tilde{u}^2(k) - \tilde{v}^2(k) \tag{50}$$

and

$$K_u = K_v = E^1 .$$

A path in augmented state-space E^4, generated by a strategy pair that satisfies the necessary conditions for optimality, will be represented by $\tilde{r}^* : k \rightarrow r^k = \tilde{r}^*(k)$, where $\tilde{r}^* \triangleq (\tilde{x}_o^*, \tilde{x}_1^*, \tilde{x}_2^*, \tilde{x}_3^*)$, and the corresponding controls will be denoted by $\tilde{u}^* : k \longrightarrow u = \tilde{u}^*(k)$, $\tilde{v}^* : k \longrightarrow v = \tilde{v}^*(k)$.

In view of Condition (c) of Theorem 1, the Hamiltonian is

$$\mathfrak{H}^{k+1} = \tilde{x}_1(k)\tilde{x}_2(k) + \tilde{u}^2(k) - \tilde{v}^2(k) + \tilde{\lambda}_1(k+1)\tilde{x}_2(k) + \tilde{\lambda}_2(k+1)(\tilde{u}(k)+\tilde{v}(k)) \tag{51}$$

and the adjoint equations relative to $\tilde{\lambda}_1$ and $\tilde{\lambda}_2$ are

$$\begin{aligned} \tilde{\lambda}_1(k+1) - \tilde{\lambda}_1(k) &= - \tilde{x}_2^*(k) \\ \tilde{\lambda}_2(k+1) - \tilde{\lambda}_2(k) &= - \tilde{x}_1^*(k) - \tilde{\lambda}_1(k+1) \end{aligned} \tag{52}$$

for $k = i, i+1, \ldots, K-1$.

From Condition (a) of Theorem 1 we obtain

$$\begin{aligned} 2\tilde{u}^*(k) + \tilde{\lambda}_2(k+1) &= 0 \\ -2\tilde{v}^*(k) + \tilde{\lambda}_2(k+1) &= 0 \end{aligned} \tag{53}$$

which implies that

$$\tilde{u}^*(k) = -\tilde{v}^*(k) \tag{54}$$

Then, from Condition (d) of Theorem 1 we obtain

$$\tilde{\lambda}_1(K) = \tilde{\lambda}_2(K) = 0 \tag{55}$$

Let

$$\begin{aligned} \tilde{x}_o^*(K) &= x_o^K = C \\ \tilde{x}_1^*(K) &= x_1^K \\ \tilde{x}_2^*(K) &= x_2^K \end{aligned} \tag{56}$$

From (49) - (55) we deduce that:

<u>at stage K-1</u>

$$\tilde{x}_2^*(K-1) = \tilde{x}_2^*(K) = x_2^K$$

$$\tilde{x}_1^*(K-1) = \tilde{x}_1^*(K) - \tilde{x}_2^*(K-1) = x_1^K - x_2^K$$

$$\tilde{x}_o^*(K-1) = C - x_2^K(x_1^K - x_2^K)$$

and

$$\tilde{\lambda}_1(K-1) = x_2^K$$

$$\tilde{\lambda}_2(K-1) = x_1^K - x_2^K$$

<u>at stage K-2</u>

$$\tilde{x}^*_2(K-2) = \tilde{x}^*_2(K-1) = x_2^K$$

$$\tilde{x}^*_1(K-2) = \tilde{x}^*_1(K-1) - x_2^K = x_1^K - 2x_2^K$$

$$\tilde{x}^*_0(K-2) = -x_2^K(x_1^K - 2x_2^K) + \tilde{x}^*_0(K-1) = C - x_2^K(2x_1^K - 3x_2^K)$$

and

$$\tilde{\lambda}_1(K-2) = x_2^K + x_2^K = 2x_2^K$$

$$\tilde{\lambda}_2(K-2) = x_1^K - 2x_2^K + x_2^K + x_1^K - x_2^K = 2(x_1^K - x_2^K)$$

More generally, one can prove readily by recursive arguments that

<u>at stage K-m</u> $0 \leq m \leq K-1$

$$\tilde{x}^*_2(K-m) = x_2^K$$

$$\tilde{x}^*_1(K-m) = x_1^K - mx_2^K$$

$$\tilde{x}^*_0(K-m) = C - mx_2^K(x_1^K - \frac{m+1}{2}\, x_2^K) \tag{57}$$

and

$$\tilde{\lambda}_1(K-m) = mx_2^K$$

$$\tilde{\lambda}_2(K-m) = m(x_1^K - x_2^K) \tag{58}$$

Hence, from (53) and the second of Eqs. (58), we have

$$\tilde{u}^*(K-m) = -\tilde{v}^*(K-m) = -\frac{m-1}{2}(x_1^K - x_2^K) \tag{59}$$

$m = 1,2,\ldots,K-i$.

The value of the game starting at stage k is

$$V^*(\tilde{x}^*(k)) = \tilde{x}^*_0(K) - \tilde{x}^*_0(k) = (K-k)x_2^K(x_1^K - \frac{K-k+1}{2}\, x_2^K) .$$

Now let

$$\tilde{x}^*_0(k) = x_0^k$$

$$\tilde{x}^*_1(k) = x_1^k$$

$$\tilde{x}^*_2(k) = x_2^k .$$

From (57) we deduce

$$x_2^k = x_2^K$$

$$x_1^k = x_1^K - (K-k)x_2^K$$

and hence, the value of the game starting at stage k may be written as

$$V^*(\bar{x}^*(k)) = (K-k)\, x_2^k \left(x_1^k + \frac{K-k-1}{2}\, x_2^k\right) \tag{60}$$

It follows that the equation of surface $\Sigma_k(C)$ is

$$x_0^k + (K-k)\, x_2^k \left(x_1^k + \frac{K-k-1}{2}\, x_2^k\right) = C \tag{61}$$

$$k = 0,1,2,\ldots,K\ .$$

Surface $\Sigma_k(C)$ is shown on figures 1-3 in the x_0-x_1-x_2 space, for $k \neq K$, $K-1$, for $k = K-1$ and $k = K$, respectively.

One can easily verify that the sets $\Omega_P(K-m)$ and $\Omega_E(K-m)$, $m = 0,1,\ldots,K-i-1$, in this example, satisfy Assumption 2.

REFERENCES

1. On the Geometry of Optimal Processes

1. A. BLAQUIERE and G. LEITMANN, On the Geometry of Optimal Processes, Parts I,II, III, Univ. of California, Berkeley, IER Repts. AM-64-10, AM-65-11, AM-66-1.

2. G. LEITMANN, Some Geometrical Aspects of Optimal Processes, J. SIAM, Ser. A: Control 3, No. 1, 1965

3. A. BLAQUIERE, Further Investigation into the Geometry of Optimal Processes, J. SIAM, Ser. A: Control 3, No. 2, 1965

4. A. BLAQUIERE and G. LEITMANN, Some Geometric Aspects of Optimal Processes, Part I: Problems with Control Constraints, Proc.Congr. Automatique Théorique, Paris 1965, Dunod Ed.

5. A. BLAQUIERE and G. LEITMANN, On the Geometry of Optimal Processes, in Topics in Optimization, Academic Press, 1967, pp. 265-371 (G. Leitmann ed.)

6. K.V. SAUNDERS and G. LEITMANN, Some Geometric Aspects of Optimal Processes, Part II: Problems with State Constraints, Proc.Congr. Automatique Théorique, Paris 1965, Dunod Ed.

7. H. HALKIN, The Principle of Optimal Evolution, in Nonlinear Differential Equations and Nonlinear Mechanics, Academic Press, 1963 (J.P. LaSalle and S. Lefschetz, eds.)

8. H. HALKIN, Mathematical Foundations of System Optimization, in Topics in Optimization, Academic Press, 1967, pp. 198-260, (G. Leitmann ed.)

9. E. ROXIN, A Geometric Interpretation of Pontryagin's Maximum Principle, in Nonlinear Differential Equations and Nonlinear Mechanics, Academic Press, 1963, (J.P. LaSalle and S. Lefschetz, eds.)

10. R.E. BELLMAN, Dynamic Programming, Princeton Univ. Press, Princeton, New Jersey, 1957

11. A. BLAQUIERE and G. LEITMANN, Further Geometric Aspects of Optimal Processes: Multiple-Stage Dynamic Systems, Mathematical Theory of Control, Academic Press, 1967

2. On the Theory of Games

12. J. von NEUMANN and O. MORGENSTERN, Theory of Games and Economic Behavior, Princeton Univ. Press, Princeton, New Jersey, 1953

13. J. MAC KINSEY, An Introduction to the Theory of Games, McGraw-Hill, 1952

14. L. ZADEH, Optimality and Nonscalar-valued Performance Criteria, IEEE Transactions on Automatic Control, Vol. Ac-8, Jan. 1963, pp. 59-60

15. R. ISAACS, Differential Games, Wiley, N.Y. 1965

16. L.D. BERKOVITZ, A Variational Approach to Differential Games, in Advances in Game Theory, Princeton Univ. Press, Princeton, 1964, pp. 127-174

17. L.D. BERKOVITZ and W.H. FLEMING, On Differential Games with Integral Payoff, in Contributions to the Theory of Games III, Princeton Univ. Press, Princeton, 1957, pp. 413-435

18. L.D. BERKOVITZ, Necessary Conditions for Optimal Strategies in a Class of Differential Games and Control Problems, J. SIAM Control, Vol. 5, No. 1, 1967, pp. 1-24

19. D.L. KELENDZHERIDZE, A Pursuit Problem, in The Mathematical Theory of Optimal Processes, Interscience, N.Y., 1962

20. L.S. PONTRYAGIN, On Some Differential Games, J. SIAM Control, Vol. 3, No.1, 1965, pp. 49-52

21. L.S. PONTRYAGIN, On the Theory of Differential Games, Uspehi Mat. Nauk 21, No. 4 (130), 1966, pp. 219-274

22. E.F. MISHCHENKO and L.S. PONTRYAGIN, Linear Differential Games, Dokl.Akad.Nauk SSSR, Tom 174, No. 1, 1967, pp. 27-29, and Soviet Math. Dokl. Vol. 8, No. 3, 1967, pp. 585-588

23. Y.C. HO, A.E. BRYSON and S. BARON, Differential Games and Optimal Pursuit-Evasion Strategies, IEEE Transactions on Automatic Control, Vol. AC-10, October 1965, pp. 385-389

24. S. BARON, Differential Games and Optimal Pursuit-Evasion Strategies, Ph.D. Thesis in Applied Mathematics, Harvard University, Cambridge, 1966

25. E.N. SIMAKOVA, Differential Games (a survey paper), Avtomatika i Telemekhanika, Vol. 27, No. 11, Nov. 1966, pp. 161-178

26. I.G. SARMA and R.K. RAGADE, Some Considerations in Formulating Optimal Control Problems as Differential Games, Int.J.Control, Vol. 4, No. 3, 1966, pp. 265-279

27. A. KAUFMANN, Graphs, Dynamic Programming, and Finite Games, Academic Press, 1967

28. G. LEITMANN and G. MON, Some Geometric Aspects of Differential Games, Journal of Astronautical Sciences, Vol. 14, No. 2, Mar.-Apr., 1967, pp. 56-65

29. G. LEITMANN and G. MON, On a Class of Differential Games, Proceed. Colloquium on Advanced Problems and Methods for Space Flight Optimization, Liege 1967, Pergamon Press 1968 (also in Kibernetika, Jan. 1968)

30. A. BLAQUIERE and G. LEITMANN, Quantitative Games, Mémorial des Sciences Mathématiques, Gauthier-Villars Ed. 1968

31. A. BLAQUIERE, Quantitative and Qualitative Games, A Geometric Approach - Part I: Quantitative Games Report, Laboratoire d'Automatique Théorique, Faculté des Sciences de Paris, 1968

32. A. BLAQUIERE, F. GERARD and G. LEITMANN, Quantitative and Qualitative Games, A Geometric Approach, Academic Press, 1969

ACKNOWLEDGEMENT

The work reported here was supported by Delegation Generale à la Recherche Scientifique et Technique under Contract 69-01-691 and by NASA (NAS 12-114).

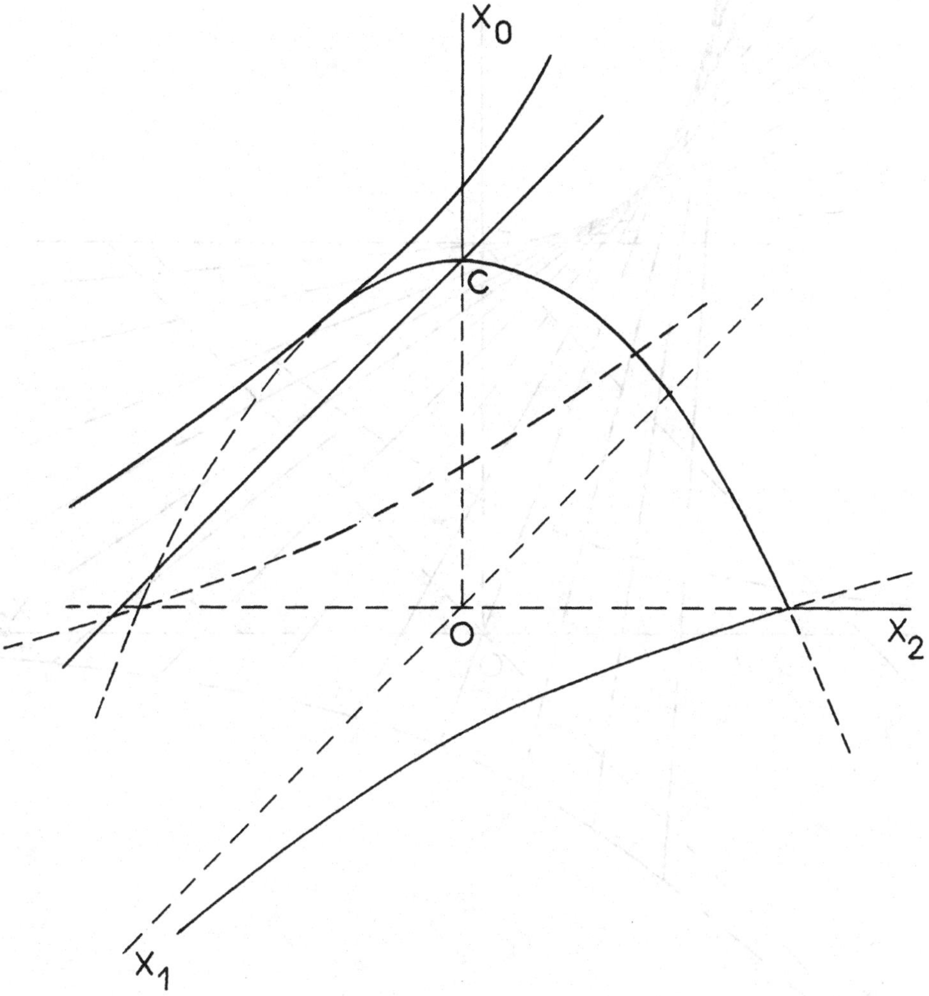

Fig. 1

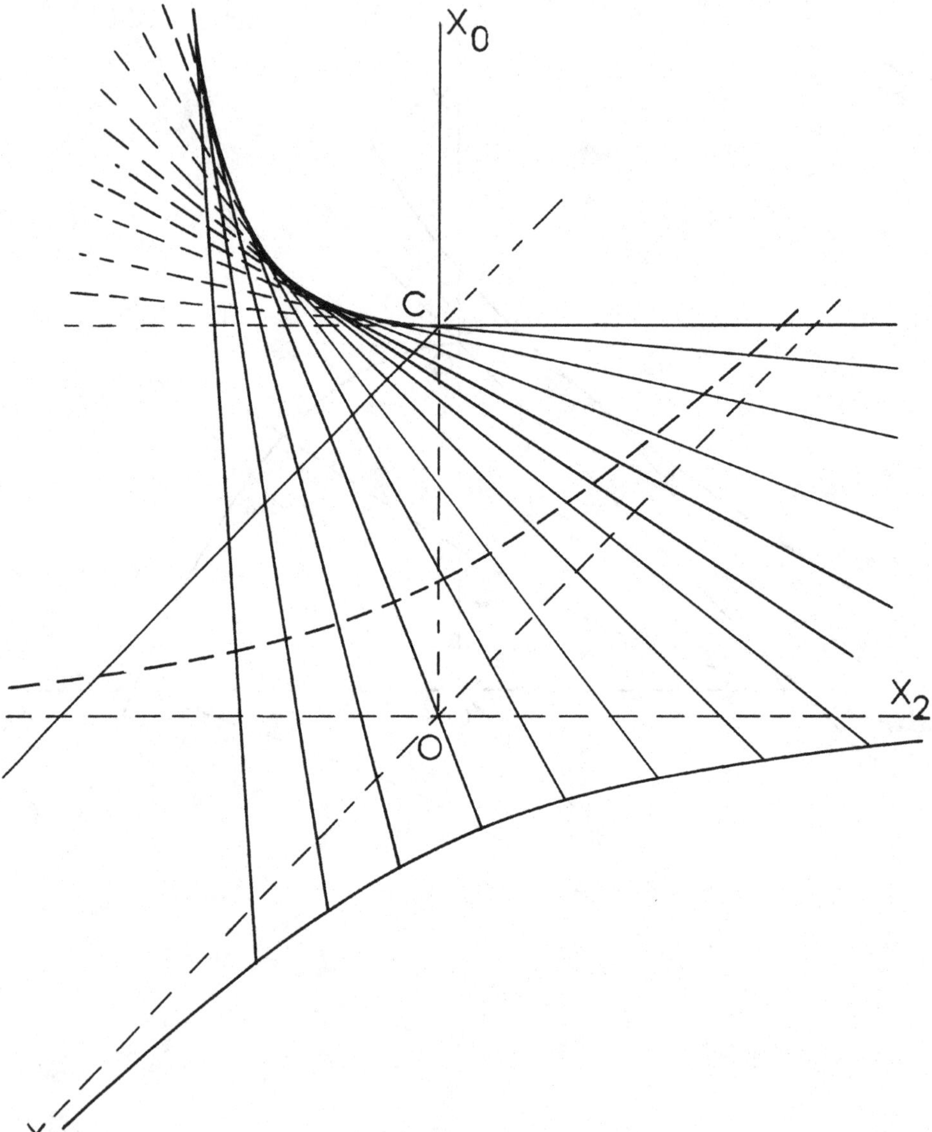

Fig. 2

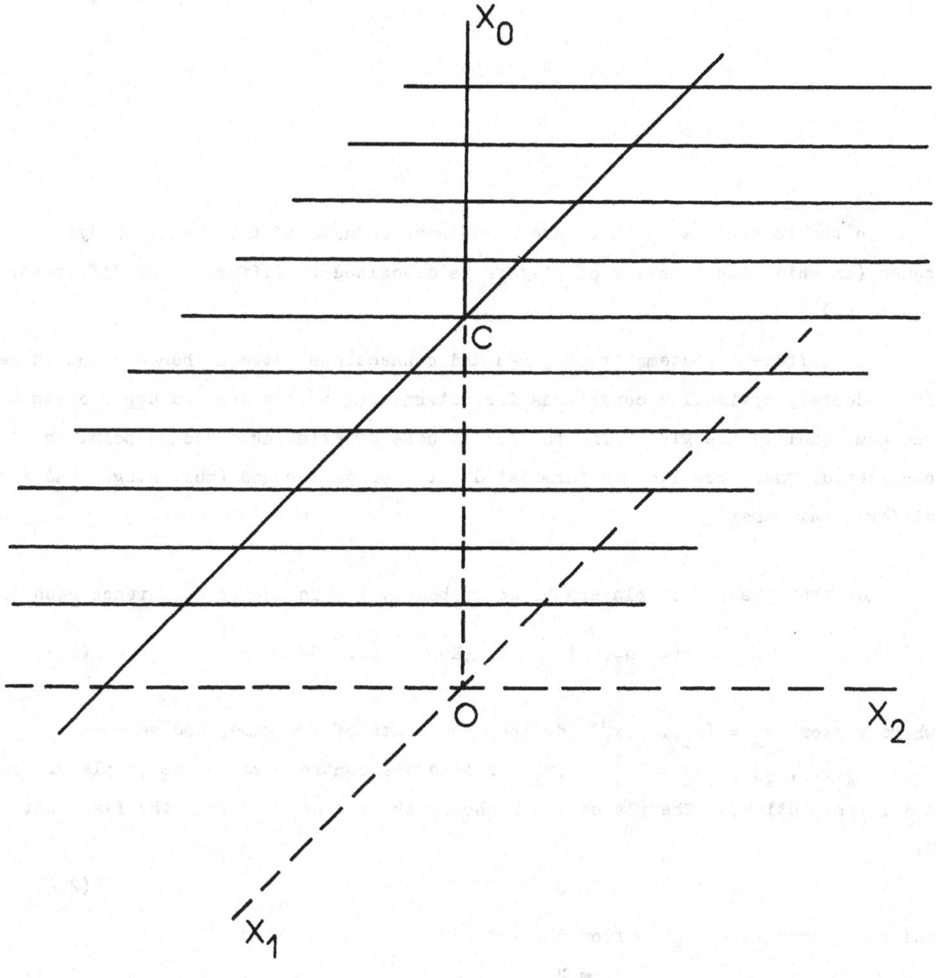

Fig. 3

ON THE THEORY OF DYNAMIC GAMES

A. Propoy

1. In the present paper there are given some results of the theory of dynamic games (in which the behavior of players is described by difference or differential equations).

The different statements of games and connections between them are considered. The necessary optimality conditions for determining of the low and upper costs of the game quality are given. The particular case of existing a saddle-point is considered. These results are formulated both for difference (multistage) and for differential games.

2. Let the behavior of players be described by the following difference equation

$$x_{K+1} = f(x_K, u_K, v_K) \qquad (K = 0,1,\dots,N-1) \tag{2.1}$$

where vector $x_K = \{x_K^1,\dots,x_K^n\}$ defines the state of the game, and vectors $u_K = \{u_K^1,\dots,u_K^r\}$, $v_K = \{v_K^1,\dots,v_K^s\}$ define the control variables of players I and II respectively. The player I can chooze the values u_K from the fixed set U:

$$u_K \in U \tag{2.2}$$

and the player II - v_K - from the set V :

$$v_K \in V \ . \tag{2.3}$$

The number of stages N is supposed to be fixed. The game quality (performance index) is following

$$I = \Phi(x_N) + \sum_{K=0}^{N-1} f_o(x_K, u_K, v_K) \tag{2.4}$$

and the player I tries to maximize the value of (2.4), but the player II - to minimize it.

Further we shall suppose that the sets U and V are compacts and the functions $\Phi(x)$, $f_i(x,u,v)$ $(i = 0,1,\dots,n)$ are continuously differentiable on $E_n \times U \times V$.

We shall call sequences $u = \{u_o,\dots,u_{N-1}\}$, $v = \{v_o,\dots,v_{N-1}\}$, $x = \{x_o,x_1,\dots,x_N\}$ the control of player I, II and the game trajectory, respectively. (This type of control is often called "the open loop control".) Obviously, $I = I(x_o,u,v)$.

Besides we introduce sequences of functions $u(K,x) = \{u_o(x),\dots,u_{N-1}(x)\}$, $v(K,x) = \{v_o(x),\dots,v_{N-1}(x)\}$ which we shall call the synthesis of the player I and II (closed loop control). These sequences also define the process: $I = I(x_o,u(K,x),v(K,x))$.

In connection with information of each players about the course of the game we shall consider the following problems.

<u>Problem 1.</u> Let's suppose that the player II may learn all the control u of the player I beforehand (as a function of time). In this situation the player I does his best if he chooses the control u to obtain

$$\max_u \min_v I(x_o,u,v) = w_1^-(x_o) \tag{2.5}$$

Choosing his control in such a way the player I guarantees himself the value of $I \geq w_1^-(x_o)$, whatever control v the player II would choose. $w_1^-(x_o)$ - is the low guaranteeing cost of the game quality for the player I in open loop control.

Similarly we define the upper cost (the case, when the player I may learn the control v of the player II beforehand):

$$\min_{v} \max_{u} I(x_o,u,v) = w_1^+(x_o)$$

Problem 2. Let's suppose that the player II may learn all the synthesis of the player I $u(K,x)$ beforehand, that is, he knows the rules by which the player I chooses his control variables u_K in dependence on the current game state x_K.

In this case the player I must choose his own synthesis to yield

$$\max_{u(K,x)} \min_{v(K,x)} I(x_o,u(K,x),v(K,x)) = w_2^-(x_o) \tag{2.6}$$

$w_2^-(x_o)$ is the low cost for the player I in closed loop control. The upper cost is defined by

$$\min_{v(K,x)} \max_{u(K,x)} I(x_o,u(K,x),v(K,x)) = w_2^+(x_o) \tag{2.7}$$

Problem 3. Now let's suppose that the player II do not know anything beforehand but he may learn the current values of state x_K and control variables u_K of his opponent. In this situation it may be proved by induction that the player I must choose his control to yield

$$\max_{u_o} \min_{v_o} \dots \max_{u_{N-1}} \min_{v_{N-1}} I(x_o,u,v) = w_3^-(x_o) . \tag{2.8}$$

If on the contrary the player I may learn the current values v_K of the player II control variables then he chooses his control to obtain

$$\min_{v_o} \max_{u_o} \dots \min_{v_{N-1}} \max_{u_{N-1}} I(x_o,u,v) = w_3^+(x_o) \quad . \tag{2.9}$$

3. Now we shall establish the correlations between the problems 1-3. For this purpose we make use of the following results in the theory of static games (see, for example, [1]):

$$\max_{x} \min_{y} \varphi(x,y) \leq \min_{y} \max_{x} \varphi(x,y) ; \tag{3.1}$$

$$\max_{x} \min_{y} \varphi(x,y) = \min_{y(x)} \max_{x} \varphi(x,y) \quad . \tag{3.2}$$

The inequality (3.1) means that the gain of the player I when his choice is known for the player II cannot be less his gain when vice versa the player I may learn the choice of player II . The equality (3.2) means that operators max and min are commutative when the information of each players about the course of the game do not change.

Using to problems 1-3 the expressions (3.1) (3.2) we may prove the following.

Theorem 3.1: For any initial state x_0 the following results are true

$$w_2^- = w_3^- \, , \quad w_2^+ = w_3^+ \, ; \tag{3.3}$$

$$w_1^- \leq w_2^- \leq w_2^+ \leq w_1^+ \, . \tag{3.4}$$

The equalities (3.3) show that the problems 2 and 3 are only the different formulations of the same problem. Therefore there is need to distinguish two main games in the theory of dynamic games. In the first game which we shall denote by $\Gamma_1(x_0,u,v)$ the players choose controls beforehand and don't use in any way the current information about the game course; it is a static game in essence.

In the second game $\Gamma_2(x_0,u(K,x),v(K,x))$ the players already define their behavior in connection with the course of the game. The results of theorem 3.1 show that the player I is able to increase the quality of his control using the closed loop control (instead the open loop control).

From the theorem 3.1 we may easily obtain

Theorem 3.2: If the game Γ_1 has a saddle-point, i.e.,

$$w_1^- = w_1^+ \quad ,$$

then the game Γ_2 has a saddle-point too, i.e.

$$w_2^- = w_2^+ \quad .$$

The inverse is not true in general.

This theorem shows that the open loop control is equivalent to the closed loop control only when the game Γ_1 has a saddle-point.

This situation is probably not frequent in real dynamic games. The more likely when w_2^- is considerably larger than w_1^- ; naturally in these cases the closed loop control is compulsory.

For more detailed analysis of these questions see paper [2].

4. Before formulating optimality conditions for problems 1-3 we define some notions.

Consider the problem of finding

$$\max_{x \in X} \min_{y \in Y} \varphi(x,y) \tag{4.1}$$

where sets X and Y are compacts, and $\varphi(x,y)$ is a continuously differentiable function on $X \times Y$.

It was proved in [3-5] that the function $\Phi(x) = \min_y \varphi(x,y)$ is only coninuous but differentiable in any direction δx and

$$\frac{\partial}{\partial} \delta\Phi(\bar{x}) = \min_{y \in Y(\bar{x})} \left[\frac{\partial\varphi(\bar{x},y)}{\partial x}\right]^T \delta x \tag{4.2}$$

where $Y(\bar{x})$ is the set of optimal responses y for $x = \bar{x}$:

$$Y(\bar{x}) = \{y \,|\, \varphi(\bar{x},y) = \min_{y \in Y} \varphi(\bar{x},y)\}$$

Besides we define an adjoint system

$$p_K = \frac{\partial f_o(x_K,u_K,v_K)}{\partial x_K} + \frac{\partial f(x_K,u_K,v_K)}{\partial x_K} p_{K+1} \qquad (K = N-1,\dots,2,1) \tag{4.3}$$

with boundary condition

$$p_N = \frac{\partial \Phi(x_N)}{\partial x_N} \quad , \tag{4.4}$$

where vectors $p_K = \{p_K^1,\dots,p_K^n\}$ and $\frac{\partial\Phi}{\partial x} = \left[\frac{\partial\Phi}{\partial x_i}\right]_i$, $\frac{\partial f_o}{\partial x} = \left[\frac{\partial f_o}{\partial x_i}\right]_i$, $\frac{\partial f}{\partial x} = \left[\frac{\partial f_j}{\partial x_i}\right]_{ij}$

$(i,j = 1,\dots,n)$.

Define also the Hamilton function

$$H(p_{K+1},x_K,u_K,v_K) = f_o(x_K,u_K,v_K) + p_{K+1}^T f(x_K,u_K,v_K) \qquad (4.5)$$

and denote by $\delta_u H$ and $\delta_v H$ the feasable differentials of this function

$$\delta_u H = \sum_{i=1}^{r} \frac{\partial H}{\partial u_i} \delta u_i, \quad \delta_v H = \sum_{i=1}^{s} \frac{\partial H}{\partial v_i} \delta v_i ,$$

where $\delta u \in K(u)$, $\delta v \in K(v)$ ($K(u)$ and $K(v)$ are the cones of feasible variations at the points $u \in U$ and $v \in V$ [6,7]).

5. Now we can formulate the optimality conditions for the problem 1. Using the notions of p. 4 and the usual reasonings in the theory of optimal discrete control (see, for example, [7]) we may prove the following:

<u>Theorem 5.1:</u> Let $\Omega(u^*)$ be the set of optimal solutions v^* of the problem

$$\{\min I;\ x_{K+1} = f(x_K,u_K^*,v_K),\ \ v_K \in V \ \ (K = 0,1,\dots,N-1)\} . \qquad (5.1)$$

Then the following inequalities

$$\min_{v \in \Omega(u^*)} \delta_u H(p_{K+1}^v,\ x_K^v,\ u_K^*,\ v_K) \leq 0 \qquad \delta u_K \in K(u_K^*) \qquad (5.2)$$

are hold for the optimal control u_K^* of the player I, where the optimal values $\{x_K^v,p_K^v\}$ are found from (2.1), (4.3) and (4.4) for $v \in \Omega(u^*)$.

We consider some particular cases now. Only for simplicity of designations we shall assume that $f_o(x,u,v) \equiv 0$, i.e. $I = \Phi(x_N)$.

<u>Theorem 5.2:</u> Let the set $f(x,U,v)$ be convex for all x and $v \in V$.

Then the equality

$$\max_{u_K \in U}\ \min_{v \in \Omega(u^*)} \left[H(p_{K+1}^v,x_K^v,u_K,v_K) - H(p_{K+1}^v,x_K^v,u_K^*,v_K)\right] = 0 \qquad (5.3)$$

is true for the optimal process.

The justice of this theorem is followed from the inequality (5.2) and the convexity of the set $f(x,U,v)$ (Compare with [4,7]).

Theorem 5.3: Let the solution of the problem (5.1) be unique. Then the following inequalities

$$\delta_u H(p^*_{K+1}, x^*_K, u^*_K, v^*_K) \leq 0 \tag{5.4}$$

$$\delta_v H(p^*_{K+1}, x^*_K, u^*_K, v^*_K) \geq 0 \tag{5.5}$$

are hold for all $\delta u_K \in K(u^*_K)$ and $\delta v_K \in K(v^*_K)$.

The inequality (5.4) is directly followed from (5.2); the inequality (5.5) is the optimality condition of control v^* .

Theorem 5.4: Let:

1. the sets $f(x,U,v)$ and $f(x,u,V)$ are convex for all x and $u \in U$, $v \in V$;
2. the solution of the problem (5.1) is unique. Then it is necessary that the Hamiltonian has a saddle-point:

$$\max_{u_K \in U} \min_{v_K \in V} H(p^*_{K+1}, x^*_K, u_K, v_K) = \min_{v_K \in V} \max_{u_K \in U} H(p^*_{K+1}, x^*_K, u_K, v_K) = H(p^*_{K+1}, x^*_K, u^*_K, v^*_K) \tag{5.6}$$

for the $\{u^*_K, v^*_K, x^*_K, p^*_K\}$ to be optimal.

The similar result is true for the general type of performance index (Com. [7]).

Proof. From the inequality (5.4) and the convexity of set $f(x,U,v)$ is followed [7]

$$\max_{u_K \in U} H(p^*_{K+1}, x^*_K, u_K, v^*_K) = H(p^*_{K+1}, x^*_K, u^*_K, v^*_K) \quad . \tag{5.7}$$

From the inequality (5.5) and the convexity of set $f(x,u,V)$ is followed

$$\min_{v_K \in V} H(p^*_{K+1}, x^*_K, u^*_K, v_K) = H(p^*_{K+1}, x^*_K, u^*_K, v^*_K) \; . \tag{5.8}$$

The equalities (5.7) and (5.8) are equivalent to (5.6) [1].

The similar results may be obtained for finding $\min_v \max_u I = w_1^+$. The more detailed considerations of the above problems may be found in [8].

6. Let's come to optimality conditions for the problems 2,3. It may be proved that the original problem 2 (or 3) is equivalent to the problems

$$\{\max_u I \; ; \; x_{K+1} = f(x_K,u_K,v_K^*(u_K,x_K)) \; , \; u_K \in U \quad (K = 0,1,\dots,N-1)\} \tag{6.1}$$

$$\{\min_v I \; ; \; x_{K+1} = f(x_K,u_K^*(x_K),v_K) \; , \; v_K \in V \quad (K = 0,1,\dots,N-1)\} \tag{6.2}$$

where $\{u_K^*(x)\}$ and $\{v_K^*(x,u)\}$ are optimal solutions of the problem 2 or 3.

Here we shall consider only a singular case, that is, we shall assume that the problems (6.1), (6.2) have unique solutions.

<u>Theorem 6.1:</u> Let:

1. the solutions $\{u_K^*(x), v_K^*(x,u)\}$ of the problems (6.1) and (6.2) are unique;
2. the sets $f(x,U,v_K^*(x,U))$ and $f(x,u_K^*,V)$ be convex for all x and $K = 0,1,\dots,N-1$.

Denote the optimal values of control variables by $u_K^* = u_K^*(x_K^*)$ and $v_K^* = v_K^*(x_K^*,u_K^*)$ $(K = 0,1,\dots,N-1)$ for a given initial state x_0.

Then

$$\max_{u_K \in U} \min_{v_K \in V} H(p_{K+1}^*,x_K^*,u_K,v_K) = H(p_{K+1}^*,x_K^*,u_K^*,v_K^*) \tag{6.3}$$

is hold at the optimal process $\{u_K^*,v_K^*,x_K^*,p_K^*\}$, where the optimal values of $\{p_K^*\}$ are found from the adjoint system (4.3) with the boundary condition (4.4).

The proof of this theorem may be found in [9].

7. Now we consider the case when the game Γ_1 or Γ_2 has a saddle-point. In this case the game problem reduces to the pair of optimal control problems.

Namely, let for example there exist such controls u^* and v^* that the game Γ_1 has a saddle-point, i.e.,

$$\max_u \min_v I = \min_v \max_u I = I^*(x_o,u^*,v^*). \tag{7.1}$$

Then the game Γ_1 is equivalent to the following optimal control problems:

$$\{\min_v I\ ,\quad x_{K+1} = f(x_K,u_K^*,v_K)\ ,\quad v_K \in V \quad (K = 0,1,\ldots,N-1)\} \tag{7.2}$$

$$\{\max_u I\ ,\quad x_{K+1} = f(x_K,u_K,v_K^*)\ ,\quad u_K \in U \quad (K = 0,1,\ldots,N-1)\} \tag{7.3}$$

Using for (7.2), (7.3) the optimality conditions for discrete processes, we obtain directly:

<u>Theorem 7.1:</u> It is necessary that

$$\delta_u H(p_{K+1}^*,x_K^*,u_K^*,v_K^*) \leq 0 \qquad \delta u_K \in K(u_K^*)\ , \tag{7.4}$$

$$\delta_v H(p_{K+1}^*,x_K^*,u_K^*,v_K^*) \geq 0 \qquad \delta v_K \in K(v_K^*) \tag{7.5}$$

for controls u^* and v^* to be saddle-point of the game Γ_1 . Here the optimal values of $\{p_K^*\}$ are found from (4.3), (4.4).

<u>Consequence.</u> Let the sets $f(x,U,v)$ and $f(x,u,V)$ are convex for all x , $u \in U$ and $v \in V$. Then if the performance index I has a saddle-point, then the Hamiltonian (4.5) has a saddle at the same point.

This is directly followed from usual in the theory of optimal discrete control reasonings [7].

Consider the game Γ_2 now.

Let such the functions $u_K^*(x)$, $v_K^*(x)$ exist, that the game Γ_2 has a saddle-point, i.e.,

$$\max_{u(K,x)} \min_{v(K,x)} I = \min_{v(K,x)} \max_{u(K,x)} I = I(x_o,u^*(K,x),v^*(K,x)) \tag{7.6}$$

Then the game Γ_2 reduces to the problems

$$\{\max_{u} I,\ x_{K+1} = f(x_K,u_K,v_K^*(x_K)),\quad u_K \in U \quad (K = 0,1,\dots,N-1)\} \tag{7.7}$$

$$\{\min_{v} I,\ x_{K+1} = f(x_K,u_K^*(x_K),v_K),\quad v_K \in V \quad (K = 0,1,\dots,N-1)\}\ . \tag{7.8}$$

Theorem 7.2: Let:

1. the solutions of the problems (7.7), (7.8) are unique;
2. the sets $f(x,U,v)$, $f(x,u,V)$ are convex for all x and $u \in U,\ v \in V$.

Then if the performance index I has a saddle-point ((7.6) is hold), the Hamiltonian (4.5) has a saddle-point,too, i.e.,

$$\max_{u_K \in U}\ \min_{v_K \in V} H(p_{K+1}^*,x_K^*,u_K,v_K) = \min_{v_K \in V}\ \max_{u_K \in U} H(p_{K+1}^*,x_K^*,u_K,v_K) = H(p_{K+1}^*,x_K^*,u_K^*,v_K^*)$$

where the optimal values $\{x_K^*,p_K^*\}$ are found from (2.1), (4.3), (4.4).

This theorem is directly followed from the theorem 6.1.

In conclusion we note that the optimal controls u and v are not coincided for the games Γ_1 and Γ_2 in general case.

8. At last we shall formulate the similar results for the differential games. In this case a game is discribed by differential equations of following type

$$\dot{x}(t) = f(x(t),u(t),v(t)),\qquad t \in [0,T] \tag{8.1}$$

with initial state $x(0) = x_o$ and time T are fixed, where
$x(t) = \{x^1(t),\dots,x^n(t)\}$, $u(t) = \{u^1(t),\dots,u^r(t)\}$, $v(t) = \{v^1(t),\dots,v^s(t)\}$
and

$$u(t) \in U\ ,\quad v(t) \in V \tag{8.2}$$

and the performance index

$$I = \Phi(x(T)) + \int_0^T f_o(x(t),u(t),v(t))\, dt\ . \tag{8.3}$$

As in the multistage games we shall distinguish two games: $\Gamma_1(x_o,u,v)$ which is connected with the problems

$$w_1^- = \max_u \min_v I \ , \tag{8.4}$$

$$w_1^+ = \min_v \max_u I \tag{8.5}$$

and $\Gamma_2(x_0, u(t,x), v(t,x))$ which is defined from the problems

$$w_2^- = \max_{u(t,x)} \min_{v(t,x)} I \tag{8.6}$$

$$w_2^+ = \min_{v(t,x)} \max_{u(t,x)} I \ . \tag{8.7}$$

These problems are completely similar to problems 1 and 2 in multistage case. As for the problem 3, we consider the following problem:

Find a control u, which yields the best value of functional (8.3) if the player II may learn the current values of $x(t)$ and $v(t)$ (it is supposed, of course, that the player II "must" do the worst for player I).

This problem is apparently similar to problem 3 of p.2 and is called the game with discrimination [10,11] or minorant (majorant) game [12]. Note, that formally this problem may be written only as a limiting case of the problem 3 (see, for example [13]).

Also by limiting transition we can establish the correlations between the differential games Γ_1 and Γ_2 similar to the theorems 3.1 and 3.2.

At last the next results may be obtained by limiting transition from multi-stage games:

Define the adjoint system

$$\dot{p} = \frac{\partial f_0(x,u,v)}{\partial x} + \frac{\partial f(x,u,v)}{\partial x} p \ , \tag{8.8}$$

with foundary condition

$$p(T) = \frac{\partial \Phi(x(T))}{\partial x(T)} \tag{8.9}$$

and the Hamilton function

$$H(p,x,u,v) = p^T f(x,u,v) \tag{8.10}$$

<u>Theorem 8.1:</u> Let u^* be an optimal control of the player I for the problem 8.4 and $\Omega(u^*)$ be the set of optimal controls of player II , i.e., the set of solutions of the problem:

$$\{\min_{v} I, \quad \dot{x} = f(x,u^*,v), \quad v \in V, \quad t \in [0,T]\} \tag{8.11}$$

Then the equality

$$\max_{u \in U} \min_{v \in \Omega(u^*)} [H(p^v,x^v,u,v) - H(p^v,x^v,u^*,v)] = 0 \tag{8.12}$$

is hold for optimal process, where the optimal values $\{x^v(t),p^v(t)\}$ are found from (8.1), (8.8), (8.9).

This theorem evidently is similar to the theorem 6.2.

<u>Theorem 8.3:</u> Let $u^*(t) = u^*(t,x^*(t))$ and $v^*(t) = v^*(t,x^*(t),u^*(t))$ be optimal controls for problem 8.6 for a given initial state x_o and let the solutions of the problems:

$$\{\max_{u} I, \quad \dot{x} = f(x,u,v^*(t,x)), \quad u \in U, \quad t \in [0,T]\}$$

$$\{\min_{v} I, \quad \dot{x} = f(x,u^*(t,x,v),v), \quad v \in V, \quad t \in [0,T]\}$$

are unique.

Then

$$\max_{u \in U} \min_{v \in V} H(p^*,x^*,u,v) = H(p^*,x^*,u^*,v^*)$$

is true for the optimal process $\{u^*(t),v^*(t),x^*(t),p^*(t)\}$ where the optimal values of $p^*(t)$ are also found from equations (8.1),(8.8),(8.9).

<u>Theorem 8.4:</u> Let the pair $\{u^*(t,x),v^*(t,x)\}$ be a saddle-point for the game $\Gamma_2(x_o,u(t,x),v(t,x))$. Then if the solutions of the following problems

$$\{\max_u I, \quad \dot{x} = f(x,u,v^*(t,x)), \ u \in U, \ t \in [0,T]\} ,$$

$$\{\min_v I, \quad \dot{x} = f(x,u^*(t,x),v), \ v \in V, \ t \in [0,T]\}$$

are unique, the Hamilton function has saddle at the same point $u^* = u^*(t,x^*(t))$, $v^* = v^*(t,x^*(t))$.

The theorem is directly followed from the theorem 8.3. Note, that a similar result was obtained in paper [14] for the case of "regular synthesis" and in paper [15] for the case of smooth (differentiable) synthesis.

In conclusion it is necessary to note that the results formulated in this item for differential games must be regarded to some extent as hypothetical ones and need accurate grounds.

9. At last we note that the results obtained in the present paper permit to develope efficient numerical methods in the theory of dynamic games.

R e f e r e n c e s

[1] Karlin, S., Mathematical methods and theory in games, programming, and economics. Pergamon Press, L.-P. 1959.

[2] Ereshko, F.I. Propoy, A., To the theory of dynamic games, Izvestia AN SSSR, Technicheskaya Kibernetica (in press).

[3] Pshenichnij, B.N., Convex programming in normed spaces, Cybernetics, No. 5, 1965.

[4] Demyanov, V.F., On the solution of certain minimax problems I,II, Cybernetics, No. 6, 1966, No. 3, 1967.

[5] Danskin, J.M., The theory of maxmin, with applications, J.SIAM Appl.Math., v.14, No.4, 1966

[6] Zojtendijk, G., Methods of feasible directions, Elsevier Co, 1962.

[7] Propoy, A., Maximum principle for discrete control systems, Automation and Remote Control, No.7, 1965.

[8] Propoy, A., Minimax control problems with a priori information, Automation and Remote Control (in press).

[9] Propoy, A., Minimax control problems with successive information, Automation and Remote Control (in press).

[10] Pontryagin, L.S., To the theory of differential games, Uspechi Math. Nauk, v.21, No.4, 1966.

[11] Krasovskii, N.N., Repin U.M., Tret'yakov, V.E., On some game situations in the theory of control, Izv. AN SSSR, Technicheskaya Kibernetica, No.4, 1965.

[12] Fleming, W.H., The convergence problem for differential games, J.Math.Anal. and Appl., No.3, 1961, p.102-116.

[13] Pshenichnij, B.N., Necessary and sufficient conditions for differential games. (to be presented to the II Int. Colloquium in Methods of Optimization, Academgorodok, July, 1968.)

[14] Berkovitz, L.D., Necessary conditions for optimal strategies in a class of differential games and control problems. SIAM J. on Control, v.5, No.1, 1967.

[15] Zelikin, M.I., Tyinyanskii, N.T., Determinate differential games, Uspechi Math. Nauk, v.20, No.4, 1965.

ON OPTIMAL ALGORITHMS FOR SEARCH

F.L. Chernousko

§1 Introduction

Many computational algorithms, processes of planning experiments and so on, can be considered as controllable processes. Thus it is natural to pose a problem of devising the optimal (that is, in some sense, the best) computational algorithm for solving of problems of some class. Let B denote a class of problems to be solved, and $\beta \in B$ is an individual problem in that class. Let A denote a class of algorithms which can be applied for solution of an arbitrary problem from the class B, and let $\alpha \in A$ denote an individual algorithm in this class. Applying an algorighm $\alpha \in A$ to a problem $\beta \in B$, we can estimate the result by means of a function $Q(\alpha,\beta)$, which is representative of the quality of the chosen algorithm α for the solution of a specific problem β. This function $Q(\alpha,\beta)$ is defined for all $\alpha \in A$ and $\beta \in B$ and can express, for example, the error of solution, the expenditure of labour or computer time for obtaining the desired accuracy and so on. Choosing the algorithm we want to minimize the estimate function Q. As before the solution the specific properties of the problem β are usually unknown, then it is reasonable to apply a minimax approach. We shall define the optimal algorithm in the class A for the solution of problems in the class B as the algorithm α^* which gives the following minimax

$$\min_{\alpha\in A} \max_{\beta\in B} Q(\alpha,\beta) \tag{1}$$

Besides the minimax approach (1) we can consider algorithms optimal "in the average". The algorithm which is optimal "in the average" can be defined by the condition

$$\min_{\alpha \in A} \int_B Q(\alpha, \beta) d\mu$$

Here μ is some measure on the set of problems B . The algorithm which is optimal in the minimax sense (1) has the following advantages: for its finding it is not necessary to introduce a measure on the set of problems B (which is often very difficult to do), and it gives the guarranteed result for the whole class of problems B .

In this paper only the minimax approach (1) is used. We shall find the optimal algorithm for the search for zeroes of functions. It is assumed that a function belongs to some class and that its value can be computed approximately at every point. The mathematical formulation and the solution of the problem are given in §2 and §3. In §4 we consider the more complicated and interesting problem of the distribution of resources throughout computation. In this paper we use the ideas of the dynamic programming method. These ideas were applied already for devising some optimal algorithms which take no account of the computational errors [1]. The results of this work are represented also in the papers [2], [3]. The obtained optimal search algorithms can be useful for carrying out of numerous and laborious computations.

§2 Definitions and formulation of the problem

Let F denote a class of funtions $f(x)$, with the following properties: 1) $f(x)$ is defined and continuous on an interval $L = [a,b]$ of the real axis; 2) for arbitrary x', x'' in L (with $x' \neq x''$), the function $f(x)$ satisfies the inequalities

$$m \leq [f(x') - f(x'')]/(x'-x'') \leq M$$

where m, M are constants $(0 < m < M < \infty)$; 3) the zero of $f(x)$ (which is unique due to condition 2) lies in L . Condition 2) implies that the derivative $f'(x)$ of $f(x)$ exists and satisfies $m \leq f'(x) \leq M$ at almost all $x \in L$.

Let us assume that in searching for the zero of a function $f \in F$, we have evaluated the latter at n points, $x_1, \ldots, x_n$, in L ; that the approximate functional values obtained are $y_1, \ldots, y_n$; and that the corresponding computational errors are known to be $\delta_1,\dots,\delta_n$. This subjects $f(x)$ to conditions

$$|f(x_i)-y_i| \le \delta_i \qquad (\delta_i \ge 0, \quad i = 1,\dots,n) \tag{2}$$

These conditions imply the existence of at least one function $f(x) \in F$ satisfying (2).

The set of points x which are zeros of those functions $f(x)$ in the class F which satisfy (2) will be called the zero localizer under conditions (2) and denoted by D . It can be shown (see ref. [3]) that D is a closed interval which can be put in the form

$$D = [a_n, b_n] = L \cap \left\{ \bigcap_{i=1}^{n} [\xi_i^+, \xi_i^-] \right\}$$

$$\xi_i^+ = \begin{cases} x_i - (y_i+\delta_i)/M & \text{if } y_i \le -\delta_i \\ x_i - (y_i+\delta_i)/m & \text{if } y_i \ge -\delta_i \end{cases} \tag{3}$$

$$\xi_i^- = \begin{cases} x_i - (y_i-\delta_i)/m & \text{if } y_i \le \delta_i \\ x_i - (y_i-\delta_i)/M & \text{if } y_i \ge \delta_i \end{cases} \qquad (i = 1,\dots,n)$$

The zero of $f(x)$ that we are looking for lies in the interval D by definition and can be anyone of its points. It follows that the length of D is representative of the accuracy with which the zero can be computed. This length can be taken as an estimate function Q in the expression (1).

The problem now is to find an optimal algorithm for determining zeros of functions in F , i.e., to devise a method for successively selecting points $x_1,\dots,x_n$ in L which among other methods would guarantee the smallest length of the interval D after the computation. The interval L , numbers m and M , integer n , and errors $\delta_i \ge 0$ for $i = 1,\dots,n$ are considered as given, whereas numbers y_i are not known beforehand and can prove to be "the worst" as far as minimizing the length of D is concerned. Therefore, for an optimal algorithm, we shall require the attainment of the minimax

$$\Delta = \min_{x_1} \max_{y_1} \min_{x_2} \max_{y_2} \dots \min_{x_n} \max_{y_n} (b_n - a_n) \tag{4}$$

The minima here are computed for $x_i \in L$, and the maxima for y_i such that there exists at least one function $f(x) \in F$ satisfying (2). Clearly, the optimal search problem is treated here as an n-step game between the "computer", who selects numbers x_i, and "nature", selecting y_i, each side informing the other about the moves. The payoff is the length of D.

§3 Optimal search algorithm

The minimax problem (4) is solved in successive steps, starting with the determination of x_n and y_n, all of the steps being similar to one another. Solving problem (4) yields the following optimal n-step search algorithm (see ref. [3]).

Setting $a_0 = a$, $b_0 = b$, let us describe the i-th step of the algorithm $(i = 1,\ldots,n)$. Let us assume that after $i-1$ steps we have found the zero localizer interval $[a_{i-1}, b_{i-1}]$, corresponding to the first $i-1$ inequalities in (2). We now let

$$x_i = (a_{i-1} + b_{i-1})/2 \qquad (i = 1,\ldots,n) \tag{5}$$

and determine y_i, i.e., an approximate value of $f(x_i)$ with a known error δ_i. Following this, we find the zero localizer after i steps

$$[a_i, b_i] = [a_{i-1}, b_{i-1}] \cap [\xi_i^+, \xi_i^-] \qquad (i = 1,\ldots,n) \tag{6}$$

Here ξ_i^+, ξ_i^- are determined from equations (3). Next we perform the subsequent step and so on. At each step, the length of the localizer is nonincreasing, and

$$b_i - a_i \leq (b_{i-1} - a_{i-1}) h\left[\frac{2\delta_i}{m(b_{i-1} - a_{i-1})}\right] \qquad (i = 1,\ldots,n) \tag{7}$$

where

$$h(z) = \begin{cases} (1-k)/2 + kz & \text{if } 0 \leq z \leq 1/2 \\ z & \text{if } 1/2 \leq z \leq 1 \\ 1 & \text{if } z \geq 1 \qquad (k = m/M) \end{cases} \tag{8}$$

Equality in (7) holds for those y_i for which minimax (4) is attained. If, in particular, $f(x)$ can be computed exactly, then $\delta_i = 0$, and from (7) and (8) we have

$$\frac{b_i - a_i}{b_{i-1} - a_{i-1}} \le \frac{1}{2}\left(1 - \frac{m}{M}\right), \quad \frac{b_n - a_n}{b - a} \le \left[\frac{1}{2}\left(1 - \frac{m}{M}\right)\right]^n \tag{9}$$

As we can see from (5), the optimal algorithm comes to halving the localizer interval. Formulas (9) demonstrate that it leads to a faster reduction of the interval than can be achieved by ordinary halving methods.

§4 Optimal distribution of resources throughout computation

Approximate values of $f(x)$ at every x are obtained as a result of some computational or experimental process. In many such a process, the error δ of the evaluation of $f(x)$ at a given point is a certain function of the resource u consumed in the computation: $\delta = R(u)$. By a resource we can mean the labour or expense involved in the measurements, the computer time consumed etc. Thus, if for every x the value of $f(x)$ is obtained as the arithmetic mean of a number of independent measurements, then $R(u) \sim u^{-1/2}$, where u is the number of measurements. If for every x the evaluation of $f(x)$ is obtained through some iteration process which converges as a geometric progression with ratio $q < 1$, then $R(u) \sim q^u$ where u is the number of iterations. If for every x computing of $f(x)$ involves numerical solution of the Cauchy problem for a system of ordinary differential equations (as in solving boundary problems through appropriate selection of missing initial conditions), then $R(u) \sim u^{-p}$. Here u is the computer time, which is inversely proportional to the integration step, and the number p depends on the order of the finite-difference approximation used.

We now formulate the problem of devising an optimal search algorithm which takes into account the distribution of resources. Suppose that $f(x) \in F$ and that we are given the interval L, the numbers m and M, the integer n, the total resource

$U \geq 0$, and the function $R(u)$ which is defined, nonnegative and nonincreasing for all $u \geq 0$. The problem is to devise a method for successively selecting points x_i in L and numbers $u_i \geq 0$ $(i = 1,\dots,n)$ which make possible the attainment of the minimax

$$\Delta' = \min_{x_1,u_1} \max_{y_1} \min_{x_2,u_2} \max_{y_2} \dots \min_{x_n,u_n} \max_{y_n} (b_n - a_n) \tag{10}$$

Here x_i and y_i vary over the same ranges as in the case of minimax (4), the interval $[a_n,b_n]$ is given by equation (3); the errors are equal to $\delta_i = R(u_i)$; and the variables u_i are subject to the restrictions

$$u_i \geq 0 \ , \quad \sum_{i=1}^{n} u_i = U \qquad (i = 1,\dots,n) \tag{11}$$

For arbitrary u_i satisfying (11) and the corresponding $\delta_i = R(u_i)$ problem (10) reduces to problem (4), whose solution is given above (see (5)-(8)). Let us deal with the problem of distribution of the resources u_i .

Let $\Phi_j(s,v)$ be the minimal length b_j-a_j of the zero localizer which can be attained through an optimal distribution of resources in a j-step searching process; here s is the initial length of the localizer and v is the total resource in j steps. Using relation (7), in which equality can be attained, we easily establish

$$\Phi_j(s,v) = \min_{0\leq u\leq v} \Phi_{j-1}\left\{h\left[\frac{2R(u)}{ms}\right]s, v-u\right\} \ , \quad \Phi_0(s,v) = s \tag{12}$$
$$(j = 1,2,\dots)$$

Through the following change of variables (with arbitrary $\alpha \geq 0$, $\beta \geq 0$)

$$R(u) = mr(u)/2 \ , \quad u = tv \ , \quad s = r(v)/\sigma$$
$$\Phi_j(\alpha,\beta) = \alpha g_j(r(\beta)/\alpha,\beta) \qquad (j = 0,1,\dots) \tag{13}$$

formulas (12) assume the form

$$g_j(\sigma,v) = \min_{0\leq t\leq 1} \left\{h\left[\sigma \frac{r(tv)}{r(v)}\right] g_{j-1}\left[\frac{\sigma r[(1-t)v]}{r(v)h[\sigma r(tv)/r(v)]}, (1-t)v\right]\right\} \qquad (14)$$

$$g_0(\sigma,v) = 1 \qquad (j = 1,2,\ldots)$$

Here g_j is the ratio of the length of the zero localizer at the end of the j-step optimal search algorithm to the localizer's initial length. Calculation of an optimal distribution of resources reduces to computations by means of the recurrence formulas (14); besides the g_j, these computations will yield the (generally not unique) value of $t = T_j(\sigma,v)$ for which a minimum is attained in (14). The function $T_j(\sigma,v)$ represents that portion of the total resource which is spent in the first step of the j-step optimal algorithm.

Suppose that the functions $g_j(\sigma,v)$ and $T_j(\sigma,v)$, for $\sigma \geq 0$, $v \geq 0$,, $j = 1,..,n$, have been evaluated. The optimal search algorithm taking account of resource distribution reduces to the following: Setting $v_0 = U$, $a_0 = a$, $b_0 = b$, let us describe the i-th step of the n-step algorithm ($i = 1,\ldots,n$). Let us assume that after i-1 steps we have found the zero localizer $[a_{i-1},b_{i-1}]$ and that we know the remaining amount $v_{i-1} \geq 0$ of the resource. Setting $s_{i-1} = b_{i-1} - a_{i-1}$; $\sigma_{i-1} = r(v_{i-1})/s_{i-1}$; $u_i = v_{i-1}T_{n-i+1}(\sigma_{i-1},v_{i-1})$; $\delta_i = R(u_i)$; $x_i = (a_{i-1} + b_{i-1})/2$, we determine y_i, i.e., an approximate value of $f(x_i)$, and spend in this the resource u_i. Then, using formulas (6) and (3) we find the interval $[a_i,b_i]$ and move on to the next step. After n steps, the length of the zero localizer interval will be bounded by the inequality

$$b_n - a_n \leq (b-a)g_n(r(U)/(b-a),U)$$

in which equality may occur.

Let us state some properties of the functions g_j and T_j. The function $g_j(\sigma,v)$ does not decrease with incresing σ and fixed j and v, and does not increase with increasing j and fixed σ and v. For $\sigma \geq 1/2$ the functions T_j are not unique and may be taken to vanish. The following holds:

$$g_1(\sigma,v) = h(\sigma)\ , \qquad T_1(\sigma,v) = 1$$

$$g_j(0,v) = [(1-k)/2]^j \qquad (j = 1,2,\ldots) \tag{15}$$

$$g_j(\sigma,v) = \sigma \quad \text{if} \quad 1/2 \leq \sigma \leq 1\ , \quad g_j(\sigma,v) = 1 \quad \text{if} \quad \sigma \geq 1$$

For $j \geq 2$ and $0 \leq \sigma \leq 1/2$, the functions g_j and T_j depend on the function $r(u)$ and on the number k occurring in formula (8) for $h(z)$; their calculation from (14) is a rather cumbersome task.

For many important computational and experimental processes we have (with A and p constant)

$$R(u) = Au^{-p}\ , \quad r(u) = (2A/m)u^{-p} \quad (A > 0,\ p > 0) \tag{16}$$

When (16) hodls, g_j and T_j depend on σ alone, which considerably facilitates their tabulation. Relations (14) in conjunction with (16) yield

$$g_j(\sigma) = \begin{cases} \min\limits_{0 \leq t \leq 1} \left\{ h(\sigma t^{-p}) g_{j-1}\left[\dfrac{(1-t)^{-p}\sigma}{h(\sigma t^{-p})}\right]\right\} \end{cases}, \quad g_o(\sigma) = 1 \quad (j = 1,2,\ldots) \tag{17}$$

For sufficiently small σ it is possible to write the following exact analytic solution of equations (17):

$$g_j(\sigma) = \left(\frac{1-k}{2}\right)^j + k\left(\frac{1-q^j}{1-q}\right)^{p+1}\sigma\ , \quad T_j(\sigma) = \frac{q^{j-1}(1-q)}{1-q^j}$$

$$q = \left(\frac{1-k}{2}\right)^{1/(p+1)} \qquad (j = 1,2,\ldots)$$

The functions g_j and T_j were computed electronically from formulas (17) in conjunction with (8) and (15). As an illustration we give some results for $k = 0.1$ and $p = 1$

σ	0	0.1	0.2	0.3	0.4	0.5	1
$g_2(\sigma)$	0.2025	0.230	0.303	0.454	0.490	0.5	1
$T_2(\sigma)$	0.40	0.40	0.33	0.33	0	0	0

The detailed proof of the results given above and also numerous data of computing of optimal algorithms are represented in the paper [3]. Here we shall only point out two qualitative properties of obtained optimal distributions of resources.

1. For the first steps of computation (when the number of steps is large) it is profitable to spend only a small portion of the total resource.

2. Let the parameters k, p, σ in the expressions (16), (17) are fixed, and j grows monotonically. Then there exists such a sufficiently large number $j_0(k,p,\sigma)$, that $T_j = 0$ when $j \geq j_0$. It means that when the initial length of the interval and the total resource are given then it is unreasonable to distribute the total resource throughout more then j_0 steps. Therefore, there exists the optimal (finite) number of steps depending on the parameters of the problem. This number is also found through the process of numerical solution of equations (17) (see ref. [3]).

References

1. R.E. Bellman, S.E. Dreyfus. Applied dynamic programming. Princeton University Press, Princeton, 1962.

2. Ф.Л. Черноусько. Об оптимальном поиске корня функции, вычисляемой приближенно. Доклады Академии наук СССР, 1967, том 177, № 1, стр. 48-51.

3. Ф.Л. Черноусько. Оптимальный алгоритм поиска корня функции, вычисляемой приближенно. Журнал вычислительной математики и математической физики, том 8, № 4, 1968, стр. 705-724.

RANDOM OPTIMIZATION AND STOCHASTIC PROGRAMMING

Ju. Ermoliev

I. Introduction

Consider the problem of minimizing the function of variables

$$F(x_1,x_2,\ldots,x_n) \tag{1}$$

subject to

$$x = (x_1,x_2,\ldots,x_n) \in D \tag{2}$$

where D is a set of a space R^n.

One can differ deterministic and stochastic methods for solving this problem.

In deterministic methods minimum search is carried out on some certain point sequence $\{x^s\}$, $s = 0,1,\ldots$ (minimizing sequence), which uniquely determined by selected method and original approach x^o.

In stochastic minimization methods the stochastic minimizing sequences $\{x^s(\omega)\}$, $s = 0,1,\ldots$ are considered. I.e. in this case from the origin point x^o the whole trajectory family outputs and searching the minimum may be realized on one of it.

Obviously the deterministic methods are partial cases of the stochastic ones.

The stochastic methods are able to solve the problems, which cannot be solved with ordinary, deterministic ones. In particular that give opportunity to solve some important problems of large volume discrete programming.

In the present paper we consider one highly general class of stochastic minimizing methods, which may be called stochastic quasigradient methods. It is shown that these methods can be naturally applied in solving nonlinear problems of mathematical programming, arising, for example, in the choice of complex systems parameters on the base of results of "playing" some "scenarios".

2. Stochastic quasi-Feyer's sequences

This concepts is introduced in reference [1]. That may be useful in the ground of random (stochastic) optimization methods and linear inequalities solution ones.

Let A be some closed set from R^n, and $C(A)$ its convex cover.

A stochastic sequence of the points $\{Z^s(\omega)\}$, $s = 0,1,\ldots$ is called a stochastic quasi-Feyer's one with respect to the set A, if conditional mathematical expectation

$$M(\|y - Z^{s+1}\|^2 / Z^0,\ldots Z^s) \leq \|y - Z^s\|^2 + g_s \qquad (3)$$

for an arbitrary $y \in A$; $s = 0,1,\ldots$; g_s are such numbers that $\sum_{s=0}^{\infty} g_s < \infty$; $M\,\|Z^0\|^2 < \infty$.

The stochastic quasi-Feyer's sequence have properties analogous to ones of the arbitrary Feyer's sequences:

a) set of the limited points $\{Z^s(\omega)\}$ is not empty almost for every ω ;

b) if $Z^*(\omega)$, $Z^{**}(\omega)$ are any limited points which do not belong to $C(A)$, then $C(A)$ lays in a plane being geometrical place of the points which are equidistant to $Z^*(\omega)$ and $Z^{**}(\omega)$ (for given ω);

c) if a limited point $Z^*(\omega)$ belongs to $C(A)$, then $Z^*(\omega)$ is the only limited one for given ω .

These properties easily follow from that the sequence $P_k(\omega) = \|y-Z^k(\omega)\|^2 + \sum_{s=k}^{\infty} g_s$, $k = 0,1,\ldots$ under (3) is a semimartingal; and hence converges almost for every ω .

3. Method of stochastic quasi-gradients

To say nothing on the local extremum suppose that function $F(x)$ is convex down and unnecessarily differentiable. A set D is convex and closed. Let $\pi(x)$ be a operator of projection on D, i.e. such that $\pi(x) \in D$,

$$\|y - \pi(x)\|^2 \leq \|y - x\|^2$$

for any $y \in D$.

Consider a stochastic sequence of the points $\{x^s\}$ defined by the following relation

$$x^{s+1} = \pi(x^s - \rho_s \gamma_s \xi^s), \quad s = 0,1,\ldots \tag{4}$$

Here x^0 is an arbitrary point for which $M\|x^0\| \leq$ const ; ρ_s is a step value; γ_s is a normalization factor; ξ^s - stochastic vector which mathematical expectation (on each component separately)

$$M(\xi^s / x^0, \ldots, x^s) = C_s \hat{F}_x(x_s) + \lambda^s \tag{5}$$

Where C_s is a nonnegative number and λ^s is a vector, depending on $x^0, \ldots, x^s$; $\hat{F}_s(x^s)$ is a generalized gradient vector satisfying the unequality

$$F(y) - F(x) \geq (\hat{F}_x(x), y - x)$$

for any points x and y.

In the case of differentiable function $F(x)$, the generalized gradient vector coincides with ordinary gradient. Generally that is directed on the normal to any plane of support (Fig.) at the point x^s of a set $\{x: F(x) \leq F(x^s)\}$.

In the deterministic case when $\xi^s(\omega) \equiv F(x^s)$, $C_s \equiv 1$, $\lambda^s \equiv 0$ the method (4) was studied in references [2] - [5]. Note that even in that case and under condition $D = R^n$ from iteration to iteration there is not observed monotone decrease of object function. In reference [5] it is shown that under certain conditions of choice of ρ_s one can get the convergence of distance squares to the minimum point with the geometrical progression speed.

The method of stochastic quasi-gradients (4) - (5) was studied in references [7]-[9].

The values C_s, λ^s can be considered as the generalized gradient extimation errors in so far as even the exact value of gradient for the differentiable function $F(x)$ is known in the rare cases.

In the following we shall suppose that there are known such constants r_s, m_s depending only on s, for which $C_s \geq r_s$, $\|\lambda^s\| \leq m_s$. Then the following assertion is true.

Theorem.

If

a) there is known the value h_s such that

$$M(\|\xi^s\|^2/x^0,\ldots,x^s) \leq h^2_s \leq M_B < \infty$$

subject to $|x^k| \leq B < \infty$ $(k = 0,1,\ldots,s)$;

b) $0 < \underline{\gamma} \leq \gamma_s h_s \leq \bar{\gamma} < \infty$;

c) $\rho_s \geq 0$, $\sum_{s=0}^{\infty} \rho_s m_s < \infty$, $\sum_{s=0}^{\infty} \rho^2_s < \infty$, $\frac{m_s}{r_s} \to 0$,

then the point sequence $\{x^s(\omega)\}$ is the stochastic quasi-Feyer's one; if moreover, $\sum_{s=0}^{\infty} \rho_s r_s = \infty$ then almost for every ω the sequence $\{x^s(\omega)\}$ converges to the solution of the problem (1) - (2). The theorem conditions are easily checked in concrete problems. In references [7], [8] there are other theorems on the convergence of the processes of the form (4) - (5) concerning the cases of presence of the restriction of the form

$$F^i(x_1,\ldots,x_n) \leq 0, \quad i = 1,2,\ldots,m$$

and also nonconvex function $F(x)$.

Now show why is the vector ξ^s estimated, satisfying (5) for the concrete problems solutions.

4. Adaptive processes of minimization.

In analysis and synthesis of the complex system there is not mostly the united analytical model describing its behavior. In general we have one or a few scenarios in accordance with that the system activity is developed. Every scenario can consist of a number of the analytical models connected with certain logical and probabilistic transitions.

Executers of that are computers and even actual objectives, men (that, for example, has concerned in professor Medov's report). In such situation one has opportunity to use only the results of separate draws of the scenarios.

It is necessary to construct such adaptive processes of the system parameters unknown values search, for which information concerned is sufficient. The method (4) - (5) allows to offer a number of such processes.

1. Let every "draw" of the "scenario" (we have the only "scenario") give us some stochastic value $y(x,\omega)$, characterizing the system efficiency under given parameters (of the plan) $x = (x_1,\dots,x_n)$. The adaption purpose is to find x for which

$$F(x) = M\,Y(x,\omega) = \min \tag{6}$$

subject to $x \in D$ (7)

Complexity of this problem is that the analytical form of the function $F(x)$ is unknown. For any x we can observe only separate realization of the value $Y(x,\omega)$. In the reference [10] for solving the problem (6) without restrictions (7) it was offered the important adaptive process (stochastic approximation method)

$$x^{s+1} = x^s - \rho_s \frac{1}{\Delta_s} \sum_{j=1}^{n} \left[Y(x^s + \Delta_s e^j, \omega^{sj}) - Y(x^s, \omega^{s0}) \right] e^j \tag{8}$$

where e^j is a point of the j-th axis; ω^{sv}, $v = 0,1,\dots,n$ are the independent tests in the s-th iteration.

It is easily seen that under the second restricted derivatives of the function $F(x)$ in the domain D

$$M\left(\frac{1}{\Delta_s}\sum_{j=1}^{n}\left[Y(x^s+\Delta_s e^j,\omega^{sj})-Y(x^s,\omega^{so})\right]e^j/x^s\right)=F_x(x^s)+W^s\Delta_s \tag{9}$$

where $\|W^s\| \leq const < \infty$. I.e. the method (8) is a special case of the process (4) under

$$\xi^s=\frac{1}{\Delta_s}\sum_{j=1}^{n}\left[Y(x^s+\Delta_s e^j,\omega^{sj})-Y(x^s,\omega^{so})\right]e^j \tag{10}$$

If still suppose the vector components dispersions are restricted in D , then in the theorem conditions a) values $h_s \equiv const$.

To estimate the vector (10) it is necessary to have $(n+1)$ observations of the values $Y(x^s,\omega)$.

Given $n = 60$ and one observation lasts 0,5 min, then one iteration time lasts 0,5 hour and therefore the method (8) can turn out unpractical. However the choice of the vector ξ^s given in the method (4) by the formula (10) is not the only possible.

One can accept, for example,

$$\xi^s=\frac{1}{\Delta_s}\sum_{k=1}^{P_s}\left[Y(x^s+\Delta_s\theta^{sk},\omega^{sk})-Y(x^s,\omega^{so})\right]\theta^{sk}$$

where $P_s \geq 1$, $\{\theta^{sk}\}$, $k = 1,\dots,P_s$ - is the independent realizations series of the stochastic vector $\theta = (\theta_1,\dots,\theta_n)$ with the components independent and evenly distributed on $[-1,1]$.

It can be easily shown, that the relation of the form (9) is valid for the vector (II) subject to the second restricted derivatives of the function $F(x)$ in the domain D. To estimate (II), $P_s + 1$ observations are needed, where $P_s \geq 1$.

2. Let there is a number of the "scenarios" $i = 1,2,\dots,m$, in every of which the value $Y_i(x,\omega)$ describes the efficiency of the plan $x = (x_1,\dots,x_n)$.

The adaptation purpose is to minimize

$$F(x) = M \max_i Y_i(x,\omega) = \min \tag{12}$$

subject to $x \in D$. (13)

Let the function $Y(x,\omega)$ be convex down for every ω , and for $x \in D$ has the second restricted derivative on x for all ω evenly.

Then one can assume that

$$\xi^s = \frac{1}{\Delta_s} \sum_{j=1}^{n} \left[Y_{i_s}(x^s + \Delta_s e^j, \omega^s) - Y_{i_s}(x^s,\omega^s)\right] e^j , \tag{14}$$

or

$$\xi^s = \frac{1}{\Delta_s} \sum_{k=1}^{P_s} \left[Y_{i_s}(x^s + \Delta_s \theta^{sk}, \omega^s) - Y_{i_s}(x^s,\omega^s)\right] \theta^{sk} \tag{15}$$

where the number i_s is defined from the condition

$$Y_{i_s}(x^s,\omega^s) = \max_i Y_i(x^s,\omega^s) .$$

It can be shown that if ξ^s is chosen in accordance with (14) or (15) and the function $F(x)$ respectively (12), then

$$M(\xi^s/x^s) = F_x(x^s) + W^s \Delta_s ,$$

where $\|W^s\| \leq \text{const} < \infty$.

In the same manner the more general game with the stochastic gain function

$$F(x) = M \max_{z \in Z} Y(x,z,\omega) = \min \tag{16}$$

is solved subject to $x \in D$. (17)

Note the function $F(x)$ defined in accordance with (12) or (16) will be undifferentiable.

5. General problems of stochastic programming.

The considered problems (6) - (7), (12) - (13), (16) - (17) are the special problems of stochastic programming, in which the objective functions and restrictions can depend on the stochastic parameters. If we consider the only case, when the plan $x = (x_1,..,x_n)$ under search must be determined, then highly general problem of stochastic programming consists in minimizing the function

$$F^o(x) = Mf^o(x,\omega) \tag{18}$$

subject to $F^i(x) = Mf^i(x,\omega) \leq 0$, (19)

$$i = 1,2,...,m$$

$$x \in D , \tag{20}$$

where $f^v(x,\omega)$ are some stochastic values, depending on $x = (x_1,...,x_n)$. In the proper choice of the values $f^v(x,\omega)$ to the problem (18)-(20) one can reduce, for example, the problems with the probabilistic restrictions of the form

$$G^i(x) = P\{Y_i(x,\omega) \leq 0\} \geq P_i$$

and the objective function

$$G^o(x) = P\{Y_o(x,\omega) \geq Y\} .$$

In such a case

$$f^o(x,\omega) = \begin{cases} 1, & Y_o(x,\omega) \geq Y , \\ 0, & Y_o(x,\omega) < Y , \end{cases}$$

$$f^i(x,\omega) + P_i = \begin{cases} 1, & Y_i(x,\omega) \leq 0 , \\ 0, & Y_i(x,\omega) > 0 , \end{cases}$$

By introducing in the proper way the functions the problem (18)-(20) can be solved with the method (4) also.

6. The problems of two-stage stochastic programming.

These very important problems of maiking decisions were for the first time examined in the reference [II]. The term "two-stage" one ought not to understand literally, i.e. there are only two stages of planing - there can be any number of them.

In the problems of two-stage stochastic programming one stage is connected with the activity plan acceptance long before the future becomes known, and the next stage consists in to correct the plan as unknown future accepts the certain values.

In such a case one tends to choose the plan so that the expenditure connected with its realization and correction to be minimal at the average.

Let the plan $x = (x_1, \dots, x_n)$ accepted ahead of some time interval must satisfy the following restrictions

$$A(\omega)x + D(\omega)y \leq b(\omega) \tag{21}$$

$$x \in D \tag{22}$$

where y is a correction vector. The plan x is acceptable till the values $A(\omega)$, $D(\omega)$, $b(\omega)$ became known. After that, the inequalities (21) are corrected with the vector (21). If $(d(\omega), y)$ is a loss on the correction then we can find a vector $y(x,\omega)$ minimizing

$$(d(\omega), y) \tag{23}$$

subject to (21), where the vector x and also $A(\omega)$, $D(\omega)$, $b(\omega)$ are fixed. The problem is to find the vector x, minimizing

$$F(x) = (c, x) + M(d(\omega), y(x,\omega)) \tag{24}$$

subject to

$$x \in D \tag{25}$$

Let along with $y(x,\omega)$ it can be obtained $u(x,\omega)$ namely the dual variables associated $y(x,\omega)$.

Then for solving the problem (24)-(25) such adaptive process can be offered. Denote by x^s approach resulted from the s-th iteration. Observe $A(\omega^s)$, $D(\omega^s)$, $b(\omega^s)$, $d(\omega^s)$ and solve under $x = x^s$, $\omega = \omega^s$ the problem (21)-(23). Obtain $y(x^s,\omega^s)$, $u(x^s,\omega^s)$.

Change x^s according to (4) subject to

$$\xi^s = C - A^T(\omega^s)u(x^s,\omega^s)$$

It can be easily shown [8], that if $F(x)$ is defined according to (24) then

$$M(\xi^s/x^s) = \hat{F}_x(x^s) ,$$

besides for the theorem conditions a) under the restricted variances of $A(\omega)$, $D(\omega)$, $b(\omega)$, $d(\omega)$ we have h_s = constant.

In reference [8] the process described above has been extended to general unlinear problems of two-stage programming. Note the problem of minimizing the undifferentiable function (24) can be considered as the stochastic problem of parameter programming and the process above as the specific method of block programming (given D has blocks).

The method stated in the reference [II] allows to solve the two-stage problems in case when only the vector $b(\omega)$ is stochastic and has the finite number of values with the prescribed probabilities.

Sometimes in terms of two-stage programming problem it can be easily formulated the synthesis problem of objective control which subjects to random disturbances.

7. Programmed control of stochastic process.

Let the behavior of an objective which subjects to random disturbances be described with the system of difference equations

$$x_i(k+1) = x_i(k) + g_i(x,y,k,\omega), \quad x_i(0) = x_i^o \tag{26}$$

$$i = 1,2,\dots,n \; ; \qquad k = 0,1,\dots,N-1$$

where the control vector $y(k) = (y_1(k),\dots,y_n(k))$ in every moment of time $k = 0,\dots,N-1$ belongs to the domain D, i.e.

$$y(k) \in D, \quad k = 0,1,\dots,N-1 \tag{27}$$

It is necessary to find the programmed control, i.e. the vector-function $y(k)$, $k = 0,\dots,N-1$, satisfying (27) for which the function

$$F(y(0),\dots,y(N-1)) = M\varphi(x(N),\omega) \tag{28}$$

under fixed N gets the minimal value.

In this case the method (4) corresponds with the following adaptive process. Let the approach $y^s(k)$, $k = 0,1,\dots,N-1$ be resulted from the s-th iteration. Make the observation ω^s and obtain from (26) a trajectory $x^s(k)$, $k = 0,1,\dots,N$. Find the solution $\lambda^s(k)$, $k = N,\dots,0$ of the following system of conjugate equations for $x(k) = x^s(k)$, $y(k) = y^s(k)$, $\omega = \omega^s$

$$\lambda_i(k) = \lambda_i(k+1) + \sum_{j=1}^{n} (\lambda_j(k+1) g_{j x_i}(x,y,k,\omega))$$

$$\lambda(N) = -\varphi_x(x(N),\omega)$$

Let

$$\xi^s(k) = \sum_{i=1}^{n} \lambda_i^s(k+1) g_{iy}(x^s,y^s,k,\omega^s)$$

and consider a set of vectors

$$\xi^s = (\xi^s(0),\dots,\xi^s(N-1))$$

The set ξ^s represents the stochastic gradient vector of the function (28) as one of variables $y(0),\dots,y(N-1)$ for which

$$M(\xi^s / y^s(0),\dots,y^s(N-1)) = \text{grad } F(y^s(0),\dots,y^s(N-1)) \ .$$

Therefore a new control $y^{s+1}(k)$, $k = 0,\dots,N-1$ can be obtained from the formula (4) and so on.

I. Ю.М.Ермольев, А.Д.Туниев. Случайные фейеровские и квазифейеровские последовательности, сб. "Теория оптимальных решения", ИК АН УССР, Киев, № 2, 1968.

2. Н.З.Шор. О структуре алгоритмов численного решения задач оптимальногоп планирования и проектирования, автореферат диссертации, ИК АН УССР, Киев, 1964.

3. Ю.М.Ермольев. Методы решения нелинейных экстремальных задач, журн. Кибернетика, № 4, К., 1966.

4. Ю.М.Ермольев, Н.З.Шор. О минимизации недифференцируемых функций, журнал Кибернетика, № 2, К., 1967.

5. Н.З.Шор. О сходимости метода обобщенного градиента, журнал Кибернетика, № 3, К., 1968.

6. Б.Т.Поляк. Один общий метод решения экстремальных задач, ДАН СССР, том 174, № I, 1967.

7. Ю.М. Ермольев, З.В.Некрылова. Метод стохастических градиентов и его применение. сб. "Теория оптимальных решений", ИК АН УССР, Киев, № I, 1967.

8. Ю.М.Ермольев, Н.З.Шор. Метод случайного поиска для двухэтапной задачи стохастического программирования и его обобщение, журнал Кибернетика, № I, К., 1968.

9. Ю.М.Ермольев, А.Д.Туниев. О некоторых прямых методах стохастического программирования, журнал "Кибернетика", № 4, К.,1968

IO. I.R.Blum. Multidimensional stochastic approximation methods. Annals of Mathematical Statistics. 1954, v. 25, no. 4.

II. I.B.Dantzig, Madansky. On the solution of two-stage linear programming problems under uncertainty. Proc. Fourth Berkeley Symposium on Mathematical Statistics and Probability, 1961, no. 1.

APPROXIMATE CALCULATION OF OPTIMAL CONTROL BY AVERAGING METHOD

Yu.G.Yevtushenko

The present paper is devoted to the solution of optimal control by means of the Kryloff-Bogoliuboff averaging method. The approximate solution obtained by this method may be sufficiently accurate, or may be used as an approximate solution for further numerical calculation. In some cases this method permits us to simplify equations and after that to use numerical calculation.

The potentialities inherent in this approach are demonstrated by applying it to two problems of optimal satellite motion programming. In the first part the problem of optimal spacecraft transfer from near-earth orbit to another orbit with specified energy is solved.

The second part is devoted to the optimal correction of satellite motion by means of low thrust engine in noncentral field of the earth. It is shown that in the first approximation the problem is reduced to the solution of a certain boundary problem for the system of four differential equations.

1. Consider the problem of optimal changing of satellite energy by means of low thrust engine. This problem arises, for example, when we try to transfer a spacecraft from a specified near-earth orbit to earth-period (24-hr) orbit. The similar problems have been considered by many authors (See [1,2], for example).

Assume that the eccentricity of orbit e remains less than unity, thrust acceleration W is small. Therefore we take as a small parameter the quantity $\epsilon = W_1 r_1^2 \mu^{-1}$ equal to the ratio of characteristic value of thrust W_1 to the Earth's gravitation acceleration at a certain characteristic distance from centre of Earth to satellite r_1 (μ- gravitational constant). Introduce the following dimensionless quantities: τ - time of motion, p - focal parameter, n - mean angular velocity, w - thrust acceleration

$$\tau = t\sqrt{\mu}\, r_1^{-\frac{3}{2}}, \quad p = Pr_1^{-1}, \quad n = (1-e^2)p^{-\frac{3}{2}}, \quad w = WW_1^{-1} \tag{1.1}$$

where $t, P,$ are dimension quantities.

The spacecraft motion equations are of the form

$$\frac{dn}{du} = -\frac{3\varepsilon}{R^2}(1-e^2)n^{-\frac{1}{3}} K\sqrt{1+2e\cos\vartheta+e^2} = Kf_1$$

$$\frac{de}{du} = \varepsilon D[2K(e+\cos v) + \frac{N}{R}(1-e^2)\sin v] = Kf_{21} + Nf_{22}$$

$$\frac{d\omega}{du} = \frac{\varepsilon D}{e}[2K\sin v - \frac{N}{R}[2e+(1+e^2)\cos v]] = Kf_{31} + Nf_{32}$$

$$\frac{d\tau}{du} = \frac{(1-e^2)^{\frac{3}{2}}}{nR^2}, \quad D = \frac{(1-e^2)^2 n^{-\frac{4}{3}}}{R^2\sqrt{1+2e\cos\vartheta+e^2}}, \quad u = \vartheta + \omega$$

$$R = 1 + e\cos\vartheta, \quad K = w\cos\gamma, \quad N = w\sin\gamma \tag{1.2}$$

Where K, N are projections of the thrust acceleration in tangential and normal directions, ω is the angular longitude of perigee, u - angular displacement of satellite from line of nodes.

The optimal control problem consists of finding control functions $K(u)$, $N(u)$, $u_0 \leq u \leq u_1$ that minimize the performance of the system

$$I = \varepsilon\int_0^T (K^2+N^2)d\tau$$

The interval $u_1 - u_0$ and $n(u_1) = u_1$ will be assumed fixed. We shall use Pontryagin's maximum principle [3]. The conjugate variables p_1, p_2, p_3 correspond to the functions n, e, ω. The Hamiltonian function is

$$H = Kf_1p_1 + (Kf_{21}+Nf_{22})p_2 + (Kf_{31}+Nf_{32})p_3 - \frac{\varepsilon(1-e^2)^{\frac{3}{2}}(K^2+N^2)}{n R^2} \tag{1.3}$$

The optimal control $\mathfrak{K}, \mathfrak{N}$ is obtained by maximizing the above expression

$$\mathfrak{K} = \frac{n R^2}{2\varepsilon(1-e^2)^{\frac{3}{2}}}[f_1p_1 + f_{21}p_2 + f_{31}p_3], \quad \mathfrak{N} = \frac{n R^2}{2\varepsilon(1-e^2)^{\frac{3}{2}}}[f_{22}p_2 + f_{32}p_3] \tag{1.4}$$

The terminal boundary conditions are

$$u = u_1 \ , \quad p_2 = p_3 = 0 \quad , \quad n = n_1 \tag{1.5}$$

Inserting (1.4) into (1.3) yields

$$H = \frac{n\,R^2}{4\epsilon(1-e^2)^{\frac{3}{2}}} [(f_1 p_1 + f_{21}p_2 + f_{31}p_3)^2 + (f_{22}p_2 + f_{32}p_3)^2] \tag{1.5'}$$

We can rewrite the system (1.2) in Hamiltonian form

$$\frac{dn}{du} = \frac{\partial \mathfrak{H}}{\partial p_1} \ , \quad \frac{de}{du} = \frac{\partial \mathfrak{H}}{\partial p_2} \quad , \quad \frac{d\omega}{du} = \frac{\partial \mathfrak{H}}{\partial p_3}$$

$$\frac{dp_1}{du} = -\frac{\partial \mathfrak{H}}{\partial n} \ , \quad \frac{dp_2}{du} = -\frac{\partial \mathfrak{H}}{\partial e} \quad , \quad \frac{dp_3}{du} = -\frac{\partial \mathfrak{H}}{\partial \omega} \tag{1.6}$$

where $\mathfrak{H}$ is given by (1.5'). Inserting (1.4) into

gives

$$I = \int_{u_0}^{u} \mathfrak{H}\, du \qquad .$$

The initial problem is reduced to the boundary problem, for further calculation we can use numerical methods. Instead of this approach we shall apply asymptotic or averaging techniques of Kryloff and Bogoliuboff [4,5].

The system (1.6) obtained above has standard form [4]. The variables $n, e, \omega, p_1, p_2, p_3$ are slowly varying functions of u .

In order to design an asymptotic solution in the first approximation it is necessary that the right-hand parts of the system be averaged with respect to τ over non-disturbed motion period.

We also can average Hamiltonian

$$U = \frac{1}{2\pi} \int_0^{2\pi} \mathfrak{H}\, du = \frac{\epsilon}{8n^{\frac{3}{2}}} [18n^2p_1^2 + 5(1-e^2)p_2^2 + \frac{5-4e^2}{e^2} \ p_3^2]$$

Inserting U into (1.6) yields a system of the first approximation

$$\frac{de}{du} = \frac{5\epsilon p_2}{4}(1-e^2)n^{-\frac{5}{3}} , \qquad \frac{d\omega}{du} = \frac{\epsilon p_3}{4e^2}(5-4e^2)n^{-\frac{5}{3}}$$

$$\frac{dn}{du} = \frac{9\epsilon}{2} n_1^{\frac{1}{3}} p_1 , \qquad \frac{dp_1}{du} = \frac{5}{3n} U - \frac{9\epsilon}{2} p_1^{2} n^{-\frac{2}{3}}$$

$$\frac{dp_2}{du} = \frac{5\epsilon e}{4n^{\frac{5}{3}}}\left[p_2^{2} + \frac{p_3^{2}}{e^4}\right] , \qquad \frac{dp_3}{du} = 0 \qquad (1.7)$$

This system has the first integrals

$$U = \text{const} , \quad p_3 = \text{const} , \quad 5(1-e^2)p_2^2 + \frac{5-4e^2}{e^2} p_3^2 = \text{const}$$

$$p_1 n = p_{11} n_1 + \frac{5}{3} H(u-u_1)$$

$$n^{\frac{5}{3}} = n_1^{\frac{5}{3}} + \frac{25}{4} \epsilon H(u-u_1)^2 + \frac{15}{2} \epsilon p_{11} n_1 (u-u_1)$$

$$\omega - \frac{\epsilon}{4} p_3 \int_0^u (5-4e^2)e^{-2} n^{-\frac{5}{3}} du = \text{const}$$

Subscript 1 denotes terminal value of function. After neglecting all the terms of $-\epsilon$ we obtain

$$I = U(u_1-u_0)$$

Recalling (1.5) we get

$$e = e_0 , \quad \omega = \omega_0 , \quad p_2 = p_3 = 0$$

$$n = \left[n_1^{\frac{5}{6}} + \frac{n_0^{\frac{5}{6}} - n_1^{\frac{5}{6}}}{u_0 - u_1} (u-u_1) \right]^{\frac{6}{5}} , \qquad p_1 = \frac{4(n_0^{\frac{5}{6}} - n_1^{\frac{5}{6}})}{15 \varepsilon n^{\frac{1}{6}} (u_0 - u_1)}$$

We get an optimal programm of thrust

$$K = \frac{2(n_0^{\frac{5}{6}} - n_1^{\frac{5}{6}})\, n_1^{-\frac{1}{6}} \sqrt{1+2e\cos\vartheta + e^2}}{5\varepsilon(u_1 - u_0)\sqrt{1-e^2}} \left[n_1^{\frac{5}{6}} + \frac{n_0^{\frac{5}{6}} - n_1^{\frac{5}{6}}}{u_0 - u_1} (u - u_1) \right]^{\frac{3}{5}}$$

The obtained above asymptotic solutions of the first order approximate an accurate solution of an optimal problem with error ε within the motion interval $u \sim \varepsilon^{-1}$.

2. In some cases it is very important to provide stable satellite motion inspite of the disturbing influence of the Earth noncentral force field. To avoid these difficulties we shall find the optimal corrective thrust law.

To discribe the motion we introduce the new functions: i-angle of inclination of the orbital plane with Earth's equatorial plane, angle Ω defines the line of nodes. We write the gravitational potential of Earth in the form of

$$V = \frac{\mu}{r} \left[1 + \sum_{n=2}^{\infty} \sum_{m=0}^{n} \left(\frac{r_3}{r}\right)^n C_{nm} P_{nm}(\sin\varphi) \cos m(\Lambda - \lambda_{mn}) \right]$$

where P_{nm} are the spherical functions, r_3 is the radius of the Earth, $\varphi = \arcsin(\sin u \sin i)$ is the geocentrical latitude, Λ is the geographical longitude referenced in the plane of equator to the East of the major axis of the equatorial ellipse

$$\Lambda = \frac{\pi}{2} + \Omega - \omega_3 t + \text{arc cos}\,\frac{\cos u}{\cos\varphi} \qquad (2.2)$$

where ω_3 is the angular velocity of the Earth's rotation round the polar axis, C_{nm} are the constant functions. The force caused by non-centrality of the gravitational field of the Earth is considerably less than basic force of the Newtonian attraction. Therefore we take a small parameter, the quantity

$$\varepsilon = -3C_{20}\left(\frac{r_3}{P_o}\right)^2 \qquad (2.3)$$

proportional to the relation of acceleration caused by the first degree of the polar oblatness to the acceleration of gravity at a certain characteristic height P_o which is equal to the initial focal parameter.

The coefficients C_{30}, C_{40}, C_{22}, C_{23} in gravitational potential are of the second order of smallness with respect to the polar oblatness. Assume therefore

$$c_{nm} = \frac{C_{nm}}{\varepsilon^2}\left(\frac{r_3}{P_o}\right)^n \qquad (2.4)$$

The thrust acceleration is of the same order, so we have

$$w = \frac{Wr_3^2}{\varepsilon^2\mu P_o^2}$$

The dimensionless quantities τ, p, n were introduced by (1.1). Now we shall put $r_1 = P_o$ in (1.1).

We shall write the potential of disturbing forces related to $\varepsilon\mu P_o^{-1}$ in the form

$$U = \frac{1-3\sin^2\varphi}{6p^3}R^3 + \frac{\varepsilon R}{p}\sum_n\sum_m \frac{R^n c_{nm}}{p^n} P_{nm}(\sin\varphi)\cos m(\Lambda-\lambda_{nm}) = \frac{R^3}{6p^3}(1-3\sin^2\varphi) + \varepsilon U_1(p,R,\varphi) + \varepsilon U_2(p,R,\Lambda)$$

where U_1, U_2 are dimensionless potentials of Earth.

If U depends on $\Omega, \omega, i, n, e, M, \tau$ we have

$$\frac{d\Omega}{d\tau} = \frac{\varepsilon n^{\frac{1}{3}}}{\sqrt{1-e^2}\sin i}\frac{\partial U}{\partial I} = f_1 \quad , \quad \frac{dn}{d\tau} = -3\varepsilon n^{\frac{4}{3}}\frac{\partial U}{\partial M}$$

$$\frac{di}{d\tau} = \frac{\varepsilon n^{\frac{1}{3}}}{\sqrt{1-e^2}\sin i}\left[\cos i\frac{\partial U}{\partial \omega} - \frac{\partial U}{\partial \Omega}\right] = f_2 \quad , \quad \frac{de}{d\tau} = \frac{\varepsilon n^{\frac{1}{3}}\sqrt{1-e^2}}{e}\left[\sqrt{1-e^2}\frac{\partial U}{\partial M} - \frac{\partial U}{\partial \omega}\right] = f_3$$

$$\frac{d\omega}{d\tau} = \frac{\varepsilon n^{\frac{1}{3}}}{e}\sqrt{1-e^2}\left[\frac{\partial U}{\partial e} - \frac{e\operatorname{ctg} i}{1-e^2}\frac{\partial U}{\partial I}\right] = f_4 \quad , \quad \frac{dM}{d\tau} = n + \frac{\varepsilon n^{\frac{1}{3}}(1-e^2)}{e}\left[\frac{3ne}{1-e^2}\frac{\partial U}{\partial n} - \frac{\partial U}{\partial e}\right]$$

$$M = 2\operatorname{arc\,tg}\sqrt{\frac{1-e}{1+e}}\operatorname{tg}\frac{\vartheta}{2} - \frac{e\sqrt{1-e^2}\sin\vartheta}{1+e\cos\vartheta} \tag{2.5}$$

Consider a resonance case. The resonance takes place if the frequencies n and ω_4 are commensurable, i.e. if the coprime number m and s exist and are such that

$$n - \frac{m}{s}\omega_4 \approx O(\varepsilon) \quad (\omega_4 = \omega_3\mu^{-\frac{1}{2}}P_o^{\frac{3}{2}}) \tag{2.5'}$$

A new variable α (the phase shift between the average satellite longitude $M + \Omega + \omega$ and the longitude of the equatorial ellipse half-axis) is introduced

$$\alpha = M + \Omega + \omega - \frac{n}{s}\tau\omega_4 \tag{2.6}$$

If we neglect U_2 then the system (2.5) will have integral

$$n^{\frac{2}{3}} - n_o^{\frac{2}{3}} + 2\varepsilon(U-U_o) = 2\varepsilon F \tag{2.7}$$

Let us take into account the effect of thrust. We have an equation for n

$$\dot{n} = -3\epsilon n^{\frac{4}{3}} \frac{\partial U}{\partial M} - \frac{3\epsilon^2 n^{\frac{2}{3}} w \cos\gamma}{\sqrt{1-e^2}} \sqrt{1+2e\cos\theta + e^2}$$

where γ is the angle that the thrust makes with the direction of velocity.

From (2.7) with error of the second order of smallness we obtain for n the following expression

$$n = n_o + 3\epsilon n_o^{\frac{1}{3}}(F+U_o-U) \tag{2.9}$$

Differentiating (2.6) (2.7) owing to (2.5) using (2.8) (2.9) and after neglecting all the terms of ϵ^2 yields the following set of differential equations

$$\frac{d\Omega}{d\tau} = f_1 , \quad \frac{di}{d\tau} = f_2 , \quad \frac{de}{d\tau} = f_3 , \quad \frac{d\omega}{d\tau} = f_4$$

$$\frac{d\alpha}{d\tau} = n - \frac{m}{s}\,\omega_1 + \frac{\epsilon\, n^{\frac{1}{3}}}{\sqrt{1-e^2}} \left[tg\,\frac{i}{2}\,\frac{\partial U}{\partial I} + 3\sqrt{1-e^2}\, n\,\frac{\partial U}{\partial n} + \frac{e(1-e^2)}{1+\sqrt{1-e^2}}\,\frac{\partial U}{\partial e} \right] = f_6$$

$$\frac{dF}{d\tau} = \epsilon\,\frac{\partial U_2}{\partial \tau} - \frac{\epsilon w n^{\frac{1}{3}}\cos\gamma}{\sqrt{1-e^2}} \sqrt{1+2e\cos\theta + e^2} = f_5 \tag{2.10}$$

where $U = U(\Omega,\omega,i,n,e,M,\tau)$.

Let us consider the problem of finding $w(\tau)$, $\gamma(\tau)$ to minimize

$$I = \epsilon \int_0^T w^2\, d\tau$$

subject to the constraints (2.10) and some condition for $\alpha(T)$, $\dot{\alpha}(T)$. The conjugate variables p_1, p_2, p_3, p_4, p_5, p_6 correspond to the functions $\Omega, i, e, \omega, F, \alpha$. Minimizing

$$H = \sum_{i=1}^{6} p_i f_i - \epsilon\, w^2$$

with respect to γ, w yields

$$\gamma = 0 , \quad w = - \frac{p_5 n^{\frac{1}{3}}}{2\sqrt{1-e^2}} \sqrt{1+2e \cos \vartheta + e^2} \tag{2.13}$$

Substituting (2.13) into the system (2.10) and conjugated system we reduce the problem of determining optimal control to the solution of a boundary problem for the system of twelve ordinary differential equations. This system has a standard form and averaging techniques of Kryloff and Bogoliuboff may be used.

Averaging $U, \frac{\partial U_2}{\partial \tau}$ yields

$$\bar{U} = \frac{n}{2\pi} \int_0^{\frac{2\pi}{n}} U \, d\tau = \frac{n}{6(1-e^2)^{\frac{3}{2}}} (1 - \frac{3}{2} \sin^2 i) + O(\varepsilon)$$

$$\tilde{U} = \lim_{T \to \infty} \frac{1}{T} \int_0^T \frac{\partial U_2}{\partial \tau} \, d\tau \tag{2.14}$$

Using angular displacement measured from the line of nodes we obtain

$$\tilde{U} = \frac{(1-e^2)^{\frac{3}{2}}}{2\pi m} \int_0^{2\pi m} \frac{\frac{\partial U_2}{\partial \tau} \, du}{[1+e \cos(u-\omega)]^2} \tag{2.15}$$

The slowly varying functions of time $i, e, \Omega, \omega, \alpha, p_1, p_2, p_3, p_4, p_5, p_6$ are assumed constant over period, only angle M is changed in (2.14) (2.15)

$$M = M_0 + n\tau$$

In the case of near circular orbits we can calculate U expanding U_2 in terms of eccentricity e .

The system of the first approximation has the form

$$\frac{d\Omega}{d\tau} = -\frac{\epsilon n}{2p^2}\cos i \quad , \quad \frac{d\omega}{d\tau} = \frac{\epsilon n}{4p^2}(5\cos^2 i - 1)$$

$$\frac{de}{d\tau} = \frac{di}{d\tau} = 0 \quad , \quad \frac{dF}{d\tau} = \epsilon\tilde{U} + \frac{\epsilon}{2} p_5 n_o^{\frac{2}{3}}$$

$$\frac{d\alpha}{d\tau} = n_o - \frac{m}{s}\omega_4 + 3\epsilon n^{\frac{1}{3}}(F+U_o) + \frac{\epsilon n}{4p^2}(3\cos^2 i - 1) \tag{2.16}$$

$$\frac{dp_5}{d\tau} = -3\epsilon n_o^{\frac{1}{3}} p_6 \quad , \quad \frac{dp_6}{d\tau} = -3p_5\frac{\partial\tilde{U}}{\partial\alpha}$$

The conjugate variables p_1, p_2, p_3, p_4 do not have influence on motion and equations for them were omitted. Integrating system (2.16) yields

$$\Omega = \Omega_o - \frac{\epsilon n\tau}{2p^2}\cos i \quad , \quad \omega = \omega_o + \frac{\epsilon n_o \tau}{4p^2}(5\cos^2 i - 1) \quad , \quad i = i_o \; , \quad e = e_o$$

The remained four equations must be integrated by numerical methods. We shall consider only one particular case. Assuming the exxentricity of the elliptical orbit is small $(e \sim \epsilon)$, putting $m = n = 1$ in (2.5') yields

$$2\frac{\partial U}{\partial\tau} = -g\Big[\sin 2(\Omega - \lambda_{22} - \omega\tau)[\sin^2 i_o + (1+\cos^2 i_o)\cos 2u] + 2\sin 2u\cos i\cos 2(\Omega - \lambda_{22} - \omega_4\tau)\Big]\left(g = \frac{2C_{22}\omega_4 P_o^2}{3\,C_{20}^2 r_3^2}\right) \tag{2.18}$$

For near circular orbit

$$u = M + \omega + O(\epsilon)$$

In this case we obtain from (2.18)

$$\frac{dp_6}{d\tau} = \frac{\varepsilon}{2} g\, p_5 (1 + \cos i)^2 \cos 2(\alpha - \lambda_{22})$$

$$\frac{dF}{d\tau} = -\frac{\varepsilon g}{4} (1 + \cos i)^2 \sin 2(\alpha - \lambda_{22}) + \frac{\varepsilon}{2} n^{\frac{2}{3}} p_5 \qquad (2.19)$$

In all other cases $(m \neq 1\ ,\ s \neq 1)$ instead of (2.19) we have

$$\dot{p}_6 = 0 \quad , \quad \dot{F} = \frac{\varepsilon n^{\frac{2}{3}}}{2} p_5 \quad .$$

I. Sherman B., Low thrust escape trajectories. Proc. IAS Sym. on Vehicle System Optim. N.Y. 1969.

2. Г.Б.Ефимов, Д.Е.Охоцимский. Об оптимальном разгоне космического аппарата в центральном поле. Космические исследования, т.3, № 6, 1965.

3. Л.С.Понтрягин, В.Г.Болтянский, Р.В.Гамкрелидзе, Е.Ф.Мищенко. Математическая теория оптимальныхпроцессов. Ф.М.Москва, 1961.

4. N.Bogoliuboff and Y.Mitropolsky, Asymptotic Methods in the Theory of Nonlinear Oscillations. Gordon Breach (1961).

5. Ю.Г.Евтушенко. Влияние касательного ускорения на движение спутника. ПММ, 1966, т. XXX, № 3, 594-598.

6. Yu.G.Yevtushenko. Asymptotic calculation of satellite motion in non-central field of the Earth. Proc. of XVIII IAC, 1965.

7. Ю.Г.Евтушенко, И.А.Крылов, Р.Ф.Мержанова, В.Г.Самойлович. Движение искусственных спутников в гравитационном поле Земли. Изд. ВЦ АН СССР, М., 1967.

ON OPTIMAL STOCHASTIC ORBITAL TRANSFER STRATEGY

C.G. Pfeiffer*

ABSTRACT

The optimal stochastic orbit transfer strategy is defined as the sequence of guidance corrections which will minimize a statistical measure of final error, subject to the constraint that the total correction capability expended be less than a specified number. The dynamic programming algorithm is employed to solve this problem. It is assumed that the state of the system at any time can be described by the correction capability remaining and the maximum likelihood estimates of the orbit parameters. The numerical difficulty of storing the many values of the optimized performance index corresponding to every discrete value of the state variables is overcome by representing the performance surface only in the neighborhood of local minima and by introducing the effect of the correction capability state variable constraint in a simplified way.

1. INTRODUCTION

An important objective of some space missions is to achieve a specified elliptical orbit around a target body (e.g. Voyager, Apollo, Lunar Orbiter). After the insertion maneuver, however, the spacecraft may reach a dispersed orbit due to orbit determination and maneuver execution errors, and a sequence of guidance corrections (possibly only two) becomes mandatory in order to accomplish the mission. These corrections are applied in the form of acceleration impulses, causing a step change in velocity, and hence orbital elements,

---

*Head of Mathematical Physics Section, Guidance and Analysis Department, TRW Systems Group, One Space Park, Redondo Beach, California.

at the correction times. The guidance strategy for such a sequence of corrections is the specification of how many corrections should be applied, when, and what they should be.

The determination of a strategy which minimizes a statistical measure of the final orbit error in the presence of orbit determination (estimation) and guidance execution errors poses an important unsolved problem. The classical optimal orbit transfer analysis, which seeks the strategy which minimizes propellant (correction capability) expenditure and does not consider random errors (Reference 1), can at best yield an approximately optimal strategy. Instead, we assume that there is a constraint that the total correction capability expended during the mission be less than a specified number (the resource initially allotted), and we seek the strategy which minimizes the expected value of a weighted sum of squares of the final orbit errors. These dispersions arise from random estimation and correction execution errors.

Several authors (References 2, 3, and 4) have dealt with a problem of this type and obtained some interesting results by analysis of simplified cases or by developing a sub-optimal guidance strategy. This paper will describe a dynamic programming technique for determining the optimal stochastic orbital transfer strategy (Reference 5). It will be assumed that the state of the system at any time can be described by the correction capability remaining and the maximum likelihood estimates of the orbit parameters. These estimates do indeed provide sufficient statistics for our problem if we assume that very many tracking data points are processed between corrections, since it can be shown that the probability density function for the estimation errors asymptotically becomes Gaussian with known covariance. A further simplification will be introduced by assuming that two modes of operation are possible at each correction opportunity: capability unlimited, where an excess of correction capability is available to complete the mission and the optimal correction can be determined without constraint, and capability limited, where there is not adequate correction capability to complete the mission optimally, and a heuristically justified simple one-correction policy and measure of performance are to be employed. Thus correction capability (c) is a state variable which is constrained to be positive, and the control policy and performance associated with transferring from the state $c > 0$ to the state variable boundary

$c = 0$ are assumed to be specified. (It is reasonable to specify these functions in a heuristic fashion, because for most applications the capability limited case rarely occurs.) In effect, then, one need only treat the capability unlimited case ($c = \infty$) and tabulate the optimal correction policy, the optimized measure of performance, and the expected correction capability required to complete the mission.

These simplifications are essential to the numerical dynamic programming solution of the problem. Nevertheless, one still encounters the well known numerical difficulty of calculating and storing the many values of the optimized performance index and guidance policy corresponding to all possible values of the state vector for all of the corrections. This problem can be partially overcome by recognizing that the only regions of the performance surface which are of real interest are the neighborhoods of the local minima which result from the guidance corrections, for these determine the "aiming points" in state space for preceeding corrections. (It can be shown that such local minima do indeed exist.) Thus for the $i^{\underline{th}}$ guidance correction we need to store as functions of uncorrected state variables the coordinates and magnitudes of the local minima of the performance surface resulting from the correction, the expected value of the correction capability required for subsequent corrections, and a quadratic approximation of the local behavior of the performance surface. Given the estimate of orbit parameters prior to the $(i-1)^{\underline{st}}$ correction, these results would be used in a real-time application to determine the coordinates of the "best" reachable local minimum point. This choice specifies the aiming point for the $(i-1)^{\underline{st}}$ correction and (implicity) the correction itself, assuming that the capability required to reach this point plus the expected value to complete the mission is less than the amount presently available. This procedure would be repeated for the next correction after tracking and estimation of the corrected orbit parameters, and all subsequent corrections would be treated similarly. If adequate correction capability is not available at any opportunity it is necessary to either choose a different local minimum which is acceptable, or apply the capability limited mode of operation, or aim for "best" local minimum, assuming it is to be followed with a capability-limited correction. The best of these alternatives should be chosen.

This algorithm has been applied to a space mission of the Voyager type, and numerical results are presented in Reference 5. In this paper we will present only a discussion of the general approach.

2. DETERMINATION OF THE OPTIMAL FINAL CORRECTION

Let the statistical measure of mission success at completion of the final (N^{th}) guidance correction be

$$J_N(\bar{x}_N, \bar{u}_N) = E\left[\left(\bar{x}_N + \Delta\bar{x}_N(\bar{u}_N) - \bar{x}_d\right)^T \frac{W}{2}\left(\bar{x}_N + \Delta\bar{x}_N(\bar{u}_N) - \bar{x}_d\right)\right] \quad (1)$$

where E [. . .] denotes the statistical expectation of [..], given the state and control; $\bar{x}_N$ is a vector composed of orbit parameters just prior to the Nth correction; $\Delta\bar{x}_N$ is the change resulting from the correction; W is an apriori specified semi-positive-definite weighting matrix; $\bar{x}_d$ is the desired orbit and $\bar{u}_N$ is the control vector. The $\bar{u}_N$ is composed of the magnitudes of the components of the step change of velocity and a variable θ (say true anomaly) which denotes the point on the orbit where the correction is applied. The correction capability expended is the square root of the sum of the squares of the velocity components, which will be denoted as $|\bar{u}_N|$. The geometry of the orbit is such that, in general, $\bar{x}_N$ cannot be made equal to $\bar{x}_d$ by performing a single correction $\bar{u}_N$, that is, for every state $\bar{x}_N$ there is a reachable set of points.

Applying assumptions usually made in orbit determination and guidance analysis, we suppose that a maximum likelihood estimate of $\bar{x}_N$ (estimate denoted by star) is available and, that the error in the estimate is approximately Gaussian with known variance, that is,

$$\bar{\epsilon}_N(\theta_N) = \left[\bar{x}_N^*(\theta_N) - \bar{x}_N(\theta_N)\right]$$

$$= \{\text{Gaussian, mean zero, covariance known}\}$$

As pointed out in the Introduction, this assumption can be justified if the estimate is based upon many data points between corrections, for it can be shown that the probability density function of the maximum likelihood estimation error asymptotically becomes Gaussian with variance which can be computed apriori. We also suppose that, for any given correction, the error in executing the correction results in a state error which is approximately Gaussian with known variance, that is,

$$\bar{\eta}_N \left(\text{given } \bar{u}_N, \ \bar{x}_N^*\right) = \left[\Delta\bar{x}_N^* - \Delta\bar{x}_N\right]$$

$$= \{\text{Gaussian, mean zero, covariance known}\}$$

where $\Delta\bar{x}_N^*$ denotes the commanded (expected) value of $\Delta\bar{x}_N$. The $\bar{\eta}_N$ depends upon $\bar{u}_N$, and $\bar{x}_N^*$, and is zero if no correction is made. Then, assuming $\bar{\eta}_N$ and $\bar{\epsilon}_N$ to be independent, we have

$$J_N = \left(\bar{x}_N^* + \Delta\bar{x}_N^* - \bar{x}_d\right)^T \frac{W}{2} \left(\bar{x}_N^* + \Delta\bar{x}_N^* - \bar{x}_d\right)$$

$$+ E\left[\bar{\epsilon}_N{}^T \frac{W}{2} \bar{\epsilon}_N\right] + E\left[\bar{\eta}_N{}^T \frac{W}{2} \bar{\eta}_N\right] \tag{2}$$

Thus the state variables for the problem are $\bar{x}_N^*$ and c_N, where c_N is the correction capability available, and the optimal final correction is then found from solving

$$p_N\left(\bar{x}_N^*, \ c_N\right) = \begin{cases} \underset{(\bar{u}_N)}{\text{minimum}} \left[J_N\left(\bar{x}_N^*, \ \bar{u}_N\right)\right] & \text{if } |\bar{u}_N| < c_N \\ f\left(\bar{x}_N^*, \ c_N\right) & \text{if } |\bar{u}_N| \geq c_N \end{cases} \tag{3}$$

where p_N is the optimized performance index, and $f\left(\bar{x}_N^*, \ c_N\right)$ is a specified penalty function reflecting the result of a single-correction policy to be applied when correction capability is not adequate to complete the mission. For example, f [. .] might be the performance resulting from expending all the correction capability to achieve a tangential velocity change at the best point on the orbit. As discussed in the Introduction, this representation assumes that two cases can apply: correction capability limited, where a heuristic performance index is adequate, and correction capability unlimited, where the optimization algorithm can be applied without regard to c_N. In the latter case the $\bar{u}_N$ which minimizes J_N must be obtained by a numerical search procedure for each $\bar{x}_N^*$, for the final state is a nonlinear function of $\bar{x}_N^*$, $\bar{u}_N$. From Equations (2 and 3) it can be seen that a minimum value of J_N is characterized by

$$\frac{\partial J_N}{\partial \bar{u}_N} = 0 = \left[\frac{\partial\left(\Delta\bar{x}_N^*\right)}{\partial \bar{u}_N}\right]^T W \left(\bar{x}_N^* + \Delta\bar{x}_N^* - \bar{x}_d\right) + \left[\frac{\partial g_N}{\partial \bar{u}_N}\right] \tag{4}$$

where $g_N\left(\bar{x}_N^*, \bar{u}_N\right)$ is the stochastic dispersion factor, composed of the sum of the last two terms on the right hand side of Equation (2). Equation (4) implicitly defines $\bar{u}_N\left(\bar{x}_N^*\right)$.

3. DETERMINATION OF THE CORRECTION SEQUENCE

The procedure for determining $p_N\left(\bar{x}_N^*, c_N\right)$ can be applied for a large number of values of c_N and sample vectors $\bar{x}_N^*$, a multidimensional table of the results can be stored, and, applying the dynamic programming algorithm, the corrections $\bar{u}_{N-1}$ and the optimized performance index $p_{N-1}\left(\bar{x}_{N-1}^*, c_{N-1}\right)$ can be determined. This is a formidable task, for the number of values to be stored is $\left(\text{number of vectors } \bar{x}_N^*\right) \times \left(\text{number of components of } \bar{x}_N^*\right) \times \left(\text{number of values of } c_N\right)$. The storage requirements can be greatly reduced if the random errors are small, however, for then we are only interested in values of the performance index and the associated expected correction capability required to complete the mission in the neighborhoods of the local minima of the performance surface. To simplify notation, let

$$p_N\left(\bar{x}_N^*, c_N = \infty\right) = p_N\left(\bar{x}_N^*\right) \tag{5}$$

Ec_N = expected value of correction capability to optimally complete the mission, given $\bar{x}_N^*$ (6)

Since we are supposing that the case $c_{N-1} + |\bar{u}_{N-1}| > Ec_N$ can be treated in a simplified way (see Introduction), we seek to determine and represent $p_N(\bar{x}_N^*)$ and Ec_N in the regions of local minima.

The $p_N(\bar{x}_N^*)$ surface must attain its absolute minimum at $\bar{x}_N^* = \bar{x}_d$, for W is semi-positive definite. The neighborhood of this point is a quadratic region for the $N^{\underline{th}}$ correction. On the other hand, there must be other local minima. From Equations (2) and (3) and the implicit relationship $\bar{u}_N(\bar{x}_N^*)$ defined by Equation (4), it follows that these points are characterized by

$$0 = \left(\frac{\partial p_N}{\partial \bar{x}_N^*}\right) = \left(\frac{\partial J_N}{\partial \bar{x}_N^*}\right) + \left(\frac{\partial J_N}{\partial \bar{u}_N}\right)\left(\frac{\partial \bar{u}_N}{\partial \bar{x}_N^*}\right)$$

$$= \left[I + \left(\frac{\partial \Delta \bar{x}_N^*}{\partial \bar{u}_N}\right)\right]^T W \left[\bar{x}_N^* + \Delta \bar{x}_N^* - \bar{x}_d\right] + \left(\frac{\partial g_N}{\partial \bar{x}_N^*}\right) \qquad (7)$$

Considering the effects of execution errors and correction capability constraints, and the fact that it may not be physically possible to attain the desired orbit with a single correction, it may be desirable to aim for an intermediate orbit (i.e., an aiming point not equal to $\bar{x}_d$) specified by the coordinates of a local minimum rather than $\bar{x}_d$. Thus we shall represent the $p_N(\bar{x}_N^*)$ the following way: let $\tilde{\bar{x}}_{Ni}$ be the coordinates of the m local minima (i = 1 . . . m); let $\tilde{p}_{Ni}$ be the corresponding m values of the optimized performance index; let $\tilde{W}_{Ni}$ be the m weighting matrices describing curvature of the $p_N(\bar{x}_N^*)$ surface in the neighborhood of the $\tilde{\bar{x}}_{Ni}$, determined from a local quadratic fit to the p_N surface, and let Ec_{Ni} be the expected correction capability required to complete the mission optimally if $\bar{x}_N^* = \tilde{\bar{x}}_{Ni}$. Then in a small neighborhood of $\tilde{\bar{x}}_{Ni}$ we have

$$p_{Ni}(\bar{x}_N^*) \cong \tilde{p}_{Ni} + \left(\bar{x}_N^* - \tilde{\bar{x}}_{Ni}\right)^T \frac{\tilde{W}_{Ni}}{2} \left(\bar{x}_N^* - \tilde{\bar{x}}_{Ni}\right) \quad \left\{\begin{matrix} \bar{x}_N^* \text{ in region } i \\ \\ c_N > Ec_{Ni} \end{matrix}\right\} \qquad (8)$$

This approximation represents $p_N(\bar{x}_N^*)$ in the important regions of $\bar{x}_N^*$ space, and requires that fewer numbers be stored. For example, if the dimension of $\bar{x}_N^*$ is 3 we have (since $\tilde{W}_{Ni}$ is a symmetric matrix)

number of stored values to represent $p_N(\bar{x}_N^*)$

$$= \left[\left(6 \text{ numbers for } \tilde{W}_{Ni}\right) + \left(3 \text{ numbers for } \tilde{\tilde{x}}_{Ni}\right) + \left(1 \text{ number for } p_{Ni}\right) + \left(1 \text{ number for } Ec_{Ni}\right)\right] \times [m]$$

$$= 11 \cdot m \text{ numbers}$$

The algorithm for finding the control $\bar{u}_{N-1}$ can now be much simplified, for applying the results of part 2, we replace J_N with $p_N(\bar{x}_N^*)$ as given by Equation (8) to obtain

$$p_{N-1}\left(\bar{x}_{N-1}^*\right) = \left\{\begin{matrix}\text{minimum}\\ \bar{u}_{N-1},\ i\end{matrix}\right\} \left[p_{Ni}\left(\bar{x}_{N-1}^*,\ \bar{u}_{N-1}\right)\right] \tag{9}$$

The correction capability limited mode of operation, where $c_{N-1} + |\bar{u}_{N-1}| > Ec_N$, is treated as described in the Introduction. This procedure can be applied to any number of earlier corrections, and hence the complete correction sequence can be determined as a function of $\bar{x}_{N-k}^*$. The significant results obtained by this analysis are the coordinates of the local minima to be aimed for as a function of $\bar{x}_{N-k}^*$, for these specify the intermediate orbits and the number of corrections to apply. For example, essentially a single correction policy would result if one always sought the minimum centered at $\bar{x}_d$; otherwise multiple corrections would be necessary. Note that the correction capability to be allotted to the mission is obtained from this analysis as the $\max_i \left[Ec_{ki}\right]$, where k denotes the initial correction.

4. CONCLUSION

The dynamic programming approach to stochastic orbit transfer described here was applied by Nishimura to a space mission of the Voyager type (Reference 5). The numerical results were obtained with a computer program which determined the optimized performance index resulting from the final correction for a large number of points in state space. The optimal performance index contours were plotted to find the local minima (Figure 1), and these coordinates were used to define the intermediate orbits, i.e., the "aiming points" for the preceeding correction. An example correction sequence is depicted in Figure 2, which is interesting to compare to the classical Hohman transfer case. Because of the stochastic dispersion factor, the Hohman transfer will yield poorer performance than the optimized strategy shown in Figure 2.

In summary, then, we have described a non-linear stochastic control problem where expenditure of correction capability is introduced as a state variable constraint, where the maximum likelihood estimates of the orbit parameters provide sufficient statistics to define the state of the system, and where the properties of the local minima of performance surface provide adequate information for a dynamic programming solution of the problem. More work is required to develop this approach to the point where it is practical for real-time guidance application. Some possibilities for simplification have recently become apparent and are being investigated.

REFERENCES

1. Lawden, D. F., "Impulsive Transfer between Elliptical Orbits," pp. 323-351 in Leitmann, G., Optimization Techniques, Academic Press, 1962

2. Orford, R. J., "Optimal Stochastic Control Systems," J. Math. Anal. Appl. 6, 419-429 (1963)

3. Rosenbloom, A., "Final Value Systems with Total Effort Constraint," Proceedings of the First International Federation of Automatic Control (Butterworth Scientific Publications, Ltd., London, 1960)

4. Pfeiffer, C. G., "A Dynamic Programming Analysis of Multiple Guidance Corrections of a Trajectory," AIAA Journal, Vol. 3, No. 9, pp. 1674-1681, Sept. 1965

5. T. Nishimura and C. G. Pfeiffer, "A Dynamic Programming Approach to Optimal Stochastic Orbital Transfer" paper 68-872 presented at AIAA Guidance Control, and Flight Dynamics Conference, Pasadena, Calif., Aug. 14, 1968.

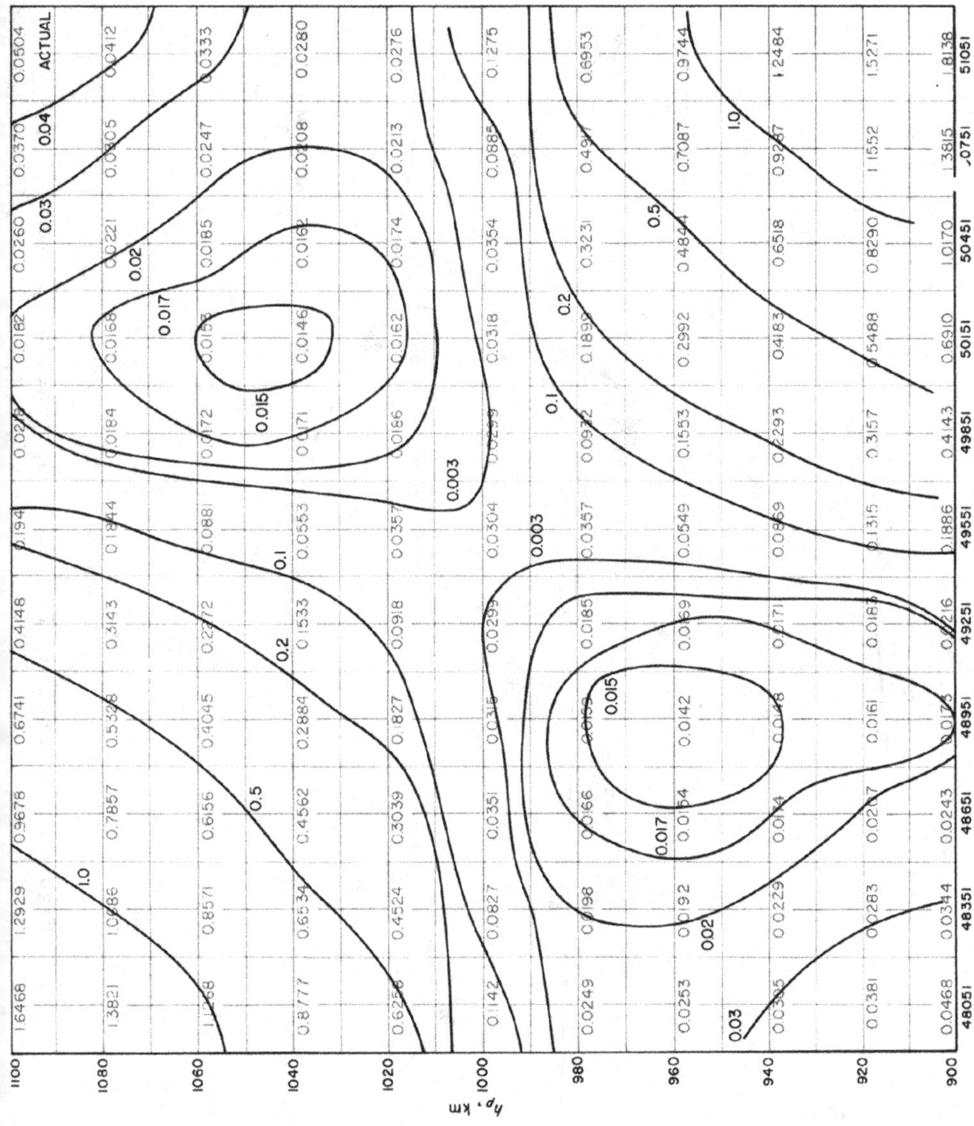

Figure 1. Performance Index Contour Map (ω = 1 deg)

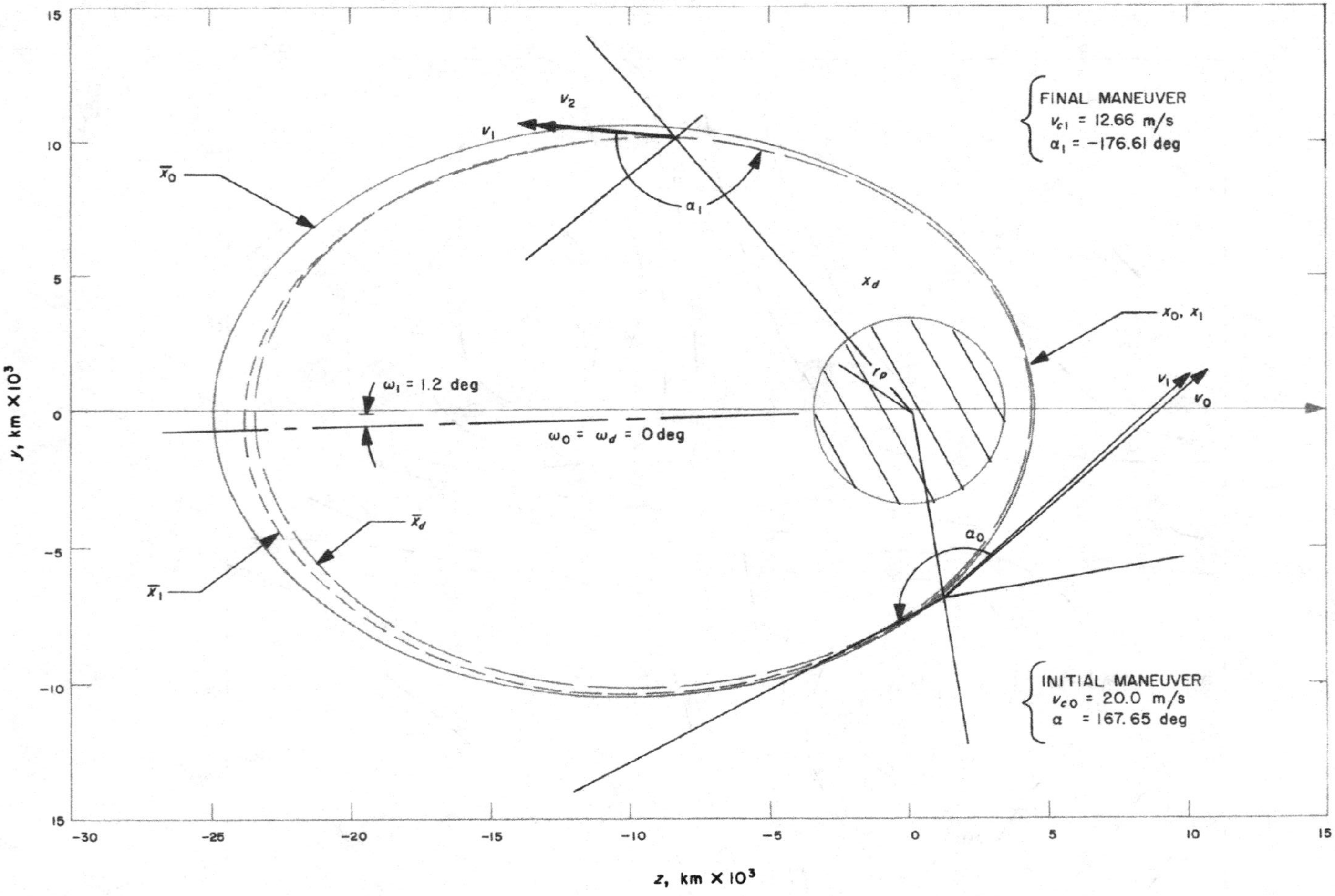

Figure 2. Example of Two-Stage Optimal Maneuvers

Conditions d'Optimalité pour les Domaines de Manoeuvrabilité à Frontière Semi-Affine

Pierre Contensou

Resumé

On recherche des conditions nécessaires d'optimalité complètant celles qui découlent du principe du maximum dans le cas de domaines de manoeuvrabilité à frontière semi-affines, c'est-à-dire de frontières ayant en commun en chaque point avec leur hyperplan tangent une sous-variété affine de cet hyperplan.

On montre d'abord la possibilité de transformer tout problème à frontière semi-affine en un problème équivalent à frontière totalement affine. Inversement on montre qu'un problème totalement affine peut-être ramené à un système restreint semi-affine, caractérisé par une dimension de moins sur la variable d'état. En experimant la convexité du domaine de manoeuvrabilité dans le cas du système restreint, on retrouve le critère d'optimalité de Kelley-Brison. Dans le cas où le système restreint est lui-même affine, on peut itérer le processus et on aboutit ainsi à une généralisation de ce critère.

Une autre généralisation apparait par l'application combinée du critère à chacun des degrés de linéarité de la frontière. Une substitution linéaire sur ces paramètres conduit à la condition qu'une certaine forme quadratique soit semi-définie positive.

La suffisance des conditions nécessaires trouvées n'est pas discutée.

1 - Introduction

L'auteur a récemment donné un apercu des propriétés générales des domaines de manoeuvrabilité à frontières semi-affines (1) qui cherchait à compléter le système des conditions d'optimisation classiques de relations supplémentaires qui soient nécessaires et, si possible, suffisantes.

Le problème est repris ici à l'aide d'une méthode nouvelle, fondée sur un changement de variable portant sur les variables d'état.

2 - Rappel du Probleme

La manoeuvrabilité d'un système représenté par un vecteur "état" $\vec{x}$ à n dimensions, élément d'un espace vectoriel (x), est définie par un domaine de manoeuvrabilité ($\mathcal{M}$), fonction donnée de x et éventuellement du temps, affecté au vecteur vitesse $\vec{u} = \frac{d\vec{x}}{dt}$ domaine qui est une partie convexe de l'espace hodographe (u).

Le problème consiste à définir, pour des conditions initiales déterminées, le domaine accessible à tout instant dans l'espace "état", et plus spécialement les trajectoires optimales qui conduisent à la frontière du domaine accessible.

Le domaine de manoeuvrabilité étant convexe possède en tout point au moins un hyperplan d'appui, c'est-à-dire un hyperplan qui laisse d'un meme coté tous les points du domaine qui ne lui appartiennent pas. Un domaine de manoeuvrabilité strictement convexe n'a qu'un point commun avec chacun de ses hyperplans d'appui. Une face semi-affine du domaine de manoeuvrabilité est une face le long de laquelle une certaine sous-variété affine des hyperplans d'appui appartient au domaine.

3 - Representation Parametrique d'une Face Semi-Affine

Il convient tout d'abord de distinguer deux cas à propos du domaine de manoeuvrabilité.

1er cas - Le domaine de manoeuvrabilité n'appartient à aucune variété à moins de n dimensions de l'espace à n dimensions (u).

(1) Le mot affine est employé ici au lieu de linéaire pour suivre l'usage qui tend à restreindre le sens de fonction linéaire à la fonction linéaire homogène.

Le domaine est alors limité par une hypersurface de l'espace (u). La frontière sépare l'espace (u) en deux parties disjointes, le domaine de manoeuvrabilité s'identifiant sans ambiguité avec celle des deux parties qui est convexe. (Si la frontière est bornée, l'une des parties seule est bornée et elle constitue le domaine).

La frontière peut être représentée par des équations paramétriques du type

$$\dot{x}_i = u_i = f_i(\vec{\nu}) \quad .$$

Le vecteur $\vec{\nu}$ étant à $n-1$ composantes. Si la frontière est semi-affine, cette dépendance sera partiellement affine. Nous dédoublerons le vecteur $\vec{\nu}$ en $\vec{\lambda}$ et $\vec{\nu}$ en écrivant

$$\dot{x}_i = u_i = a_i^k(\vec{\nu})\,\lambda_k + b_i(\vec{\nu}). \tag{1}$$

K variant de 1 à p, et, le vecteur $\vec{\nu}$ étant à q composantes avec

$$p + q = n - 1 \; .$$

Les coefficients a_i^k et b_i, bien que nous ne le rappelions pas dans l'écriture, sont bien entendu des fonctions de $\vec{x}$ et de t.

<u>2ème cas</u> - <u>Le domaine de manoeuvrabilité est contenu dans une variété à m dimensions $(\bar{u})$ de l'espace (u) (avec $m < n$),</u> (et dans aucune variété à moindre nombre de dimensions).

<u>La variété $(\bar{u})$ est nécessairement affine</u>. En effet, l'intersection du domaine par une variété affine à $(m-1)$ dimensions est un domaine convexe unidimentionnel, c'est-à-dire un segment de droite.

Le domaine de manoeuvrabilité est alors confondu avec sa frontière relative à l'espace (u). Mais il possède aussi une frontière relative au sous-espace $(\bar{u})$ (qu'on pourra appeler sa sous-frontière) qui est une hypersurface de cet espace et définit sans ambiguité l'intérieur du domaine relativement à $(\bar{u})$.

Cet intérieur peut être représenté par les équations même de la variété affine, de type

$$u_i = C_i^k\,\bar{u}_k + d_i \tag{2}$$

les u_k représentent les composantes d'un vecteur à m dimensions et les coefficients C_i^k et d_i sont maintenant des constantes (abstraction faite de leur dépendance par rapport à $\vec{x}$ et t).

Les $\bar{u}_k$ qu'on peut considérer comme des coordonnées de l'espace $(\bar{u})$ peuvent être choisis arbitrairement à l'intérieur d'un domaine convexe de $(\bar{u})$ dont la frontière (qui est la sous-frontière du domaine de manoeuvrabilité) peut elle-même présenter des faces semi-affines, représentables par des équations du type (1) dans lesquelles u_i est remplacé par $\bar{u}_i$ et où la somme p+q est maintenant égale à m-1. En transportant ces valeurs de $\bar{u}_i$ dans (2), on trouvera encore pour u_i des expressions du type (1), avec comme seule différence par rapport au 1° cas que la somme p+q des dimensions de $\vec{\lambda}$ et $\vec{\varphi}$ est maintenant inférieure à n-1.

Dans tous les cas, une face semi-affine du domaine de manoeuvrabilité pourra donc être représentée par des équations du type

$$\dot{x}_i = u_i = a_i^k(\vec{\varphi})\ \lambda_k + b_i(\vec{\varphi}) \qquad (3)$$

le vecteur $\vec{\lambda}$ étant à p composantes et le vecteur $\vec{\varphi}$ à q composantes, avec $p+q = m-1 \leq n-1$.

Les fonctions $a_i^k(\vec{\varphi})$ et $b_i(\vec{\varphi})$ ne sont pas quelconques: leur structure doit être compatible avec la convexité de la face. En particulier si ces fonctions sont dérivables par rapport à $\vec{\varphi}$, il existe en tout point un hyperplan de l'espace tangent à la face (dont tout prolongement affine à (m-1) dimensions dans (u) est un hyperplan d'appui) et cet hyperplan doit être indépendant de la valeur de $\vec{\lambda}$. Si nous repérons la direction d'un hyperplan de (u) par un vecteur $\vec{\xi}$ de l'espace dual, défini à une homothétie près, ceci revient à dire que les équations

$$\xi_i\ a_i^k = 0 \qquad \forall k \text{ compris entre 1 et p}$$

$$(4) \qquad \xi_i \frac{\partial a_i^k}{\partial \varphi_l} = 0 \qquad \forall l \quad \text{"} \quad 1 \text{ et q}$$

$$\xi_i \frac{\partial b_i}{\partial \varphi_l} = 0$$

sont vérifiées pour une variété linéaire à n-m dimensions de l'espace des vecteurs $\vec{\xi}$.

Une trajectoire qui utilise un point intérieur à la face définie par les équations (3) est dite trajectoire intermédiaire. Le problème qui nous occupe est celui de l'optimalité de telles trajectoires. Une condition nécessaire de cette optimalité est qu'un certain vecteur $\vec{\xi}(t)$, satisfaisant aux relations (4) satisfasse aussi aux équations (5) du système adjoint

$$\dot{\xi}_i = -(a^k_{j,i}\ \lambda_k + b_{j,i})\ \xi_j \tag{5}$$

(en notant $a^k_{j,i}$ la dérivée partielle $\frac{\partial a^k_j}{\partial x_i}$).

4 - Frontieres Affines et Reduction du Cas Semi-Affine au Cas Affine

Un cas particulier de frontière semi-affine est celui de la frontière affine. Les équations (3) prennent alors la forme (6) où le vecteur $\vec{\mathcal{V}}$ a disparu

$$u_i = a^k_i\ \lambda_k + b_i \qquad (6) \text{ avec } k \text{ variant de } 1 \text{ à } p \text{ et } p \leq n-1 .$$

Ce cas englobe en particulier, mais non exclusivement celui, déjà signalé, où le domaine de manoeuvrabilité lui-même se réduit à une portion de la variété affine définie par (6).

Il est important de remarquer que tout problème dans lequel la frontière est semi-affine peut être ramené à un problème dans lequel la frontière est affine.

Considérons en effet, à côté du système initial S défini par un point de l'espace état (x), un système élargi S' défini comme un point de l'espace produit (x) x $(\mathcal{V})$, (ce qui revient à dire que nous introduisons les composantes $\mathcal{V}_i$ du vecteur $\vec{\mathcal{V}}$ parmi les variables d'état) et proposons-nous d'optimiser les trajectoires du système S' . Le nouveau problème et l'ancien ne sont pas réellement distincts. Les solutions du problème initial se déduisent des solutions du problème élargi par restriction des résultats au sous-espace (x) de l'espace état (x) x (ψ). Inversement, on passe des solutions du problème initial aux solutions du problème élargi en conservant les mêmes trajectoires et en attribuant à l'instant initial et à l'instant final au vecteur ψ la valeur qu'impose d'une part la donnée des conditions initiales, d'autre part le critère d'optimalité sur les valeurs finales de l'état - ce qui est toujours possible puisque la loi de variation de ψ est complètement arbitraire.

La formulation analytique du problème élargi peut se traduire par les équations suivantes

(7,1) $\dot{x}_i = a_i^k \lambda_k + b_i$ pour i variant de 1 à

(7,2) $\dot{\psi}_l = \lambda_{p+l}$ pour l variant de 1 à q .

Les équations (7,2) où les λ_{p+l} sont de nouveaux paramètres arbitraires non bornés expriment bien la possibilité pour le vecteur $\vec{\psi}$ de varier suivant une loi quelconque du temps. L'ensemble des équations (7), où les a_i^k et b_i ne dépendant plus que de l'état et du temps sont bien de la forme (6) pour l'ensemble des variables x_i et ψ_l, ce qui montre que, par rapport au système élargi, la frontière de manoeuvrabilité est bien devenue affine.

Nous pourrons donc sans diminuer la généralité de nos résultats, traiter uniquement le problème d'un domaine de manoeuvrabilité à frontière affine.

5 - Essai de Reduction du Cas de la Frontière Affine au Cas de la Frontière Semi-Affine

Nous allons maintenant examiner la possibilité de l'opération inverse. Un problème à manoeuvrabilité affine étant donné, est-il possible de la ramener à un problème semi-affine portant sur un nombre moindre de variables d'état?

Reprenons un système d'équations de la forme (6) et isolons un des paramètres λ_k , par exemple λ_1 , en écrivant:

$$\dot{x}_i = u_i = a_i \lambda_1 + b_i \, . \tag{8}$$

Il est entendu que, dans cette écriture, b_i représente non pas un terme constant, mais une forme linéaire des paramètres λ_k autres que λ_1. Les coefficients a_i sont des fonctions des x_i et du temps.

Nous utiliserons un <u>changement de variable sur les variables d'état</u>, suggéré par les considérations suivantes:

Supposons que la face affine étudiée s'étende à l'infini. Le paramètre λ_1 en particulier peut être pris arbitrairement grand. Ceci signifie que les composantes u_i de la vitesse peuvent recevoir des valeurs arbitrairement grandes, proportionnelles aux coefficients a_i. Il existe donc une ligne de l'espace état <u>instantanément accessible</u>. On la déterminera en intégrant le système différentiel

$$\frac{dx_1}{a_1} = \frac{dx_2}{a_n} = \ldots = \frac{dx_n}{a_n} = d\tau \,. \qquad (9)$$

Dans ce système t figure comme paramètre constant. (Il est loisible de se représenter la variable auxiliare τ comme une sorte de temps accéléré, s'écoulant infiniment plus vite que le temps réel.)

L'intégration du système (9) donne des solutions de la forme

$$x_i = f_i(X_1, X_2 \ldots X_{n-1}, X_n^* + \tau, t) \qquad (10)$$

où $X_1 \ldots X_n^*$ sont des constantes d'intégration dont l'une X_n^* s'ajoute simplement à τ, qui ne figure pas dans (9) autrement que par $d\tau$.

En résolvant les équations (10), par rapport aux constantes, on trouve l'expression des intégrales premières du système (9) sous forme

$$\begin{aligned} X_i &= X_i(\bar{x}, t) \qquad \text{pour } i < n \\ X_n^* + \tau &= X_n(\bar{x}, t) \,. \end{aligned} \qquad (11)$$

Les $X_1 \ldots X_n^*$ sont des constantes lorsque le point x décrit une ligne instantanément accessible, ce qui implique

$$\begin{aligned} a_j X_{i,j} &= 0 \qquad \text{pour } i < n \\ a_j X_{n,j} &= 1 \,. \end{aligned} \qquad (12)$$

Utilisons pour la représentation de l'état du système les nouvelles variables y_i définies par

$$y_i = X_i(\vec{x},t) . \qquad (13)$$

Il s'agit donc d'un <u>changement de variable fonction du temps</u>. A ce changement de variable sur l'état correspond un changement de variable sur la vitesse obtenu en dérivant par rapport au temps les équations (13). Il s'exprime donc par

$$\dot{y}_i = v_i = X_{i,j} \, u_j + X_i' . \qquad (14)$$

En notant $X_i' = \dfrac{\partial X_i}{\partial t}$.

La frontière du domaine de manoeuvrabilité dans le nouvel espace hodographe a donc pour équation

$$\dot{y}_i = X_{i,j}(a_j \lambda_1 + b_j) + X_i'$$

ce qui, compte tenu de (12), s'écrit

$$\dot{y}_i = b_j X_{i,j} + X_i' \quad \text{pour} \quad i < n$$

$$\dot{y}_n = \lambda_1 + b_j \, X_{n,j} + X_n' . \qquad (15)$$

La propriété fondamentale du nouveau système réside dans le fait que le paramètre λ_1 ne figure que dans la dernière équation. Comme λ_1 peut être choisi arbitrairement grand, on peut choisir pour y_n une fonction arbitraire du temps, sans aucune répercussion sur le choix des autres variables λ_k . Il est donc possible de considérer y_n comme une variable de contrôle et de ramener le problème, par le processus inverse de celui qui a été utilisé précédemment, à l'étude du système restreint aux variables d'état $y_1, y_2 \ldots y_{n-1}$, avec un domaine de manoeuvrabilité défini par des équations du type

$$\dot{y}_i = v_i = b_j \, X_{i,j} + X_i' \; (i = 1,2 \ldots n-1) . \qquad (16)$$

Les seconds nombres dépendent linéairement des λ_k (avec k variant de 2 à p) par les formes linéaires b_j, et de la variable arbitraire y_n (en général non linéairement) à la fois par les coefficients b_j et X_i.

6 - Application du Changement de Variables Précédent a L'Etude des Conditions D'Optimalité

6.1. Considérons d'abord un changement de variable quelconque défini par des équations (13)

$$y_i = X_i(\vec{x},t).$$

Soient:

$$\dot{x}_i = u_i(\vec{x},\vec{\lambda},t)$$

les équations de la frontière du domaine de manoeuvrabilité relatif au vecteur représentatif $\vec{x}$, où la dépendance de u_i par rapport au paramètre $\vec{\lambda}$ est supposée absolument quelconque.

Avec le vecteur représentatif $\vec{y}$, les équations correspondantes s'écrivent:

$$\dot{y}_i = v_i = X_{i,k}\, u_k + X'_i \; . \qquad (17)$$

A la première représentation, est associé un vecteur adjoint $\vec{\xi}$ qui satisfait aux relations

$$\xi_i \frac{\partial u_i}{\partial \lambda_k} = 0 \qquad \forall K \qquad (18,1)$$

et

$$\dot{\xi}_i = -u_{j,i}\, \xi_j \; . \qquad (18,2)$$

A la deuxième représentation, est associé un vecteur adjoint $\vec{\eta}$ qui satisfait à

$$\eta_i \frac{\partial v_i}{\partial \lambda_k} = 0 \qquad \forall K \qquad (19,1)$$

$$\dot{\eta}_i = -\frac{\partial v_j}{\partial y_i}\, \eta_j \; . \qquad (19,2)$$

(Nous explicitons ici le symbole de la dérivation par rapport aux y_i, l'indice après la virgule représentant toujours une dérivation par rapport aux x_i.)

Entre les différentielles $d\vec{x}$ et $d\vec{y}$ à temps donné, existe la relation

$$dy_i = X_{i,j}\, dx_i \; . \qquad (20)$$

L'interprétation géométrique du vecteur adjoint qui définit la direction du plan tangeant au domaine accessible suggère qu'entre ξ et η doit exister la relation contragrédiente de (19) soit

$$\xi_i = X_{j,i}\, \eta_j \, . \tag{21}$$

On vérifie en effet que si on définit $\vec{\xi}$ par (21) et si $\vec{\eta}$ vérifie les équations (19), $\vec{\xi}$ vérifie lui-même les équations (18).

La vérification est immédiate en ce qui concerne (18,1), (compte tenu du fait que dans (16) u_k est le seul terme dépendant des λ_k).

Vérifions (18,2), en dérivant totalement (21)

$$\dot{\xi}_i = X'_{j,i}\, \eta_j + X_{j,ik}\, \eta_j u_k - X_{k,i} \frac{\partial v_j}{\partial y_k}\, \eta_j \, .$$

Mais $\displaystyle \frac{\partial v_j}{\partial y_k} X_{k,i} = v_{j,i} = X_{j,ik}\, u_k + X_{j,k}\, u_{k,i} + X'_{j,i} \, .$

D'où finalement

$$\dot{\xi}_i = -X_{j,k}\, u_{k,i}\, \eta_j = -u_{k,i}\, \xi_k \, .$$

Nous pouvons donc affirmer que dans tout changement de variable relatif à un problème d'optimisation, le vecteur adjoint sera transformé par la formule (21) (1).

6.2. Revenons maintenant au problème qui nous intéresse.

Dans la formulation (8) du problème (état représenté par le vecteur x), la condition portant sur le vecteur ξ intéressant la commande λ_1 s'écrit

$$\frac{\partial H}{\partial \lambda_1} = \xi_i a_i = 0 \, . \tag{22}$$

Exprimons cette condition en fonction du vecteur adjoint η associé aux variables y_i.

---

(1) On peut noter que cette transformation n'implique pas l'invariance du Hamiltonien si le changement de variable dépend du temps.

Il vient d'après (21)

$$\xi_i a_i = X_{j,i}\ \eta_j a_j$$

soit compte tenu de (12)

$$\xi_i a_i = \eta_n \ . \qquad (23)$$

$\xi_i a_i = 0$ se traduit donc dans le deuxième système par $\eta_n = 0$ ce que donne bien évidemment par ailleurs le traitement direct de ce système.

Mais la considération du système restreint (16) permet d'aller plus loin. Puisque dans ce système y_n est devenu une commande, nous devons d'abord avoir

$$\frac{\partial H'}{\partial y_n} = \eta_i \frac{\partial v_i}{\partial y_n} = 0 \qquad (24)$$

or, d'après la condition (19,2) imposée au vecteur $\vec{\eta}$, on a

$$\eta_i \frac{\partial v_i}{\partial y_n} = -\dot{\eta}_n = -\frac{d}{dt}(a_i \xi_i) \ .$$

La condition (24) n'apporte donc rien de plus que la dérivation totale par rapport au temps de la condition $a_i \xi_i = 0$. Il faut remarquer que le premier membre de cette équation ne comporte pas de terme en λ_1. (Cette circonstance, évidente sur la forme (24), se vérifie d'ailleurs aussi bien en explicitant $\frac{d}{dt}(a_i \xi_i) = 0$.) En revanche, c'est une forme affine des autres paramètres $\lambda_k (k \geq 2)$.

Nous devons ensuite exprimer que la valeur stationnaire du Hamiltonien définie par la relation (24) constitue un maximum - ou en d'autres termes exprimer la convexité du domaine de manoeuvrabilité relatif au système restreint. Ceci implique entre autres l'inégalité

$$\eta_i \frac{\partial^2 v_i}{\partial y_n^2} < 0 \ . \qquad (25)\ (*)$$

Or, si on calcule $\frac{d^2}{dt^2}(a_i \xi_i) = -\frac{d}{dt}(\eta_i \frac{\partial v_i}{\partial y_n})$, on trouve, compte tenu de $\dot{y}_n = \lambda_1$

(*) Le signe de cette inégalité suppose évidemment que nous choisissons pour les vecteurs adjoints (définis à une homothétie près) un sens dirigé vers l'extérieur du domaine de manoeuvrabilité.

$$\frac{d^2}{dt^2}(a_1\xi_1) = -\eta_1 \frac{\partial^2 v}{\partial y_n^2}\lambda_1 + B \qquad (26)$$

où B représente un ensemble non explicité de termes indépendants de λ_1.

La condition (25) se traduit donc, par rapport aux variables initiales, par la condition que le terme en λ_1, de l'expression $\frac{d^2}{dt^2}(a_1\xi_1)$ soit positif, ce qui peut encore s'écrire, H étant le Hamiltonien sans ce système

$$\frac{\partial}{\partial\lambda_1}\left[\frac{d^2}{dt^2}\left(\frac{\partial H}{\partial\lambda_1}\right)\right] > 0 . \qquad (27)$$

Nous retrouvons donc, par une méthode tout à fait indépendante, une condition d'optimalité donnée par Kelley (6) sous la forme condensée suggérée à l'auteur par Bryson.

7 - Généralisation au Cas où le Système Restreint est Lui-Même Linéaire

Le critérium précédent se trouve en défaut si la dérivée seconde de $a_1\xi_1$ par rapport au temps se trouve dépourvue de terme en λ_1. Ceci se produira, en particulier, dans un cas pratiquement fréquent, celui où le système restreint présente lui-même une frontière de manoeuvrabilité affine. Les v_i sont alors des formes affines du paramètre y_n et l'expression $\eta_i \frac{\partial^2 v_i}{\partial y_n}$ est identiquement nulle.

Supposons donc que le système restreint se présente sous la forme

$$v_i = \bar{a}_i y_n + \bar{b}_i \qquad \text{pour } i = 1, 2 \ldots, n-1 .$$

La relation précédemment établie

$$\eta_i \frac{\partial v_i}{\partial y_n} = -\frac{d}{dt}(a_i\xi_i)$$

s'écrit maintenant

$$\eta_i\bar{a}_i = -\frac{d}{dt}(\xi_i a_i). \qquad (28)$$

Or, nous pouvons évidemment itérer l'opération précédente, c'est-à-dire traiter le système restreint, par rapport au paramètre y_n, comme nous avons traité le système initial par rapport au paramètre λ_1. Le role qu'a joué l'expression $\xi_1 a_1$ sera maintenant dévolu à l'expression $\eta_1 \bar{a}_1$. L'opération pourra être répétée autant de fois qu'il sera nécessaire pour aboutir à un système non totalement affine, une relation du type (27) reliant les expressions du type $\xi_1 a_1$ de deux systèmes successifs.

Pour traiter le cas général, affectons l'indice 0 aux grandeurs caractéristiques du premier système non strictement affine rencontré, les systèmes antérieurs étant affectés des indices successifs dans l'ordre inverse de leur introduction. En passant de l'indice $\varkappa$ à l'indice $\varkappa + 1$, le système s'enrichit d'une variable d'état supplémentaire, constituée par le paramètre λ_1^1 intervenant linéairement dans le précédent, et d'une composante supplémentaire du vecteur adjoint. Les λ_k autres que λ_1, sont communs à tous les systèmes.

Nous pouvons écrire deux relations récurrentes liant chaque système au suivant. La dernière relation (15) donne d'une part

$$\frac{d}{dt} \lambda_1^{\varkappa-1} = \lambda_1^{\varkappa} + \mathcal{C}_{\varkappa-1} \tag{29}$$

et la relation (28)

$$\xi_1^{\varkappa-1} a_1^{\varkappa-1} = - \frac{d}{dt}(\xi_1^{\varkappa} a_1^{\varkappa}) \ . \tag{30}$$

On peut démontrer dans ces conditions les relations successives

$$\begin{aligned} \frac{d^2}{dt^2} (a_1^1 \xi_1^1) &= - A_0 \lambda_1^1 + B_0 \\ \frac{d^4}{dt^4} (a_1^2 \xi_1^2) &= A_0 \lambda_2^2 + B_1 \\ \cdots\cdots & \quad \cdots\cdots \\ \frac{d^{2\varkappa}}{dt^{2\varkappa}} (a_1^{\varkappa} \xi_1^{\varkappa}) &= (-1)^{\varkappa} A_0 \lambda_1^{\varkappa} + B_{\varkappa-1} \ . \end{aligned} \tag{31}$$

L'indice $\varkappa$ affecté dans ces formules aux coefficients A B C signifie que ces coefficients ne renferment pas de variables appartenant à un système de rang supérieur à $\varkappa$. Les coefficients A_0 et B peuvent contenir aussi les paramètres λ_k autres que les $\lambda_1^{\varkappa}$, et leurs dérivées par rapport au temps.

La première des relations (31) n'est que la transcription de (26). Les autres se démontrent par récurrence.

Supposons les relations (31) vraies jusqu'au rang $\varkappa$-1.
On a par dérivation de (30)

$$\frac{d^{2\varkappa-1}}{dt^{2\varkappa-1}} (\xi_1^\varkappa a_1^\varkappa) = - \frac{d^{2(\varkappa-1)}}{dt^{2(\varkappa-1)}} (\xi_1^{\varkappa-1} a_1^{\varkappa-1}) = (-1)^\varkappa A_0 \lambda_1^{\varkappa-1} + B_{\varkappa-1}. \qquad (32)$$

En dérivant totalement par rapport au temps, on trouve bien au second membre un terme

$$(-1)^\varkappa A_0 \frac{d\lambda_1^{\varkappa-1}}{dt} = (-1)^\varkappa A_0(\lambda_1^\varkappa + \mathcal{C}_{\varkappa-1})$$

accompagné de termes ne contenant que des variables a_1 jusqu'à l'ordre $\varkappa$-1, ce qui démontre la formule (31) pour le rang $2\varkappa$.

La condition d'optimalité $A_0 < 0$ peut aussi bien s'exprimer, quel que soit $\varkappa$ par

$$(-1)^\varkappa \frac{\partial}{\partial\lambda_1^\varkappa} \quad \frac{d^{2\varkappa}}{dt^{2\varkappa}} \quad (a_1^\varkappa \xi_1^\varkappa) < 0.$$

En revenant à nos notations initiales et au système initial, on peut donc énoncer la règle suivante relative aux conditions d'optimalité.

On forme la quantité

$$a_1 \xi_1 = \frac{\partial H}{\partial \lambda_1} \quad .$$

On prend ses dérivées successives. Soit $2\varkappa$ le rang de la première dérivée qui contient un terme en λ_1. Ce rang est nécessairement pair en vertu de la formule (32) qui montre que la dérivée de rang $(2\varkappa-1)$ commence par un terme en $\lambda_{\varkappa-1}$ comme la dérivée de rang $(2\varkappa-2)$.

Une condition nécessaire d'optimalité est donc

$$(-1)^\varkappa \frac{\partial}{\partial\lambda_1} \left[\frac{d^{2\varkappa}}{dt^{2\varkappa}} \left(\frac{\partial H}{\partial\lambda_1}\right) \right] < 0. \qquad (33)$$

Naturellement la nullité des dérivées successives de $\frac{\partial H}{\partial \lambda_1}$ sont aussi des conditions nécessaires d'optimalité, la dérivée d'ordre 2κ est la première qui, égalée à zéro, fournit une condition où intervienne λ_1.

Exemple d'application

Considérons, à titre d'exemple, le système à 3 variables d'état x, y, z dont la manoeuvrabilité est définie par

$$\begin{aligned} \dot{x} &= f(y) \\ \dot{y} &= z \\ \dot{z} &= \lambda \quad . \end{aligned}$$

Il a pour système adjoint (vecteur adjoint ξ, η, ζ)

$$\begin{aligned} \dot{\xi} &= 0 \\ \dot{\eta} &= -f'_y \cdot \xi \\ \dot{\zeta} &= \eta \quad . \end{aligned}$$

La méthode précédente s'applique ici sans changement de variable. On restreint le système en supprimant la 3°, puis la 2° équation de chaque groupe, les commandes successives étant λ, puis z, puis y.

Le dernier système obtenu

$$\dot{x} = f(y) \qquad \dot{\xi} = 0$$

se résout évidemment par les conditions

$$f'_y = 0 \qquad \xi \cdot f''_{y^2} < 0 \; .$$

On vérifie aisément que cette condition se retrouve aussi bien à partir du système à deux variables par la condition

$$\frac{\partial}{\partial z}\left[\frac{d^2}{dt^2}\left(\frac{\partial H_2}{\partial z}\right)\right] < 0 \qquad \text{ou} \quad \frac{\partial H_z}{\partial z} = \eta$$

ou à partir du système à 3 variables par la condition

$$\frac{\partial}{\partial \lambda}\left[\frac{d^4}{dt^4}\left(\frac{\partial H_3}{\partial \lambda}\right)\right] > 0 \qquad \text{ou} \quad \frac{\partial H_3}{\partial \lambda} = \zeta \; .$$

8 - Introduction de L'Ensemble des Paramètres λ_k

Nous avons montré la possibilité d'itérer en chaîne la méthodes de changement de variable proposée lorsqu'elle conduisait à un système restreint lui-même complètement affine. On peut se demander s'il est possible de l'itérer "en parallèle" en traitant successivement les paramètres λ_k. Le pas suivant consisterait en partant du système (15) à aboutir à un système où, λ_1, ne figurant toujours que dans la n° ligne, λ_2 ne figurerait que dans la (n - 1)°. On vérifie facilement que c'est généralement impossible, le traitement du système (14) par rapport à λ_2 selon la méthode utilisée pour le système initial par rapport à λ_1, ayant pour effet de réintroduire λ_1, dans toutes les équations. La recherche directe d'un changement de variable réalisant d'un seul coup le cantonnement respectif de λ_1 et λ_2 dans deux des équations aboutit d'ailleurs à des conditions généralement incompatibles.

Ceci nous conduit aux considérations suivantes. Si (le nombre de dimensions n étant quelconque), il y a un seul paramètre λ (domaine de manoeuvrabilité constitué d'une droite indéfinie), la région de l'espace instantanément accessible à partir d'un point donné est constitué d'une courbe. On pourrait penser que s'il y a deux paramètres λ (domaine de manoeuvrabilité constitué d'un plan indéfini), la région de l'espace instantanément accessible sera constitué d'une surface. S'il en était ainsi, un changement de variable approprié devrait précisément aboutir à des équations du type $\dot{x}_1 = \lambda_1$, $\dot{x}_2 = \lambda_2$, les autres $\dot{x}_i$ ne dépendant pas de λ_1, et λ_2. Cette réduction étant impossible, l'hypothèse elle-même est à rejeter, et il faut probablement en conclure (sous réserve d'une étude plus poussée) que dès que le domaine de manoeuvrabilité comprend une variété affine indéfinie, au moins bi-dimensionnelle, l'espace état tout entier est instantanément accessible (ou tout au moins une portion de l'espace qui ne se réduit pas à une variété à moins de n dimensions).

Cette circonstance ne supprime pas l'intérêt de l'étude des faces multi-affines des domaines de manoeuvrabilité bornés (pour lesquels il n'y a évidemment pas de domaine instantanément accessible) mais empêche la généralisation de la méthode de réduction employée (qui épuise la question dans le cas d'un seul paramètre).

Il reste évidemment la possibilité d'appliquer le critérium (27) ou (33) à chacun des paramètres λ_k. Mais, on peut faire plus. On remarque en effet qu'une substitution linéaire quelconque sur le vecteur $\vec{\lambda}$ laisse subsister la forme des équations (6). Soit λ^* l'une des nouvelles variables substituées aux λ_k. Le critérium (27) doit s'appliquer à λ^*. Or on a :

$$\frac{\partial}{\partial \lambda^*} = \frac{\partial}{\partial \lambda_k} \frac{\partial \lambda_k}{\partial \lambda^*}$$

et on peut prendre pour $\frac{\partial \lambda_k}{\partial \lambda}$ des constantes α_k quelconques.

La relation (27) s'écrit alors

$$\alpha_1 \alpha_k \frac{\partial}{\partial \lambda_1} \quad \frac{d^2}{dt^2} \quad \frac{\partial H}{\partial \lambda_k} \quad > 0 \qquad \text{pour tout vecteur } \vec{\alpha} .$$

Ceci exprime que la forme quadratique Q associée à la forme bilinéaire de terme général $\frac{\partial}{\partial \lambda_1} \left[\frac{d^2}{dt^2} \left(\frac{\partial H}{\partial \lambda_k}\right)\right]$ doit être semi-définie positive (critérium 34).

9 - Retour sur Quelques Hypotheses

9.1 - Le système donné au départ répondait aux équations $\dot{x}_i = a_i^k \lambda_k + b_i$ avec a_i et b_i fonction de $\vec{\varphi}$. Nous avons dû pour le rendre affine le compléter des relations

$$\dot{\varphi}_1 = \lambda_{p+1} \quad .$$

L'application du critère (27) à un paramètre du type λ_{p+1} ou du critère (34) à l'ensemble des paramètres du type λ_{p+1} n'exprime pas autre chose, comme on le vérifie facilement, que la convexité du domaine de manoeuvrabilité du système initial assurée par hypothèse. Il est probable en revanche que l'application du critère (34) à l'ensemble des λ_k et λ_{p+1} apporte des conditions supplémentaires par rapport à l'application du critère (34) à l'ensemble λ_k seul. Il est donc prudent de traiter le système rendu totalement affine et non le système initial.

9.2 - Nous n'avons démontré le caractère nécessaire des critères trouvés que dans le cas où la face affine du domaine de manoeuvrabilité est indéfinie. Ce caractère reste acquis cependant pour une face bornée. Si le critère n'est pas vérifié, le point de fonctionnement utilisé P n'appartient pas à la frontière du domaine de manoeuvrabilité convexisé du système restreint. Non seulement la trajectoire n'est pas optimale, mais elle est surclassée par une infinité de trajectoires utilisant alternativement des points de la génératrice infiniment voisins de P et situés de part et d'autre, trajectoires qui sont permises à un système dont le domaine de manoeuvrabilité ne comprend qu'un tronçon, si petit soit-il, de la génératrice à condition qu'il entoure le point P.

Dans le cas de la face indéfinie, nous aurions pu écrire des conditions nécessaires plus restrictives. Il est certain par exemple que les relations (24) et (25)

$$\eta_1 \frac{\partial v_1}{\partial y_n} = 0 \qquad\qquad \eta_1 \frac{\partial^2 v_1}{\partial y_n^2} < 0$$

n'expriment que l'existence d'un maximum local du Hamiltonien, en fonction de y_n, alors qu'un maximum absolu est indispensable si y_n est réellement arbitraire. Mais si $\dot{y}_n = \lambda_1$ est limité, le maximum local peut correspondre à un optimum pour un temps de trajet assez court. On le verra clairement sur l'exemple suivant. Un ballon libre, maître de sa vitesse verticale et devant se déplacer horizontalement suivant Ox dans une atmosphère où règne un vent fonction de l'altitude, choisira l'altitude correspondant au maximum absolu de la vitesse du vent si sa vitesse verticale n'est pas bornée. Mais si sa vitesse verticale est bornée et pour un trajet assez court, il peut avoir intérêt à rallier l'altitude d'un maximum local de vitesse, plus proche de sa propre altitude de départ.

10 - Conclusion

Nous avons montré la possibilité de traiter les problèmes de manoeuvrabilité semi-affine par les processus suivants:

- passage d'un problème semi-affine à un problème totalement affine par extension de l'espace état,

- possibilité inverse, à partir d'un système totalement affine, de réduction d'une unité des dimensions de la variable d'état par un changement de variable approprié. L'application du principe du maximum au système restreint permet de retrouver le critère nécessaire d'optimalité de Kelley-Bryson. Une généralisation de ce critère est obtenue par iteration du processus lorsque le système restreint est lui-même totalement affine. Une autre généralisation est obtenue par la considération successive des divers paramètres intervenant dans la représentation affine de la frontière.

Le problème du caractère suffisant des conditions d'optimalité trouvées n'a pas été abordé.

References

(1) - CONTENSOU P. - "Note sur la cinématique générale du mobile dirigé" - Communication à l'Association Technique Maritime et Aéronautique, Juin 1946, vol. 45, n° 836.

(2) - CONTENSOU P. - "Application des méthodes de la mécanique du mobile dirigé à la théorie du vol plané" - Communication à l'Association Technique Maritime et Aéronautique, Session 1950, vol. 49, n° 958.

(3) - CONTENSOU P. - "Etude théorique des trajectoires optimales dans un champ de gravitation. Application au cas d'un centre d'attraction unique". - Astronautica Acta, 8, 134 - 150 (1962). Symposium de l'Académie 1961.

(4) - PONTRYAGIN L.S., BOLTYANSKII V.G., GAMKRELIDZE R.V., MISHCHENKO E.F. - The Mathematical Theory of Optimal Processes. (Interscience Publishers, John Wiley and Sons, Inc. New-York, 1962).

(5) - ROZONOER L.I. - "Principe du Maximum de L.S. Pontryagin dans la théorie des systèmes optimaux". Automatismes et Télémécanique. Tome 20, n° 10 - 11 - 12 (1959).

(6) - KELLEY H.J. - "A Second Variation Test for Singular Extremals" AIAA Journal, vol. 2 - n° 8.

(7) - LAWDEN D.E. - "Optimal Intermediate - Thrust Arc in a Gravitational Field". - Astronautica Acta, VIII - 106 (1962).

(8) - MARCHAL C. - "Généralisation tridimensionnelle et étude de l'optimalité des arcs à poussée intermédiaire de Lawden. Astronautica Acta (à paraître).

CONTRIBUTION A L'ETUDE DES RENDEZ-VOUS MULTI-IMPULSIONNELS, OPTIMAUX, DE DUREE MOYENNE, ENTRE ORBITES QUASI-CIRCULAIRES, PROCHES, NON COPLANAIRES.

Jean-Pierre Marec

RÉSUMÉ

Cette étude analytique aborde le problème des rendez-vous optimaux (recherche de la consommation de masse minimale pour un propulseur capable de fournir des impulsions) entre orbites proches, de faible excentricité, non coplanaires, lorsque la durée est suffisante : l'erreur relative sur la consommation est de l'ordre de l'inverse du carré de l'angle de transfert.

Les zones du domaine accessible (dans l'espace des variations des éléments orbitaux) correspondant à certaines solutions hexa - et penta - impulsionnelles sont étudiées en détail.

Вклад в исследование много-импульсных, оптимальных встреч, средней продолжительности между квази-круговыми, соседними, некомпланарными орбитами.

КРАТКОЕ ИЗЛОЖЕНИЕ

Это аналитическое исследование приступает к рассмотрению вопроса оптимальных встреч (изыскание минимального массового расхода для двигателя могущего дать импульсы) между соседними, некомпланарными орбитами малого эксцентриситета, в том случае, если продолжительность достаточна – относительная ошибка в области расхода порядка обратной величины квадрата угла перехода.

Зоны доступной области (в пространстве изменений орбитальных элементов), соответствующие некоторым шести – или пятиимпульсным решениям, подробно изучены.

CONTRIBUTION TO THE STUDY OF THE MULTI-IMPULSE OPTIMAL RENDEZ-VOUS OF MEDIUM DURATION, BETWEEN NEAR-CIRCULAR, NON COPLANAR, CLOSE ORBITS

SUMMARY

This analytical study is a first approach to the problem of optimal rendez-vous (minimum fuel consumption for a thrustor capable of delivering impulses) between non coplanar, close orbits of small excentricity, when the duration is sufficiently large : the relative error on the consumption is of the order of the inverse of the square of the transfer angle.

The zones of the reachable domain (in the space of the variations of the orbital elements) corresponding to certain six - and five - impulse solutions are studied in detail.

Beitrag zur Untersuchung der multiimpulsionellen, optimalen Begegnungen mittlerer Dauer zwischen quasi-kreisförmigen, benachbarten, nicht komplanar Umlaufsbahnen.

Zusammenfassung

Diese analytische Untersuchung erörtert die Frage der optimalen Begegnungen (Untersuchung des minimalen Massenverbrauches eines Impulsionsmotors zwischen benachbarten, nicht komplanaren Umlaufsbahnen kleiner Exzentrizität, wenn die Dauer genügend ist, d. h. wenn der relative Fehler des Verbrauches von der Grössenordnung des inversen Quadrates des Überführungswinkels ist.

Die Bereiche des zugänglichen Gebietes (im Raume der Bahnelementenvariation), die gewissen hexa- und pentaimpulsionellen Lösungen entsprechen, werden ausführlich untersucht.

1 - INTRODUCTION

L'étude analytique des rendez-vous multi-impulsionnels optimaux, entre orbites quasi-circulaires, proches, coplanaires ou non, dans un champ de gravitation central est particulièrement importante, car les orbites rencontrées dans la pratique ont souvent une excentricité faible et le problème de la légère modification ("correction") des six éléments orbitaux à la fois, à l'aide d'un propulseur capable de fournir des impulsions, se pose très fréquemment.

La présente étude complète, par l'analyse de certains rendez-vous hexa - et penta - impulsionnels, de durée moyenne, entre orbites quasi-circulaires, proches, quelconques, les résultats déjà obtenus relatifs aux transferts simples [1, 2], aux rendez-vous de longue durée [3], aux rendez-vous de durée moyenne entre orbites coplanaires [4], ainsi que ceux relatifs aux rendez-vous de durée quelconque entre orbites circulaires, coplanaires [5].

La résolution du problème est basée sur l'application du Principe du Maximum de PONTRYAGIN [6], l'utilisation de la notion de "domaine accessible" de CONTENSOU [7] et l'étude de l'évolution au cours du temps de la longueur du "vecteur efficacité" [8] ou "Primer Vector" [9].

2 - GÉNÉRALITÉS

Il est supposé que, au cours du rendez-vous, l'orbite osculatrice (O) s'écarte peu d'une orbite nominale $(\bar{O})$ (écart maximal de l'ordre de $\varepsilon \ll 1$) et que son excentricité est faible ($e \lesssim$ ordre $\varepsilon \ll 1$).

Il est alors possible de retenir, en première approximation, la solution obtenue par linéarisation autour d'une orbite nominale circulaire ($\bar{e} = 0$), acceptant ainsi une erreur relative d'ordre ε sur la solution.

L'orbite nominale devant jouer un rôle privilégié, afin de simplifier les calculs, les grandeurs suivantes sont prises pour unités :

Unité de longueur : $\bar{a}$ = demi-grand axe de $(\bar{O})$
Unité de temps : $\bar{T}/2\pi$ = [période de révolution sur l'orbite $(\bar{O})$]$/2\pi$

L'unité de vitesse est alors la vitesse circulaire $V_c(\bar{a})$ à la distance $\bar{a}$ du centre d'attraction O, et l'unité d'accélération, l'accélération de gravitation g $(\bar{a})$ à la distance $\bar{a}$ de O.

De plus, le plan de $(\bar{O})$ est pris comme plan de référence (donc $\bar{i} = 0$, fig. 1).

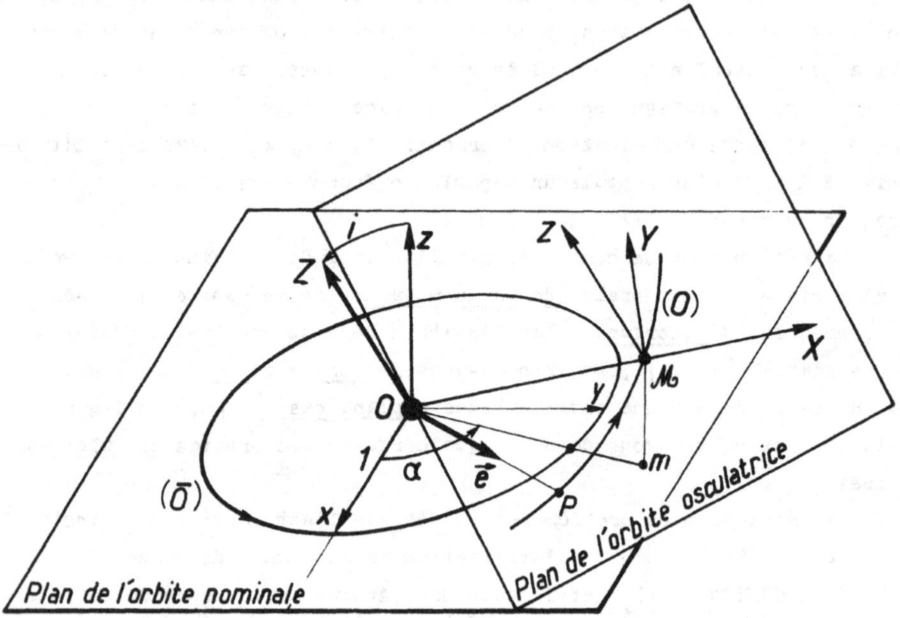

Fig. 1 - Notations

Le propulseur, à vitesse d'éjection imposée, est supposé capable de fournir des impulsions. Un rendez-vous est réalisé de façon optimale si la vitesse caractéristique Δ C correspondante est minimale ($\Delta C = \int_{t_0}^{t_f} |\vec{\gamma}|\, dt$ où $\vec{\gamma}$ est l'accélération de poussée).

Dans cette étude, il est commode de prendre comme variable de description, non pas le temps, mais l'ascension droite $\alpha = (\overrightarrow{Ox}, \overrightarrow{Om})$ du mobile $\mathcal{M}$ (fig. 1).

Un rendez-vous, dans lequel l'ascension droite α du mobile varie de α_o à α_f (fixés), peut être défini, dans cette étude linéarisée, par la seule donnée des 6 variations suivantes :

$\Delta j_x = -\Delta Z_y$ et $\Delta j_y = \Delta Z_x$, composantes sur les axes fixes $\overrightarrow{Ox}$, $\overrightarrow{Oy}$ de la rotation infinitésimale $\overrightarrow{\Delta j} = \vec{z} \wedge \overrightarrow{\Delta Z}$

du plan de l'orbite (fig. 2);

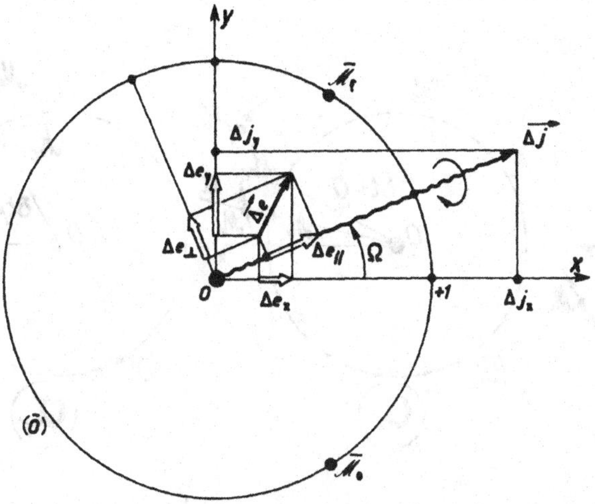

Fig. 2 - Notations

$\frac{\Delta a}{\bar{a}} = \Delta a$, "dilatation" relative du demi-grand axe;

Δe_x et Δe_y, composantes de la variation $\vec{\Delta e}$ du vecteur périgée $\vec{e}$ dirigé vers le périgée et de longueur e (excentricité);

$\Delta\tau$, variation du paramètre de rendez-vous :

(1) $$\tau = -\delta t + \frac{3}{2}\alpha\,\delta a + 2\vec{Y}.\vec{e}$$

qui n'a pas ici tout à fait la même définition que celle donnée en [8] .

Dans(1), $\vec{Y}$ = vecteur unitaire de l'horizontale locale (fig. 1),

$\delta a = a - \bar{a}$,

$\delta t = t - \bar{t}$ = retard temporel, sur l'orbite nominale $(\bar{O})$, du mobile nominal $\bar{\mathcal{M}}$ de même ascension droite α que le mobile réel $\mathcal{M}$, par rapport au mobile cible nominal $\bar{\mathcal{M}}_c$ (fig. 3). Le mobile cible nominal $\bar{\mathcal{M}}_c$

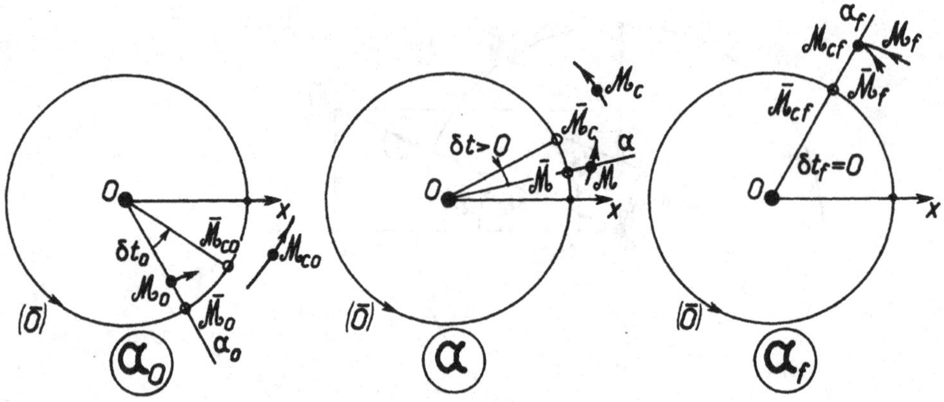

Fig. 3 - Définition du retard δt

décrit l'orbite nominale circulaire ($\bar{O}$) d'un mouvement képlérien uniforme de façon à avoir la même ascension droite finale α_f que le mobile cible $\mathcal{M}_{cf}$ (qui coïncide alors, évidemment, avec le mobile réel $\mathcal{M}_f$).

Choisissant alors pour axe de référence $\overrightarrow{Ox}$ l'axe de symétrie de l'arc de transfert $\widehat{\alpha_o \alpha_f}$ (si bien que $\alpha_o = -\Delta\alpha/2$, $\alpha_f = +\Delta\alpha/2$, où $\Delta\alpha = \alpha_f - \alpha_o$ est l'angle de transfert) et pour demi-grand axe de l'orbite nominale la moyenne arithmétique du demi-grand axe initial et du demi-grand axe final [$\bar{a} = (a_o + a_f)/2$], la variation du paramètre de rendez-vous s'écrit :

(2) $$\Delta\tau = \delta t_o + 2\,\overrightarrow{Y_f}\cdot\overrightarrow{e_f} - 2\,\overrightarrow{Y_o}\cdot\overrightarrow{e_o}$$

Comme $\bar{n} = \frac{2\pi}{T} = 1$, δt_o peut être interprété comme le retard <u>angulaire</u> initial du mobile réel $\mathcal{M}_o$ par rapport au mobile cible nominal $\bar{\mathcal{M}}_{co}$.

Il est important de remarquer que la variation $\Delta\tau$ peut être d'ordre $\varepsilon\Delta\alpha$ et non pas seulement d'ordre ε, car le rattrapage angulaire est dû à l'effet cumulatif dans le temps d'un écart sur la période, écart qui, par définition même de ε, peut, au cours du rendez-vous, être dans certains cas d'ordre ε.

Aussi, dans l'étude des rendez-vous de longue durée, est-il souvent préférable d'utiliser la variation :

$$(3) \qquad \Delta\tilde{\tau} = \frac{4}{3}\,\frac{\Delta\tau}{\Delta\alpha} = \frac{4}{3}\,\frac{\delta t_o}{\Delta\alpha} + \text{ordre } \varepsilon/\Delta\alpha$$

($\Delta\tilde{\tau}$ est, pour $\Delta\alpha$ grand, égal à l'opposé du paramètre $\Delta\mathcal{T}$ introduit dans [3]).

Les équations linéarisées donnant les variations des éléments orbitaux et la consommation s'écrivent :

$$(4)\begin{cases}
\dfrac{dj_x}{d\alpha} = j'_x = -Z'_y = \gamma_Z\cos\alpha \\[2mm]
\dfrac{dj_y}{d\alpha} = j'_y = Z'_x = \gamma_Z\sin\alpha \\[2mm]
\dfrac{da}{d\alpha} = a' = 2\gamma_Y \\[2mm]
\dfrac{de_x}{d\alpha} = e'_x = \gamma_X\sin\alpha + 2\gamma_Y\cos\alpha \\[2mm]
\dfrac{de_y}{d\alpha} = e'_y = -\gamma_X\cos\alpha + 2\gamma_Y\sin\alpha \\[2mm]
\dfrac{d\tau}{d\alpha} = \tau' = -2\gamma_X + 3\alpha\gamma_Y \\[2mm]
\dfrac{dC}{d\alpha} = C' = \gamma = \sqrt{\gamma_X^2 + \gamma_Y^2 + \gamma_Z^2}
\end{cases}$$

où γ_X, γ_Y, γ_Z sont les composantes de l'accélération de poussée $\vec{\gamma}$ dans les axes tournants $MXYZ$ ($\overrightarrow{MX}$: radial; $\overrightarrow{MY}$: circonférentiel, ou encore "horizontal"; $\overrightarrow{MZ}$: normal au plan de l'orbite, fig. 1).

Le but ultime d'une telle étude serait de définir, pour un angle de transfert $\Delta\alpha$ et une consommation ΔC donnés, le "domaine accessible" [7] dans l'hyperespace à 6 dimensions des variations Δj_x, Δj_y, Δa, Δe_x, Δe_y, $\Delta\tilde{\tau}$ (Il est à noter que le contour apparent de ce domaine dans le plan Δa, $\Delta\tilde{\tau}$ a déjà été obtenu [10] ; la connaissance de ce contour apparent suffit pour résoudre le problème du rendez-vous optimal "en ascension droite" ou acquisition de l'ascension droite et de la période correctes).

Il est commode de définir le domaine accessible par ses sections $\Omega = \operatorname{arc\,tg}(\Delta j_y/\Delta j_x)$ = constante, Δa = constante dans l'hyperespace à 4 dimensions $\Delta e_{/\!/}$, $\Delta e_{\perp}$, $\Delta j = |\overrightarrow{\Delta j}|$, $\Delta\tilde{\tau}$ où $\Delta e_{/\!/}$ est la composante de $\overrightarrow{\Delta e}$ parallèle à $\overrightarrow{\Delta j}$ et $\Delta e_{\perp}$ la composante de $\overrightarrow{\Delta e}$ perpendiculaire à $\overrightarrow{\Delta j}$ (fig. 2).

Plus exactement, comme les solutions optimales (correspondant à des points de la surface convexe du domaine accessible) sont seules intéressantes, il suffit de calculer en chaque point de la section (Ω, Δa) dans l'espace à 3 dimensions $\Delta e_{/\!/}$, $\Delta e_{\perp}$, Δj (point obligatoirement situé à l'intérieur du domaine accessible relatif aux transferts simples, déjà obtenu en [2] pour un angle de transfert supérieur à 2π) les valeurs minimale ($\Delta\tilde{\tau}_{min}$) et maximale ($\Delta\tilde{\tau}_{max}$) de $\Delta\tilde{\tau}$ et d'associer également à ce point le type de solution optimale qui doit être utilisée dans chacun de ces deux cas (nombre d'impulsions, périodes d'attente etc....).

Dans cette perspective, il est utile d'étudier au préalable les symétries du domaine accessible. Le tableau ci-après indique les modifications des paramètres Ω, Δa, $\Delta e_{/\!/}$, $\Delta e_{\perp}$, Δj, $\Delta\tilde{\tau}$ dues à des modifications, identiques pour toutes les poussées, du signe des paramètres α (point d'application) et X, Y, Z (cosinus directeurs de la direction de poussée) :

(0)	α	X	Y	Z	Δj_x	Δj_y	Δa	Δe_x	Δe_y	$\Delta\tilde{\tau}$	Ω	$\Delta e_{\parallel}$	$\Delta e_{\perp}$	Δj
(1)	α	X	Y	$-Z$	$-\Delta j_x$	$-\Delta j_y$	Δa	Δe_x	Δe_y	$\Delta\tilde{\tau}$	$\Omega+\pi$	$-\Delta e_{\parallel}$	$-\Delta e_{\perp}$	Δj
(2)	$-\alpha$	$-X$	Y	Z	Δj_x	$-\Delta j_y$	Δa	Δe_x	$-\Delta e_y$	$-\Delta\tilde{\tau}$	$-\Omega$	$\Delta e_{\parallel}$	$-\Delta e_{\perp}$	Δj
(3)	$-\alpha$	X	$-Y$	Z	Δj_x	$-\Delta j_y$	$-\Delta a$	$-\Delta e_x$	Δe_y	$\Delta\tilde{\tau}$	$-\Omega$	$-\Delta e_{\parallel}$	$\Delta e_{\perp}$	Δj
(1) x (2)	$-\alpha$	$-X$	Y	$-Z$	$-\Delta j_x$	Δj_y	Δa	Δe_x	$-\Delta e_y$	$-\Delta\tilde{\tau}$	$-\Omega-\pi$	$-\Delta e_{\parallel}$	$\Delta e_{\perp}$	Δj
(1) x (3)	$-\alpha$	X	$-Y$	$-Z$	$-\Delta j_x$	Δj_y	$-\Delta a$	$-\Delta e_x$	Δe_y	$\Delta\tilde{\tau}$	$-\Omega-\pi$	$\Delta e_{\parallel}$	$-\Delta e_{\perp}$	Δj
(2) x (3)	α	$-X$	$-Y$	Z	Δj_x	Δj_y	$-\Delta a$	$-\Delta e_x$	$-\Delta e_y$	$-\Delta\tilde{\tau}$	Ω	$-\Delta e_{\parallel}$	$-\Delta e_{\perp}$	Δj
(1) x (2) x (3)	α	$-X$	$-Y$	$-Z$	$-\Delta j_x$	$-\Delta j_y$	$-\Delta a$	$-\Delta e_x$	$-\Delta e_y$	$-\Delta\tilde{\tau}$	$\Omega+\pi$	$\Delta e_{\parallel}$	$\Delta e_{\perp}$	Δj

1) Une section (Ω, Δa) quelconque ne présente aucune symétrie particulière dans l'espace $\Delta e_{\parallel}$, $\Delta e_{\perp}$, Δj.

2) Il est possible de se borner à considérer des valeurs de Ω dans un intervalle d'amplitude π, la section ($\Omega+\pi$, Δa) se déduisant de la section (Ω, Δa) par symétrie par rapport à l'axe $\overrightarrow{O\Delta j}$ [voir changement (1)]

Il est même possible de limiter l'intervalle de variation de Ω à $[0, \pi/2]$ car la valeur $\Delta\tilde{\tau}_{max}$ au point ($\Delta e_{\parallel}$, $\Delta e_{\perp}$, Δj) de la section (Ω, Δa) est la même qu'au point ($-\Delta e_{\parallel}$, $\Delta e_{\perp}$, Δj) de la section ($-\Omega$, $-\Delta a$), symétrique du premier par rapport au plan $\Delta e_{\parallel}=0$ [voir changement (3)].

3) Il est possible de se borner à la recherche de la valeur maximale $\Delta\tilde{\tau}_{max}$ de $\Delta\tilde{\tau}$ au point ($\Delta e_{\parallel}$, $\Delta e_{\perp}$, Δj) de la section (Ω, Δa), la valeur minimale $\Delta\tilde{\tau}_{min}$ en ce point étant l'opposée de la valeur maximale au point ($-\Delta e_{\parallel}$, $-\Delta e_{\perp}$, Δj) de la section (Ω , $-\Delta a$), point symétrique du premier par rapport à $\overrightarrow{O\Delta j}$. [voir changement (2) x (3)].

L'application du Principe du Maximum de PONTRYAGIN [6] au problème d'optimisation ainsi défini conduit à introduire le vecteur adjoint

$$(5)\qquad \vec{P} = \left[p_{z_x} = p_z \cos\alpha_z \,;\, p_{z_y} = p_z \sin\alpha_z \,;\, p_a \,;\, p_{e_x} = p_e \cos\alpha_e \,;\, p_{e_y} = p_e \sin\alpha_e \,;\, p_t \geqslant 0 \,;\, p_c = -1 \right]$$

qui est constant, car les seconds membres des équations (4) ne contiennent pas de paramètres d'état.

Le vecteur efficacité $\vec{P_V}$ indiquant la direction optimale de la poussée [8] a pour composantes :

$$(6)\qquad \vec{P_V} \begin{cases} X = p_e \sin(\alpha - \alpha_e) - 2p_t = \mp p_e \cos\beta - 2p_t \\ Y = 2p_a + 3\alpha p_t + 2p_e \cos(\alpha - \alpha_e) = \lambda + 3\beta p_t \pm 2p_e \sin\beta \\ Z = p_z \sin(\alpha - \alpha_z) = \mu \cos\beta + \nu \sin\beta \end{cases}$$

en posant(*) :

$$(7)\qquad \begin{cases} \beta = \alpha - \left(\alpha_e \mp \frac{\pi}{2}\right) \\ \lambda = 2p_a + 3p_t\left(\alpha_e \mp \frac{\pi}{2}\right) \\ \mu = \mp p_z \cos(\alpha_z - \alpha_e) \\ \nu = \mp p_z \sin(\alpha_z - \alpha_e) \end{cases}$$

(*) Le signe $\pm$ sert à assurer pour les solutions "symétriques", la continuité de β , λ , μ , ν lorsque p_e passe par la valeur zéro, donc lorsque α_e subit une discontinuité de $-\pi$: ce cas se présente pour les solutions de type IV construites à partir de deux solutions de type I ⑤ ($\Delta e_\perp$ = 0) (voir réf. [2]).

Le module p_V du vecteur efficacité doit être inférieur à l'unité sur l'arc de transfert $\widehat{\alpha_o \alpha_f}$, sauf aux points d'application des impulsions où il est égal à un (fig. 4) [8, 9] .

Or, lorsque β croît de 2π , X et Z reprennent la même valeur, mais Y augmente de $6\pi\, p_t \geqslant 0$.

Donc, si p_τ n'est pas nul, les impulsions ne peuvent être appliquées que dans le premier et (ou) dans le dernier tour. D'autre part, $p_\tau \leqslant$ ordre $1/\Delta\alpha$, puisque $|Y| \leqslant 1$ (car $p_V \leqslant 1$) sur $\widehat{\alpha_0\alpha_f}$.

Le premier et le dernier tour jouant des rôles privilégiés, il est commode de poser :

$$(8)\quad \begin{cases} \alpha = \pm\, P(N)\pi + \hat{\alpha} & \text{selon que } \alpha \gtrless 0 \\ \beta = \pm\, P(N)\pi + \hat{\beta} & \text{selon que } \beta \gtrless 0 \end{cases}$$

où P (N) représente la partie paire du nombre N de tours complets. Donc :

$$(9)\quad 0 \leqslant \hat{\alpha}_f = -\hat{\alpha}_0 \leqslant 2\pi$$

Remarque : L'intervalle de variation de $\hat{\alpha}_f$ peut être réduit à $[0, \pi]$, car le changement de α en $\alpha \pm \pi$ et de Z en - Z, pour toutes les poussées laisse Ω , Δa et Δj inchangés, en changeant $\Delta e_{/\!/}$ en $-\Delta e_{/\!/}$, $\Delta e_\perp$ en $-\Delta e_\perp$ et $\Delta\tilde{\tau}_{max}$ en $\Delta\tilde{\tau}_{max}$ (1 + ordre $1/\Delta\alpha^2$) (= $\Delta\tilde{\tau}_{max}$ à l'ordre considéré dans cette étude). La section (Ω , Δa^-) du domaine accessible $\hat{\alpha}_f + \pi$ se déduit donc de la section (Ω , Δa) du domaine accessible $\hat{\alpha}_f$, par symétrie par rapport à l'axe $\overrightarrow{O\Delta j}$.

Dans toute la suite, le nombre de tours N est supposé suffisamment grand. Plus exactement, dans l'étude des rendez-vous de "longue durée" [3] , on se contentait de définir la solution de façon que la consommation correspondante soit égale à la consommation minimale pour le rendez-vous donné à l'ordre relatif : max (ε , $1/\Delta\alpha$) près. Nous nous proposons ici d'améliorer cette étude de façon que l'erreur relative sur la consommation soit, le plus souvent, seulement d'ordre : max (ε , $1/\Delta\alpha^2$).

La solution pourra alors être utilisée, en première approximation, pour des rendez-vous de "durée moyenne" (quelques révolutions) qui sont très fréquents dans la pratique (par exemple : mise en place précise d'un satellite).

Comme $p_\tau \leqslant$ ordre $1/\Delta\alpha$, posons :

$$p_\tau = \delta p_\tau = \frac{K}{\Delta\alpha}\,(1+\xi) > 0 \quad \text{où} \quad \xi \underset{\Delta\alpha\to\infty}{\longrightarrow} 0$$

Le vecteur efficacité (6) a alors pour composantes :

$$
(10) \quad \overrightarrow{P_V} \left\{ \begin{array}{l} X = \mp p_e \cos\hat{\beta} + \text{ordre } 1/\Delta\alpha \\ Y = \lambda + \frac{3K}{\Delta\alpha}(1+\xi)\left[\textcircled{\pm}\left(\frac{\Delta\alpha}{2} - \hat{\alpha}_f\right) + \hat{\beta}\right] \pm 2p_e \sin\hat{\beta} = \\ \qquad \lambda \textcircled{\pm} \frac{3}{2}K \pm 2p_e \sin\hat{\beta} + \text{ordre } \max(|\xi|, 1/\Delta\alpha) \\ Z = \mu \cos\hat{\beta} + \nu \sin\hat{\beta} \end{array} \right.
$$

d'où :

$$
(11) \quad \overrightarrow{P_V}^2 = X^2 + Y^2 + Z^2 = p_e^2 \cos^2\hat{\beta} + \left(\lambda \textcircled{\pm} \frac{3}{2}K \pm 2p_e \sin\hat{\beta}\right)^2 + \left(\mu\cos\hat{\beta} + \nu\sin\hat{\beta}\right)^2 + \text{ordre } \max(|\xi|, 1/\Delta\alpha) \leq 1 \text{ sur } \widehat{\alpha_0 \alpha_f}
$$

Si on admet, tout d'abord, sur la consommation une erreur relative d'ordre $1/\Delta\alpha$, l'expression (11) de l'efficacité suffit pour définir les solutions optimales (car, il sera montré plus loin que $|\xi| \leq$ ordre $1/\Delta\alpha$). Ce problème a déjà été résolu [3] ; il conduit soit aux solutions quadri-impulsionnelles de <u>type IV, réductibles à des solutions tri-impulsionnelles</u> ou, quelquefois même, bi-impulsionnelles, et qui contiennent comme cas limites les solutions optimales de types I, II et III, soit aux solutions singulières quasi-planes de <u>type IV bis</u> , contenant comme cas limites les solutions optimales de type I bis et II ou III (plan).

Les solutions de <u>type IV bis</u> s'obtiennent pour p_e et $|\lambda|$ d'ordre $1/\Delta\alpha$ et $\sqrt{\mu^2+\nu^2}$ d'ordre $1/\sqrt{\Delta\alpha}$. Alors, si K = 2/3 , $\overrightarrow{P_V}^2 = 1 +$ ordre $1/\Delta\alpha$ ce qui montre qu'à l'ordre conservé, la position du point d'application de la poussée peut être choisie de façon arbitraire au premier ou au dernier tour (dégénérescence).

Pour les solutions de <u>type IV</u>, les 4 points d'application des impulsions sont symétriques par rapport à $\hat{\beta} = 0$ (donc, $\lambda = \nu = 0$); ce sont (à 2π près) $-\hat{\beta}_f$, $\hat{\beta}_f + \pi$ au premier tour et $\hat{\beta}_f$, $-\hat{\beta}_f - \pi$ au dernier tour (figs. 5 et 6).

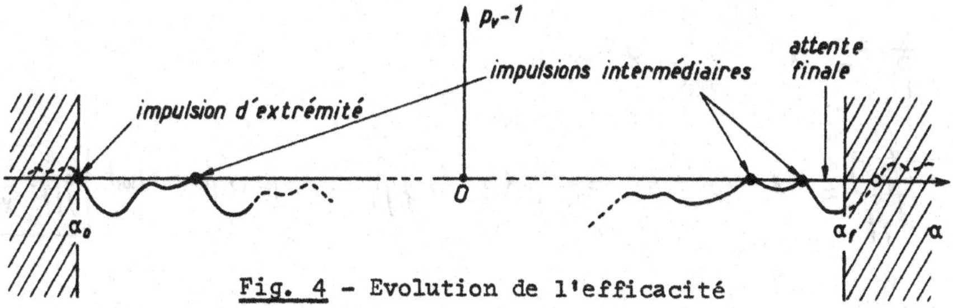

Fig. 4 - Evolution de l'efficacité

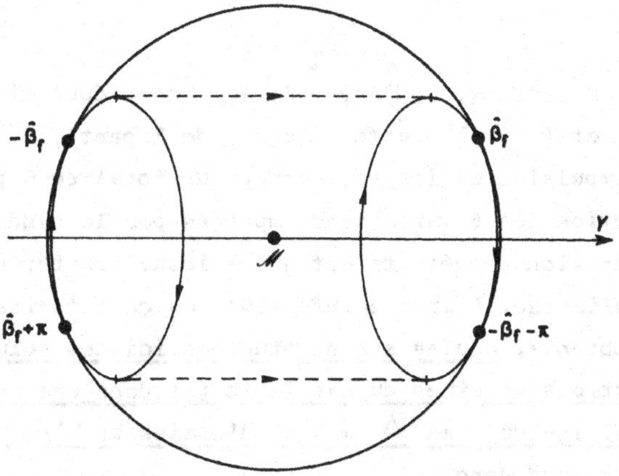

Fig. 5 - Courbe d'efficacité - Solutions de type IV
(Etude à l'ordre $1/\Delta\alpha$ près)

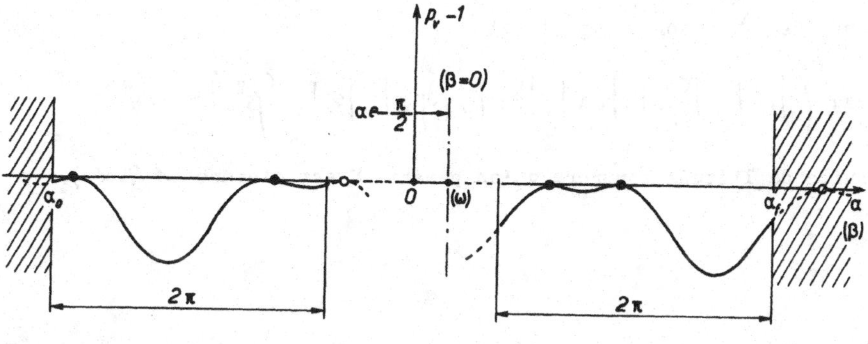

Fig. 6 - Evolution de l'efficacité - Solutions de type IV
(Etude à l'ordre $1/\Delta\alpha$ près)

L'angle $\hat{\beta}_f$ est tel que :

(12) $$\overrightarrow{p_v}^2 = X^2 + Y^2 + Z^2 = 1$$

(13) $$\frac{1}{2}\frac{d(\overrightarrow{p_v}^2)}{d\hat{\beta}_f} = XX' + YY' + ZZ' = -\left[(X^2+Z^2)\,tg\,\hat{\beta}_f + 2XY\right] = 0 \text{ (contact) soit: } \boxed{tg\,\hat{\beta}_f = \frac{-2XY}{X^2+Z^2}}$$

(14) $$\frac{1}{2}\frac{d^2(\overrightarrow{p_v}^2)}{d\hat{\beta}_f^2} = 3X^2 - Z^2 \leq 0 \quad \left(\overrightarrow{p_v}^2 \text{ maximum ; condition d'"angle utile"}\right)$$

Les 5 paramètres α_e , $\hat{\beta}_f$, X, Y, Z sont donc liés par les deux relations (12) et (13). Il reste 3 degrés de liberté, qui ajoutés aux 3 grandeurs d'impulsion efficaces, forment un total de 6 paramètres calculables en fonction des 6 variations imposées par le rendez-vous.

Si la précision précédente est jugée insuffisante, une étude à l'ordre supérieur s'impose. L'indice inférieur (0) caractérisera la solution précédemment obtenue. Seules seront étudiées ici des solutions voisines du cas IV; les solutions voisines du cas IVbis (et donc une portion du domaine accessible voisine du plan $\Delta j = 0$ et d'épaisseur $1/\sqrt{\Delta\alpha}$) seront laissées de côté. Posons donc :

(15) $$\lambda = \lambda_{(0)} + \delta\lambda = \delta\lambda \; ; \; p_e = p_{e(0)} + \delta p_e \; ; \; \mu = \mu_{(0)} + \delta\mu \; ; \; \nu = \nu_{(0)} + \delta\nu = \delta\nu \; ; \; \hat{\beta} = \hat{\beta}_{(0)} + \delta\beta$$

où tous les δ sont $\ll$ 1, et

$$M = \max\left(|\delta\lambda|, |\delta p_e|, |\delta\mu|, |\delta\nu|, |\delta p_e|, |\xi|, \delta\beta^2\right)$$

(la suite de l'étude montrera qu'en général M est d'ordre $1/\Delta\alpha$).

Le vecteur efficacité (6) a alors pour composantes :

$$(16)\quad \vec{p_v}\begin{cases} X = X_{(0)} - X_{(0)}\,\mathrm{tg}\,\hat{\beta}_{(0)}\delta\beta \mp \delta p_e \cos\hat{\beta}_{(0)} - 2\delta p_c - \frac{X_{(0)}}{2}\delta\beta^2 + \text{ordre } M^{3/2} \\ Y = Y_{(0)} - 2X_{(0)}\delta\beta + \delta\lambda \pm 2\delta p_e \sin\hat{\beta}_{(0)} \circledcirc{\pm} \frac{3}{2}K\xi + 3\delta p_c\left(\hat{\beta}_{(0)} \circledcirc{\mp} \hat{\alpha}_f\right) + X_{(0)}\,\mathrm{tg}\,\hat{\beta}_{(0)}\delta\beta^2 + \text{ordre } M^{3/2} \\ Z = Z_{(0)} - Z_{(0)}\,\mathrm{tg}\,\hat{\beta}_{(0)}\delta\beta + \delta\mu\cos\hat{\beta}_{(0)} + \delta\nu\sin\hat{\beta}_{(0)} - \frac{Z_{(0)}}{2}\delta\beta^2 + \text{ordre } M^{3/2} \end{cases}$$

avec :

$$(17)\qquad K = \frac{2Y_{(0)}\left(3X_{(0)}^2 - Z_{(0)}^2\right)}{3\left(1 - Y_{(0)}^2\right)}$$

d'où :

$$(18)\qquad \vec{p_v}^{\,2} - 1 = \underbrace{\boxed{2Y_{(0)}\delta\lambda + 2Z_{(0)}\sin\hat{\beta}_{(0)}\delta\nu + 2\delta p_c\left(3Y_{(0)}\hat{\beta}_{(0)} - 2X_{(0)}\right) + \left(3X_{(0)}^2 - Z_{(0)}^2\right)\delta\beta^2}}_{f}$$
$$\circledcirc{\mp}\, 6Y_{(0)}\hat{\alpha}_f\delta p_c + 2\delta p_e\left(\mp X_{(0)}\cos\hat{\beta}_{(0)} \pm 2Y_{(0)}\sin\hat{\beta}_{(0)}\right) + 2Z_{(0)}\cos\hat{\beta}_{(0)}\delta\mu \circledcirc{\pm} \frac{3}{2}KY_{(0)}\xi + \text{ordre } M^{3/2} \leq 0 \text{ sur } \widehat{\alpha_0\alpha_f}$$

Il est important de remarquer que cette expression ne contient évidemment pas de terme en $\delta\beta$ à cause de la condition de maximum (13).

Les valeurs de $\hat{\beta}_{f(0)}$ à utiliser sont indiquées sur les figures 7 et 8, à l'ordre M près.

Appelant β_0 et β_f les extrémités de l'arc $\widehat{\beta_0\beta_f}$ sur lequel $p_v \leq 1$, il faut distinguer deux familles de solutions selon que les solutions de base (solutions (0)) sont "symétriques" ou "antisymétriques" :

1) <u>Solutions "symétriques"</u> (fig. 7 et 9)

Les points $\beta_{f(0)}$ et $\beta_{0(0)}$ sont symétriques par rapport à $\beta = 0$ c'est-à-dire :

$$(19)\qquad \boxed{\beta_{f(0)} = -\beta_{0(0)}}$$

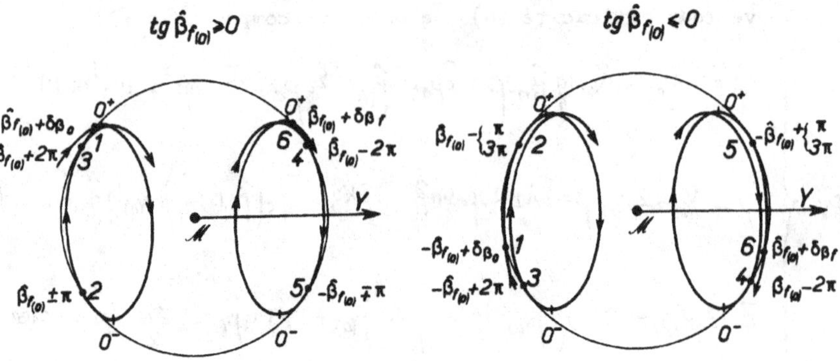

$0 \leq \hat{\beta}_{f(o)} \leq \frac{\pi}{2}$	$\beta = \alpha - (\alpha_e - \frac{\pi}{2})$, origine en 0^+	$\frac{\pi}{2} \leq \hat{\beta}_{f(o)} \leq \pi$
$\pi \leq \hat{\beta}_{f(o)} \leq \frac{3\pi}{2}$	$\beta = \alpha - (\alpha_e + \frac{\pi}{2})$, origine en 0^-	$\frac{3\pi}{2} \leq \hat{\beta}_{f(o)} \leq 2\pi$

Fig. 7 - Courbe d'efficacité des solutions "symétriques"
(Le contact en 2, 3, 4, 5 est facultatif)

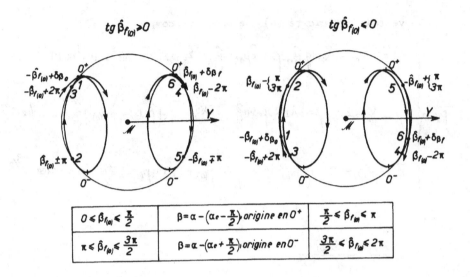

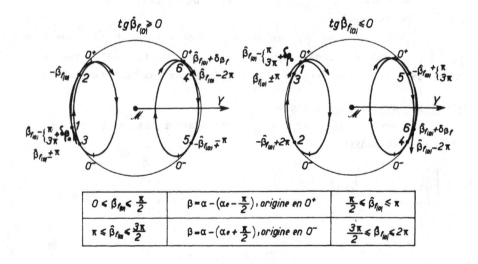

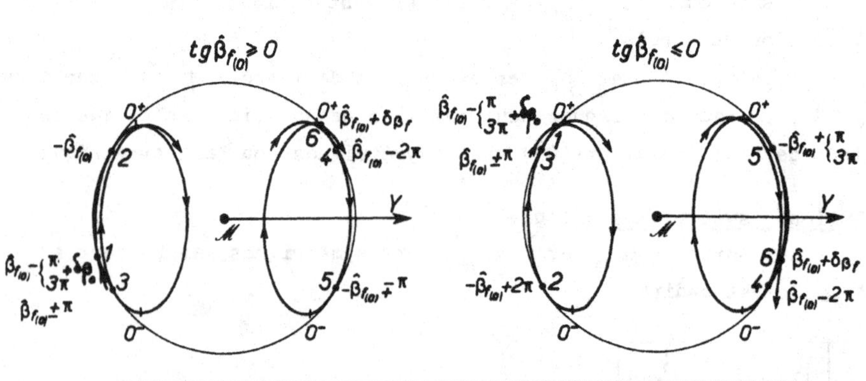

$0 \leq \hat{\beta}_{f(o)} \leq \frac{\pi}{2}$	$\beta = \alpha - (\alpha_e - \frac{\pi}{2})$, origine en 0^+	$\frac{\pi}{2} \leq \hat{\beta}_{f(o)} \leq \pi$
$\pi \leq \hat{\beta}_{f(o)} \leq \frac{3\pi}{2}$	$\beta = \alpha - (\alpha_e + \frac{\pi}{2})$, origine en 0^-	$\frac{3\pi}{2} \leq \beta_{f(o)} \leq 2\pi$

Fig. 8 - Courbe d'efficacité des solutions "antisymétriques"
(Le contact en 2, 3, 4, 5 est facultatif)

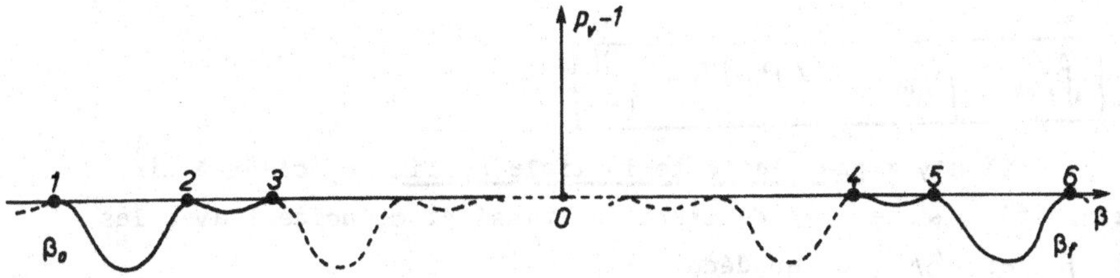

Fig. 9 - Solution de base symétrique

ou, encore :

(20) $\hat{\beta}_{f(c)} = - \hat{\beta}_{o(c)}$

2) Solutions "antisymétriques" (fig. 8 et 10)

Les points $\beta_{f(c)}$ et $\beta_{o(c)}$ ne sont pas symétriques par rapport à $\beta = 0$, mais on a :

(21) $\hat{\beta}_{o(c)} = \hat{\beta}_{f(c)} - \begin{cases} \pi \\ 3\pi \end{cases}$

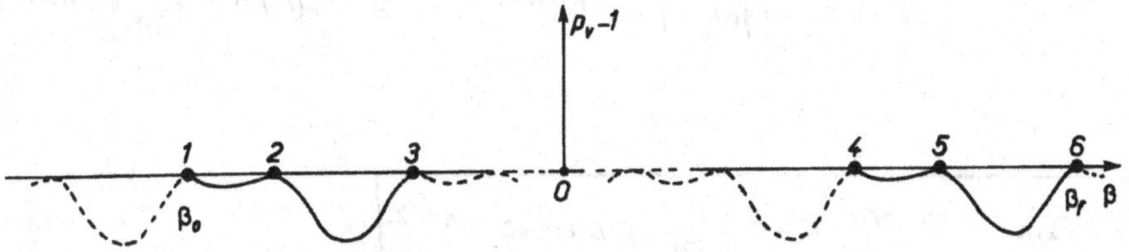

Fig. 10 - Solution de base antisymétrique

c'est-à-dire :

$$\beta_{f(0)} - \beta_{o(0)} = 2P(N)\pi + \begin{cases}\pi \\ 3\pi\end{cases} \tag{22}$$

S'il n'y a pas d'attente initiale ou finale, c'est-à-dire si les extrémités α_o et α_f de l'arc de transfert coincident avec les points β_o et β_f, on en déduit :

1) Solutions "symétriques"

$$\hat{\beta}_f = \hat{\beta}_{f(0)} + \delta\beta_f = \hat{\alpha}_f - \left(\alpha_e \mp \frac{\pi}{2}\right) \tag{23}$$

$$\hat{\beta}_o = \hat{\beta}_{o(0)} + \delta\beta_o = -\hat{\beta}_{f(0)} + \delta\beta_o = \hat{\alpha}_o - \left(\alpha_e \mp \frac{\pi}{2}\right) = -\hat{\alpha}_f - \left(\alpha_e \mp \frac{\pi}{2}\right) \tag{24}$$

d'où :

$$\alpha_e = \pm \frac{\pi}{2} + \text{ordre } M^{1/2} \tag{25}$$

et :

$$\beta = \alpha + \text{ordre } M^{1/2} \tag{26}$$

2) Solutions "antisymétriques"

$$\alpha_f - \alpha_o = \beta_f - \beta_o = \beta_{f(0)} - \beta_{o(0)} + \text{ordre } M^{1/2} = 2P(N)\pi + \begin{cases}\pi \\ 3\pi\end{cases} + \text{ordre } M^{1/2} \tag{27}$$

soit :

$$\Delta\alpha = 2P(N)\pi + \begin{cases}\pi \\ 3\pi\end{cases} + \text{ordre } M^{1/2} \tag{28}$$

Donc, s'il n'y a pas d'attente initiale ou finale, les solutions "antisymétriques" ne peuvent se présenter que si l'angle de transfert est très voisin d'un nombre impair de demi-tours.

D'autre part, l'équation (24) est remplacée par :

$$\hat{\beta}_0 = \hat{\beta}_{\alpha(0)} + \delta\beta_0 = \hat{\beta}_{f(0)} - \left\{\begin{matrix}\pi\\ 3\pi\end{matrix}\right. + \delta\beta_0 = -\hat{\alpha}_f - \left(\alpha_e \mp \frac{\pi}{2}\right) \tag{29}$$

d'où :

$$\boxed{\hat{\beta}_{f(0)} = \pi - \alpha_e + \text{ordre } M^{1/2}} \tag{30}$$

Lorsqu'on passe d'un des points 1, 2, 3, 4, 5, 6 à l'autre, seuls les termes (f), encadrés dans (18), sont susceptibles de varier. Plus précisément :

(31) (sym)

$$\left\{\begin{aligned} f_{1s} &= -2Y_{(0)}\delta\lambda - 2Z_{(0)}\sin\hat{\beta}_{f(0)}\delta\nu + 2\delta p_\tau\left(3Y_{(0)}\hat{\beta}_{f(0)} - 2X_{(0)}\right) + \left(3X_{(0)}^2 - Z_{(0)}^2\right)\delta\beta_0^2 + \text{ordre } M^{3/2}\\ f_{2s} &= -2X_{(0)}\delta\lambda + 2Z_{(0)}\sin\hat{\beta}_{f(0)}\delta\nu + 2\delta p_\tau\left[-3Y_{(0)}\left(\hat{\beta}_{f(0)} \pm \pi\right) + 2X_{(0)}\right] + \text{ordre } M^2 \text{ selon que } \operatorname{tg}\hat{\beta}_{f(0)} \gtrless 0\\ f_{3s} &= -2X_{(0)}\delta\lambda - 2Z_{(0)}\sin\hat{\beta}_{f(0)}\delta\nu + 2\delta p_\tau\left[-3Y_{(0)}\left(-\hat{\beta}_{f(0)} + 2\pi\right) - 2X_{(0)}\right] + \text{ordre } M^2 \end{aligned}\right.$$

(32) (antisym)

$$\left\{\begin{aligned} f_{1a} &= -2Y_{(0)}\delta\lambda + 2Z_{(0)}\sin\hat{\beta}_{f(0)}\delta\nu + 2\delta p_\tau\left[-3Y_{(0)}\left(\hat{\beta}_{f(0)} - \pi\right) + 2X_{(0)}\right] + \left(3X_{(0)}^2 - Z_{(0)}^2\right)\delta\beta_0^2 + \text{ordre } M^{3/2}\\ f_{2a} &= -2Y_{(0)}\delta\lambda - 2Z_{(0)}\sin\hat{\beta}_{f(0)}\delta\nu + 2\delta p_\tau\left[-3Y_{(0)}\left(-\hat{\beta}_{f(0)} + \left\{\begin{matrix}0\\ 2\pi\end{matrix}\right.\right) - 2X_{(0)}\right] + \text{ordre } M^2 \text{ selon que } \operatorname{tg}\hat{\beta}_{f(0)} \gtrless 0\\ f_{3a} &= -2Y_{(0)}\delta\lambda + 2Z_{(0)}\sin\hat{\beta}_{f(0)}\delta\nu + 2\delta p_\tau\left[-3Y_{(0)}\left(\hat{\beta}_{f(0)} + \pi\right) + 2X_{(0)}\right] + \text{ordre } M^2 \end{aligned}\right.$$

(33)

$$\left\{\begin{aligned} f_4 &= 2Y_{(0)}\delta\lambda + 2Z_{(0)}\sin\hat{\beta}_{f(0)}\delta\nu + 2\delta p_\tau\left[3Y_{(0)}\left(\hat{\beta}_{f(0)} - 2\pi\right) - 2X_{(0)}\right] + \text{ordre } M^2\\ f_5 &= 2Y_{(0)}\delta\lambda - 2Z_{(0)}\sin\hat{\beta}_{f(0)}\delta\nu + 2\delta p_\tau\left[-3Y_{(0)}\left(\hat{\beta}_{f(0)} \pm \pi\right) + 2X_{(0)}\right] + \text{ordre } M^2 \text{ selon que } \operatorname{tg}\hat{\beta}_{f(0)} \gtrless 0\\ f_6 &= 2Y_{(0)}\delta\lambda + 2Z_{(0)}\sin\hat{\beta}_{f(0)}\delta\nu + 2\delta p_\tau\left(3Y_{(0)}\hat{\beta}_{f(0)} - 2X_{(0)}\right) + \left(3X_{(0)}^2 - Z_{(0)}^2\right)\delta\beta_f^2 + \text{ordre } M^{3/2} \end{aligned}\right.$$

où X, Y, Z sont relatifs au point 6 et en se bornant au cas $0 \leq \hat{\beta}_{f(0)} \leq \pi$ qui sera seul à se présenter dans la suite.

3 - ETUDE DES SOLUTIONS HEXA-IMPULSIONNELLES ET DES SOLUTIONS QUI EN DERIVENT -

3.1. - Solutions hexa-impulsionnelles

3.1.1. - Solutions hexa-impulsionnelles symétriques

L'égalisation des niveaux d'efficacité ($\dot{f}$) aux points 1, 2, 3, 4, 5, 6 (fig. 11), à l'ordre M^2 près pour les niveaux 2, 3, 4, 5 et seulement à l'ordre $M^{3/2}$ près pour les niveaux 1 et 6, conduit à un système de 5 équations linéaires à 4 inconnues : $\delta\lambda, \delta\nu, \delta\beta_0^2, \delta\beta_f^2$.

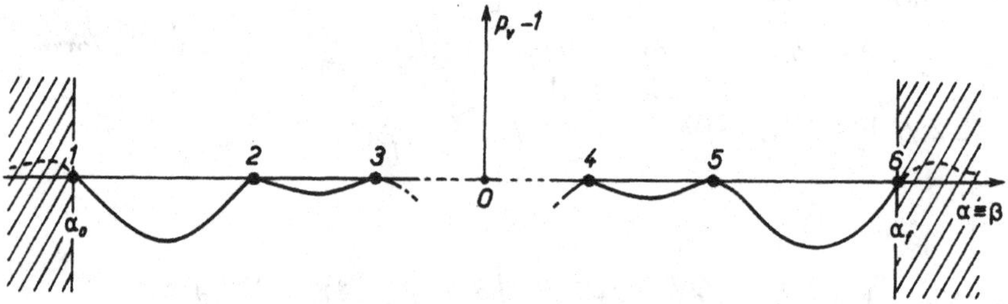

Fig. 11 - Solutions hexa-impulsionnelles symétriques

Il y a donc une condition de compatibilité :

(34) $$3Y_{(0)}\left(\hat{\beta}_{f(0)}-\begin{Bmatrix}\pi/2\\3\pi/2\end{Bmatrix}\right)-2X_{(0)}=\text{ordre } M \text{, selon que } \operatorname{tg}\hat{\beta}_{f(0)} \gtrless 0$$

et alors :

(35) $$\delta\lambda = \text{ordre } M^2$$

(36) $$\delta\nu = \text{ordre } M^2$$

(37) $$\delta\beta_0 = 2\sqrt{\frac{3\pi\, Y_{(0)}\,\delta p_\tau}{Z_{(0)}^2-3X_{(0)}^2}} + \text{ordre } M = 2\sqrt{\frac{1}{N}\,\frac{Y_{(0)}^2}{1-Y_{(0)}^2}} + \text{ordre } M$$

(38) $$\delta\beta_f = -2\sqrt{\frac{3\pi\, Y_{(0)}\,\delta p_\tau}{Z_{(0)}^2-3X_{(0)}^2}} + \text{ordre } M = -2\sqrt{\frac{1}{N}\,\frac{Y_{(0)}^2}{1-Y_{(0)}^2}} + \text{ordre } M$$

Remarque : Les deux dernières équations supposent que $1-Y_{(0)}^2 \neq 0$ c'est-à-dire que la solution considérée n'est pas voisine du type IVbis plan comme il a été mentionné précédemment.

Comme il n'y a pas d'attente initiale ou finale, les équations (25) et (26) sont utilisables et comme $\delta\beta_f = -\delta\beta_0$ + ordre M, elles sont vraies au moins à l'ordre M près et non pas à l'ordre $M^{1/2}$ près.

Une étude aux ordres supérieurs montrerait sûrement que l'on a plus précisément :

(39) $$\delta\lambda = \delta\nu = 0$$

(40) $$\delta\beta_0 = -\delta\beta_f$$

(41) $$\alpha_e = \pm\,\pi/2$$

(42) $$\beta \equiv \alpha$$

c'est-à-dire que la solution est rigoureusement symétrique.

Comme $\hat{\beta} = \hat{\alpha}$, on peut ne s'intéresser qu'aux valeurs de $\hat{\beta}_{f(0)}$ de l'intervalle $[0,\pi]$. Ainsi, les 3 cosinus directeurs $X_{(0)}$ $Y_{(0)}$ $Z_{(0)}$ sont calculables en fonction de l'angle $\hat{\beta}_{f(0)}$, à l'ordre M près, négligeable dans la suite, par les 3 équations (12),(13) et (34). A l'ordre conservé, on a donc :

(43) $$Y_{(0)} = \frac{1}{\sqrt{1-3\left(\hat{\beta}_{f(0)} - \begin{cases}\pi/2\\ 3\pi/2\end{cases}\right)\operatorname{cotg}\hat{\beta}_{f(0)}}} \quad \text{, selon que } \operatorname{tg}\hat{\beta}_{f(0)} \gtrless 0$$

(44) $$X_{(0)} = \frac{3}{2}\left(\hat{\beta}_{f(0)} - \begin{cases}\pi/2\\ 3\pi/2\end{cases}\right) Y_{(0)}$$

(45) $$|Z_{(0)}| = \sqrt{1 - X_{(0)}^2 - Y_{(0)}^2}$$

avec les contraintes d'inégalité :

(46) $$Y_{(0)}^2 \geqslant 0 \quad \Longrightarrow \quad \operatorname{cotg}\hat{\beta}_{f(0)}\left[\operatorname{tg}\hat{\beta}_{f(0)} - 3\left(\hat{\beta}_{f(0)} - \begin{cases}\pi/2\\ 3\pi/2\end{cases}\right)\right] \geqslant 0$$

(47) $$3X_{(0)}^2 \leqslant Z_{(0)}^2 = 1 - X_{(0)}^2 - Y_{(0)}^2 \quad \Longrightarrow \quad -3\left(\hat{\beta}_{f(0)} - \begin{cases}\pi/2\\ 3\pi/2\end{cases}\right) \leqslant \operatorname{cotg}\hat{\beta}_{f(0)}$$

Dans l'intervalle $[0,\pi]$, ces contraintes sont satisfaites pour :

(48) $$0 \leqslant \hat{\beta}_{f(0)} \leqslant \hat{\beta}^* = 14°08' \quad \left[\text{où } \operatorname{cotg}\hat{\beta}^* = 3\left(\frac{\pi}{2} - \hat{\beta}^*\right)\right]$$

Une valeur de $\hat{\beta}_{f(0)}$ étant donnée dans l'intervalle $[0,\hat{\beta}^*]$ il est aisé de calculer les cosinus directeurs $X_{(0)}$, $Y_{(0)}$, $Z_{(0)}$ (fig. 12)

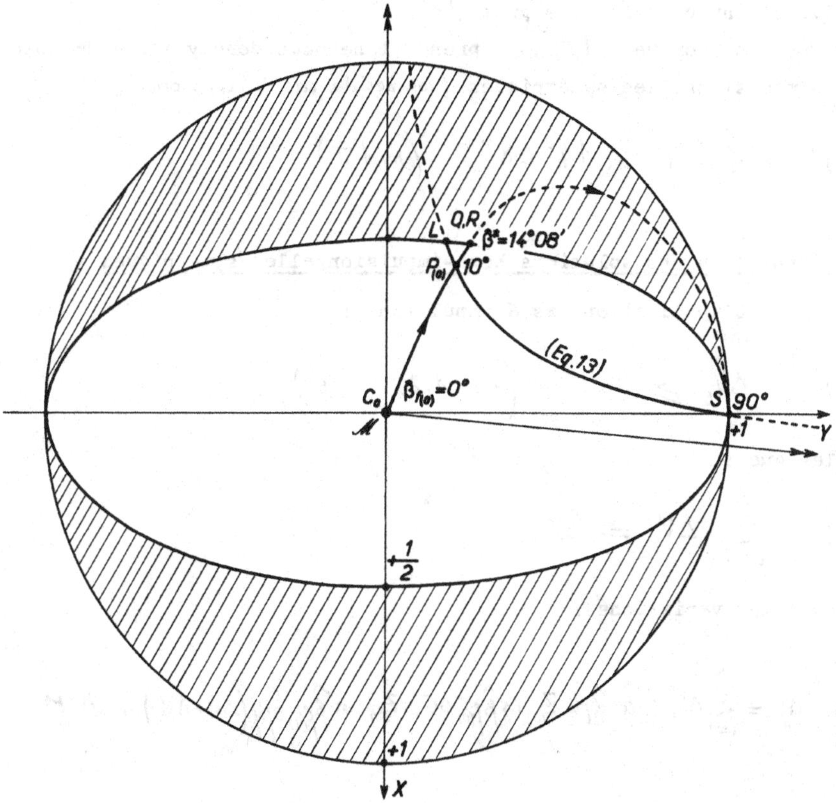

Fig. 12 - Direction optimale de poussée en fonction de l'angle de transfert.

ainsi que l'angle $\hat{\alpha}_f = \hat{\beta}_{f(0)} + \delta\beta_f$ (si le nombre de tours N est fixé), et inversement (cas pratique).

A l'ordre $1/\sqrt{\Delta\alpha}$ près, il ne peut donc y avoir de solutions hexa-impulsionnelles symétriques (voisines du cas IV) que pour :

$$(49) \qquad 0 \leq \hat{\alpha}_f \leq 14°08' \quad (\mathrm{mod}\ \pi)$$

Représentation des solutions hexa-impulsionnelles symétriques -

L'application des 6 impulsions :

$$(50) \qquad \Delta V_i \geq 0 \qquad (i = 1, 2, \ldots, 6)$$

telles que :

$$(51) \qquad \sum_{i=1}^{6} \Delta V_i = \Delta C$$

produit les variations :

$$(52) \begin{cases} \Delta j_x = \sum_{i=1}^{6} \Delta V_i Z_i \cos\hat{\alpha}_i = Z_{(0)} \cos\hat{\beta}_{f(0)} \Delta C - Z_{(0)} \sin\hat{\beta}_{f(0)} \delta\beta_f (\Delta V_4 + \Delta V_6) + \text{ordre } M \\ \Delta j_y = \sum_{i=1}^{6} \Delta V_i Z_i \sin\hat{\alpha}_i = Z_{(0)} \sin\hat{\beta}_{f(0)} (-\Delta V_1 + \Delta V_2 - \Delta V_3 + \Delta V_4 - \Delta V_5 + \Delta V_6) - Z_{(0)} \cos\hat{\beta}_{f(0)} \delta\beta_f (\Delta V_1 - \Delta V_6) + \text{ordre } M \\ \Delta a = \sum_{i=1}^{6} 2\Delta V_i Y_i = 2Y_{(0)} (-\Delta V_1 - \Delta V_2 - \Delta V_3 + \Delta V_4 + \Delta V_5 + \Delta V_6) + \text{ordre } M \\ \Delta e_x = \sum_{i=1}^{6} \Delta V_i (X_i \sin\hat{\alpha}_i + 2Y_i \cos\hat{\alpha}_i) = (X_{(0)} \sin\hat{\beta}_{f(0)} + 2Y_{(0)} \cos\hat{\beta}_{f(0)})(-\Delta V_1 + \Delta V_2 - \Delta V_3 + \Delta V_4 - \Delta V_5 + \Delta V_6) \\ \qquad + (-X_{(0)} \cos\hat{\beta}_{f(0)} + 2Y_{(0)} \sin\hat{\beta}_{f(0)}) \delta\beta_f (\Delta V_1 - \Delta V_6) + \text{ordre } M \\ \Delta e_y = \sum_{i=1}^{6} \Delta V_i (-X_i \cos\hat{\alpha}_i + 2Y_i \sin\hat{\alpha}_i) = (-X_{(0)} \cos\hat{\beta}_{f(0)} + 2Y_{(0)} \sin\hat{\beta}_{f(0)}) \Delta C \\ \qquad + (X_{(0)} \sin\hat{\beta}_{f(0)} + 2Y_{(0)} \cos\hat{\beta}_{f(0)}) \delta\beta_f (\Delta V_1 + \Delta V_6) + \text{ordre } M \end{cases}$$

Dans l'étude de la section $(\Omega, \Delta a)$ du domaine accessible relatif à la valeur $\hat{\alpha}_f$ [satisfaisant à (49)] dans l'espace $\Delta e_{\parallel}, \Delta e_{\perp}, \Delta j$ nous négligerons tout d'abord les détails d'ordre $M^{1/2}$. Il est alors possible de négliger dans les équations (52) les termes en $\delta\beta_f$. Posant alors, comme dans [3] :

$$(53) \quad \begin{cases} \Delta V_1 + \Delta V_3 = I_0' \, \Delta C \\ \Delta V_2 = I_0'' \, \Delta C \\ \Delta V_4 + \Delta V_6 = I_N' \, \Delta C \\ \Delta V_5 = I_N'' \, \Delta C \end{cases}$$

avec :

$$(54) \quad I \geqslant 0$$

$$(55) \quad I_0' + I_0'' + I_N' + I_N'' = 1$$

il vient, à l'ordre conservé :

$$(56) \quad \begin{cases} \Delta j_x = Z_{(0)} \cos \hat{\beta}_{f(0)} \, \Delta C \\ \Delta j_y = Z_{(0)} \sin \hat{\beta}_{f(0)} \left[2(I_N' + I_0'') - 1 \right] \Delta C \\ \Delta a = 2 Y_{(0)} \left[1 - 2(I_0' + I_0'') \right] \Delta C \\ \Delta e_x = \left(X_{(0)} \sin \hat{\beta}_{f(0)} + 2 Y_{(0)} \cos \hat{\beta}_{f(0)} \right) \left[2(I_N' + I_0'') - 1 \right] \Delta C \\ \Delta e_y = \left(- X_{(0)} \cos \hat{\beta}_{f(0)} + 2 Y_{(0)} \sin \hat{\beta}_{f(0)} \right) \Delta C \end{cases}$$

Les 6 grandeurs d'impulsion n'interviennent dans (56) que par les 2 groupements $I'_N + I''_0$ et $I'_0 + I''_0$. A l'ordre conservé, les solutions hexa-impulsionnelles symétriques de la section (Ω, Δa) (2 relations) sont donc représentées par un seul point $P_{(0)}$ (fig. 13).

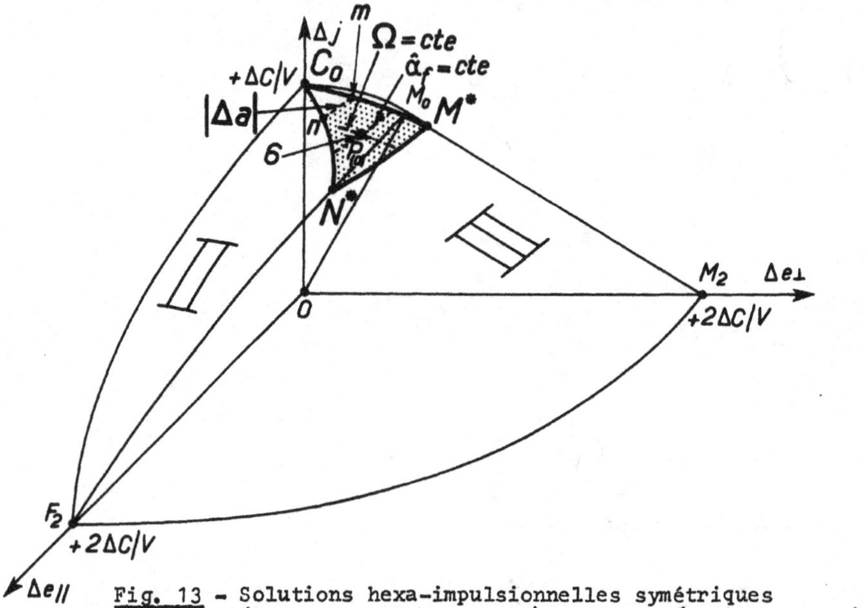

Fig. 13 - Solutions hexa-impulsionnelles symétriques (zone accessible pour $\hat{\alpha}_f$, Ω , Δa quelconques)

Le groupement $I'_0 + I''_0$ n'intervenant que dans Δa , on peut même ajouter que le point est le même pour toutes les sections (Ω fixé, Δa variable).

L'élimination des paramètres $I'_N + I''_0$, $I'_0 + I''_0$ dans les équations (56) fournit la position de ce point :

$$
(57) \quad P_{(0)} \begin{cases} \Delta e_{/\!/} = \Delta e_x \cos\Omega + \Delta e_y \sin\Omega = E \sin\Omega \, \Delta C \\ \Delta e_{\perp} = - \Delta e_x \sin\Omega + \Delta e_y \cos\Omega = \left(E \cos\Omega - \dfrac{F}{\cos\Omega}\right) \Delta C \\ \Delta j = \dfrac{G}{\cos\Omega} \Delta C \qquad \left(0 \leq \Omega \leq \dfrac{\pi}{2}\right) \end{cases}
$$

où E, F, G ne dépendent que de $\hat{\beta}_{f(0)}$ $\left[= \hat{\alpha}_f \text{ (à l'ordre conservé)}\right]$:

$$(58)\quad \begin{cases} E = \dfrac{2}{\cos u \sqrt{1 + 3u \operatorname{tg} u}} \\ F = \dfrac{\sin u \, (4 \operatorname{tg} u - 3u)}{2\sqrt{1 + 3u \operatorname{tg} u}} \\ G = \dfrac{\sqrt{3}}{2} \sin u \sqrt{\dfrac{u(4 \operatorname{tg} u - 3u)}{1 + 3u \operatorname{tg} u}} \end{cases} \qquad \left(u^* = 75°52' \leq u = 90° - \hat{\beta}_{f(0)} \leq 90°\right)$$

Lorsque $\hat{\beta}_{f(0)}$ et Ω varient (à Δa fixé) de façon que les inégalités (48) et :

$$(59)\qquad -1 \leq 2\left(I'_N + I''_0\right) - 1 = \operatorname{tg} \Omega \operatorname{cotg} \hat{\beta}_{f(0)} \leq +1$$

$$(60)\qquad -1 \leq 1 - 2\left(I'_0 + I''_0\right) = \frac{\Delta a}{2\, Y_{(0)}\, \Delta C} \leq +1$$

soient satisfaites, le point hexa-impulsionnel $P_{(0)}$ décrit une portion de surface (S) limitée dans l'octant $\Delta e_{/\!/} \geq 0$, $\Delta e_\perp \geq 0$, $\Delta j \geq 0$ (fig. 13) par :

1) La ligne $\widehat{N^* M^*}$ $\left(\hat{\beta}_{f(0)} = \hat{\beta}^* = 14°08'\right)$ tracée sur le cylindre :

$$(61)\qquad \Delta e_{/\!/}^2 + \left(\Delta e_\perp + \sqrt{3}\, \Delta j\right)^2 = 4\, \Delta C^2$$

correspondant aux transferts optimaux de type III. En effet, on a alors :

$$(62)\quad \begin{cases} E^* = 2 \\ F^* = \dfrac{3}{2} \sin^2 u^* = \sqrt{3}\, G^* = 1{,}41 \\ G^* = \dfrac{\sqrt{3}}{2} \sin^2 u^* = 0{,}814 \end{cases}$$

donc :

$$(63)\quad \begin{cases} \Delta e_{/\!/} = 2 \sin \Omega \\ \Delta e_\perp = 2 \cos \Omega - \sqrt{3}\, \Delta j \end{cases}$$

2) La ligne $\widehat{mn}$ $\left(Y_{(0)} = \frac{1}{\sqrt{1+3u\,tg\,u}} = \frac{|\Delta a|}{2\Delta C} \right)$ qui est également une ligne iso $\hat{\beta}_{f(0)}$ particulière.

3) La ligne $\widehat{nN^*}$ $\left(\hat{\beta}_{f(0)} - \Omega = 0 \right)$ tracée sur l'ellipsoïde de révolution :

(64) $$\frac{\Delta e_{/\!/}^2}{4} + \Delta e_{\perp}^2 + \Delta j^2 = \Delta C^2$$

correspondant aux transferts optimaux de type II. En effet, on a alors :

(65) $$\begin{cases} \Delta e_{/\!/} = 2Y\,\Delta C \\ \Delta e_{\perp} = -X\,\Delta C \\ \Delta j = |Z|\,\Delta C \end{cases}$$

Les lignes iso $\hat{\beta}_{f(0)}$ de la surface (S) se projettent sur le plan $\Delta e_{/\!/} = 0$ suivant des arcs d'hyperboles :

(66) $$\Delta j \left(G\,\Delta e_{\perp} + F\,\Delta j \right) - E\,G^2 \Delta C^2 = 0$$

et sur les plans $\Delta e_{\perp} = 0$ et $\Delta j = 0$ suivant des arcs de quartiques :

(67) $$\Delta e_{/\!/}^2\,\Delta j^2 + E^2 \left(G^2 \Delta C^2 - \Delta j^2 \right) \Delta C^2 = 0$$

(68) $$\left[\Delta e_{/\!/}^2 + \Delta e_{\perp}^2 + E(2F-E)\Delta C^2 \right] \left[E^2 \Delta C^2 - \Delta e_{/\!/}^2 \right] - E^2 F^2 \Delta C^4 = 0$$

Il est important de remarquer qu'à l'ordre retenu et dans les intervalles de variations envisagés, il ne peut exister de solutions hexa-impulsionnelles symétriques que si :

(69) $$\begin{cases} 0 \leq \hat{\alpha}_f \leq \hat{\beta}^* = 14°08' \\ 0 \leq \Omega \leq \hat{\beta}^* = 14°08' \\ 0 \leq \frac{|\Delta a|}{2\Delta C} \leq 2Y^*_{(0)} = 2\cos u^* = 0{,}488 \end{cases}$$

Les solutions hexa-impulsionnelles symétriques (voisines du type IV) ne peuvent donc se présenter que si l'angle de transfert est voisin d'un nombre entier de révolution, si l'axe de rotation est peu incliné sur l'axe de symétrie de l'arc de transfert et si la "dilatation" du demi-grand axe est faible.

Retenant maintenant, dans l'étude de la section ($\Omega, \Delta a$) du domaine accessible, les détails d'ordre $M^{1/2}$ et donc les termes en $\delta\beta_f$ dans (52), il vient :

$$(70)\quad \begin{cases} \Delta e_{/\!/} = \Delta e_{/\!/(0)} + E\cos\Omega\,(\Delta V_1 - \Delta V_6)\,\delta\beta_f + \text{ordre } M \\ \Delta e_{\perp} = \Delta e_{\perp(0)} + \left[-E\sin\Omega\,(\Delta V_1 - \Delta V_6) + \dfrac{F\,tg\,\hat{\beta}_{f(0)}}{\cos\Omega}(\Delta V_1 + \Delta V_6)\right]\delta\beta_f + \text{ordre } M \\ \Delta j = \Delta j_{(0)} - \dfrac{G\,tg\,\hat{\beta}_{f(0)}}{\cos\Omega}(\Delta V_1 + \Delta V_6)\,\delta\beta_f + \text{ordre } M \end{cases}$$

Les impulsions ΔV_1 et ΔV_6 étant arbitraires, il y a 2 degrés de liberté, les paramètres $\Delta V_1 - \Delta V_6$ et $\Delta V_1 + \Delta V_6$. Les solutions hexa-impulsionnelles ne sont plus représentées par un point $P_{(0)}$ mais par une petite surface plane au voisinage de ce point dont les dimensions sont de l'ordre de $\delta\beta_f$ c.a.d. de $1/\sqrt{\Delta\alpha}$ et dont il resterait à étudier les frontières.

Dans toute la suite, nous négligerons les détails d'ordre $1/\sqrt{\Delta\alpha}$.

3.1.2. - <u>Solution hexa-impulsionnelles antisymétriques</u>

La condition de compatibilité (34) est ici remplacée par :

$$(71)\quad 3\,Y_{(0)}\left(\hat{\beta}_{f(0)} - \begin{Bmatrix} 0 \\ \pi \end{Bmatrix}\right) - 2\,X_{(0)} = 0 \quad \text{selon que } tg\,\hat{\beta}_{f(0)} \gtrless 0$$

Il est aisé de voir que cette condition est incompatible avec la condition (13).

<u>Il n'existe donc pas de solutions hexa-impulsionnelles "antisymétriques"</u> .

Nous nous limiterons, dans ce qui suit, à l'étude d'un cas très particulier correspondant aux valeurs suivantes :

$$(71)\quad \begin{cases} \hat{\alpha}_f = 10^\circ \\ \Omega = 5^\circ \\ \Delta a = 0 \end{cases}$$

Alors :

$$(72)\quad \begin{cases} E = 2{,}31 \\ F = 1{,}83 \\ G = 0{,}87 \end{cases}$$

et le point représentatif $P_{(0)}$ des solutions hexa-impulsionnelles a pour coordonnées :

$$(73) \qquad P_{(0)} \begin{cases} \Delta e_{\parallel} = 0,201 \Delta C \\ \Delta e_{\perp} = 0,470 \Delta C \\ \Delta j = 0,873 \Delta C \end{cases}$$

3.2. - Solutions dérivant des solutions hexa-impulsionnelles (symétriques).

Nous appellerons ainsi les solutions dont l'évolution de l'efficacité a la même allure que celle des solutions hexa-impulsionnelles symétriques de la figure 11. Nous nous bornerons à l'étude des solutions dérivées caractéristiques d'une évolution hexa-impulsionnelle (par exemple, les solutions tri-impulsionnelles avec attente initiale $^+3$ ou finale 3^+ qui peuvent être obtenues également avec d'autres types d'évolution, seront laissées de côté).

3.2.1. - Solutions penta-impulsionnelles avec attente -

Toutes les solutions penta-impulsionnelles avec attente finale 5^+ (ou initiale $^+5$) dérivent de solutions hexa-impulsionnelles correspondant à une amplitude $\Delta\beta$ supérieure à l'angle de transfert $\Delta\alpha$ (fig. 14).

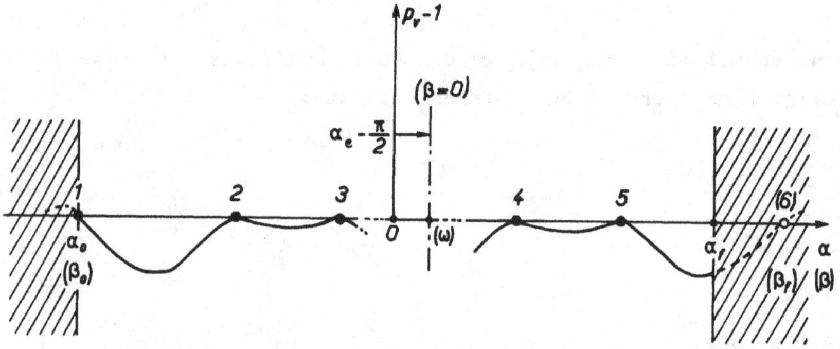

Fig. 14 - Solutions 5^+ - Evolution de l'efficacité

Pour une telle solution (fig. 15) :

$$\hat{\beta}_{f(0)} \pm \Omega' = \hat{\alpha}_f \pm \Omega \qquad \left(\hat{\beta}_{f(0)} > \hat{\alpha}_f\right) \tag{74}$$

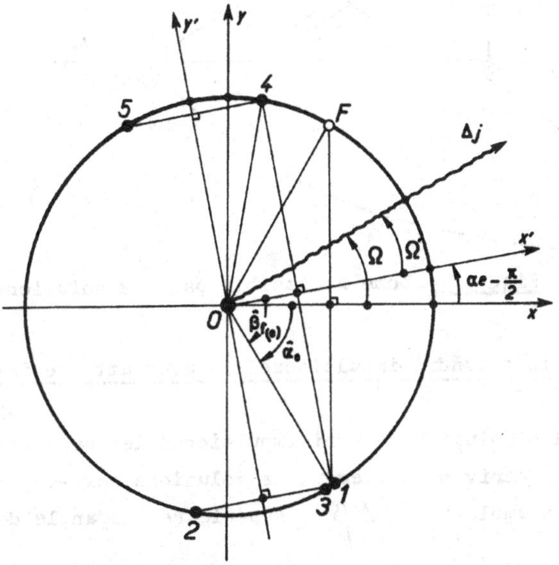

Fig. 15 - Solutions 5^+

Les impulsions 4 et 6 (1 et 3) jouant, à l'ordre conservé, des rôles équivalents et n'intervenant que par le groupement $\Delta V_4 + \Delta V_6 = I'_N \Delta C$ $\left(\Delta V_1 + \Delta V_3 = I'_0 \Delta C\right)$ où ΔV_6 (ΔV_1) peut être pris égal à zéro, les équations (56) peuvent être utilisées et le point représentatif d'une solution 5^+ $(^+5)$ décrit la ligne $\hat{\beta}_{f(0)} \pm \Omega$ = constante de la surface (S), passant par le point $(\hat{\alpha}_f, \Omega)$ jusqu'au point d'intersection Q (R) avec la frontière $\widehat{M^* N^*}$ (fig. 16).

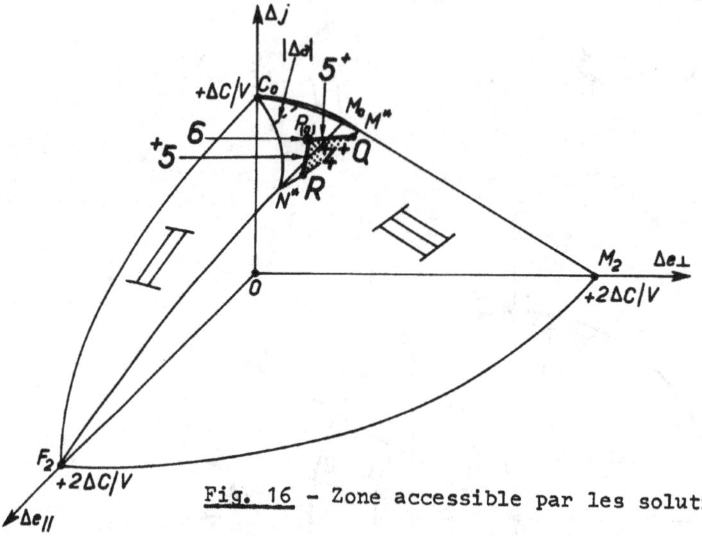

Fig. 16 - Zone accessible par les solutions 5^+, $^+5$ et $^+4^+$

3.2.2. Solutions quadri-impulsionnelles avec attente initiale et finale -

Toutes les solutions quadri-impulsionnelles avec attente initiale et finale $^+4^+$ dérivent également de solutions hexa-impulsionnelles correspondant à une amplitude $\Delta\beta$ supérieure à l'angle de transfert $\Delta\alpha$ (fig. 17)

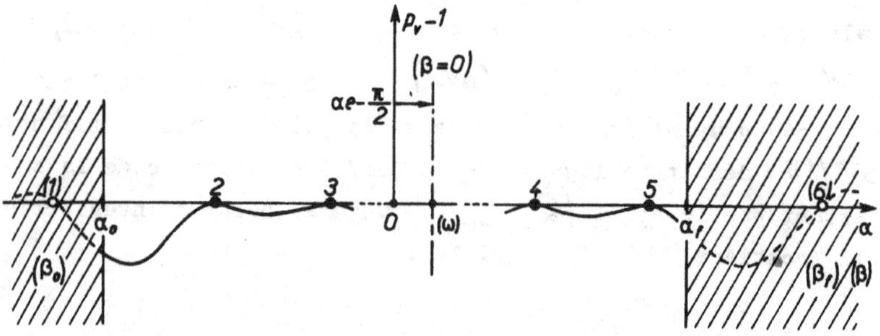

Fig. 17 - Solutions $^+4^+$. Evolution de l'efficacité

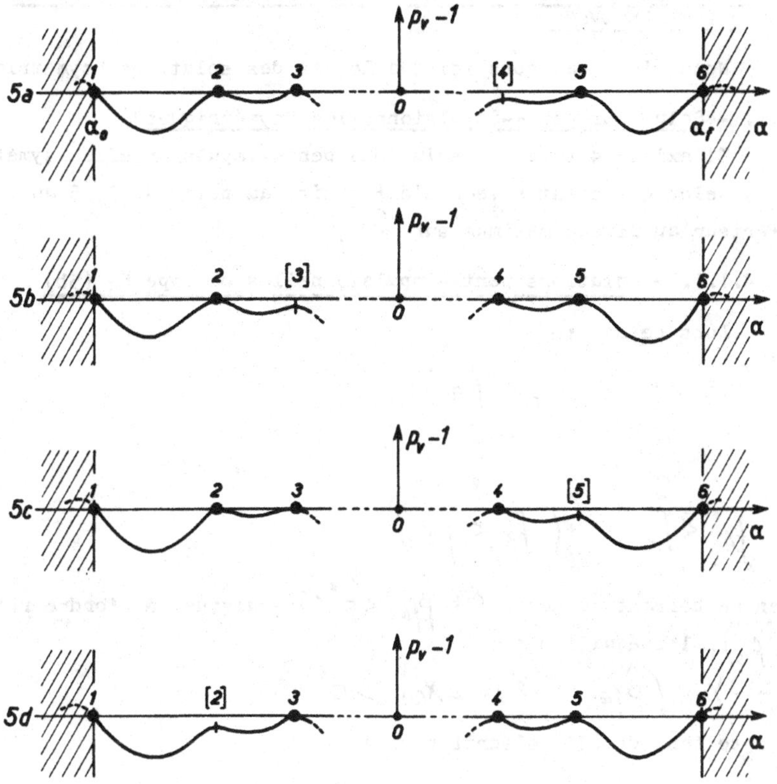

Fig. 18 - Solutions penta-impulsionnelles "symétriques"

Le point représentatif d'une telle solution est évidemment situé dans le triangle curviligne $P_{(o)}QR$ de la surface (S) (fig. 16).

4 - ETUDE DES SOLUTIONS PENTA-IMPULSIONNELLES ET DES SOLUTIONS QUI EN DERIVENT -

Nous n'envisagerons ici que le cas des solutions "symétriques".

4.1. - Solutions penta-impulsionnelles "symétriques"

Il existe 4 types de solutions penta-impulsionnelles symétriques (fig. 18) selon que c'est niveau d'efficacité au point 4, 3, 5 ou 2 qui est inférieur au niveau maximum sur $\widehat{\alpha_o \alpha_f}$.

4.1.1. - Solutions penta-impulsionnelles de type 5a (5b)

Dans ce cas :

(75) $$f_2 = f_3 \qquad (f_4 = f_5)$$

et :

(76) $$f_4 < f_5 \qquad (f_3 < f_2)$$

d'où, en se bornant au cas $0 \leq \hat{\beta}_{f(o)} \leq \hat{\beta}^* < \pi$ (puisque, à l'ordre retenu, $\hat{\beta}_{f(o)} = \hat{\alpha}_f$) l'inéquation :

(77) $$-3 Y_{(o)} \left(\hat{\beta}_{f(o)} - \pi/2 \right) + 2 X_{(o)} > 0$$

qui, compte tenu de (34) s'écrit encore :

(78) $$Y_{(o)} > Y_{(o)6}$$

où $Y_{(o)6}$ est relatif à la solution hexa-impulsionnelle (fig. 12).

Dans la solution hexa-impulsionnelle, les impulsions 4 et 6 (1 et 3) jouent, à l'ordre conservé, des rôles équivalents et il est possible en particulier d'annuler ΔV_4 (ΔV_3) tout en conservant la somme $\Delta V_4 + \Delta V_6 = I'_N \Delta C$ ($\Delta V_1 + \Delta V_3 = I'_o \Delta C$) inchangée.

Il n'y a alors, à l'ordre conservé, aucune différence entre cette solution hexa-impulsionnelle particulière et une solution penta-impulsionnelle de type 5a (5b) (en particulier les équations (56) sont valables dans les deux cas) si ce n'est que $Y_{(0)}$ est fixé dans le premier cas par (43), tandis que dans le second cas il constitue un degré de liberté soumis à la limitation (78). Le point, représentatif des solutions penta-impulsionnelles 5a (5b) décrit donc une courbe dans l'espace $\Delta e_{/\!/}, \Delta e_{\perp}, \Delta j$.

L'élimination des grandeurs d'impulsion $I'_N + I''_0$ et $I'_0 + I''_0$ dans les équations (56) conduit à :

$$(79) \quad \begin{cases} X_{(0)} = \dfrac{\cos^2\Omega - \cos^2\hat{\beta}_{f(0)}}{\sin\Omega \cos\hat{\beta}_{f(0)}} \dfrac{\Delta e_{/\!/}}{\Delta C} - \dfrac{\cos\Omega}{\cos\hat{\beta}_{f(0)}} \dfrac{\Delta e_{\perp}}{\Delta C} \\ Y_{(0)} = \dfrac{\sin\hat{\beta}_{f(0)}}{2\sin\Omega} \dfrac{\Delta e_{/\!/}}{\Delta C} \\ |Z_{(0)}| = \dfrac{\cos\Omega}{|\cos\hat{\beta}_{f(0)}|} \dfrac{\Delta j'}{\Delta C} \end{cases}$$

où $\hat{\beta}_{f(0)}$ et Ω sont évidemment fixés.

Portant ces valeurs dans les équations (12) et (13) on obtient finalement :

$$(80) \quad \Delta e_{/\!/}\left[\Delta e_{/\!/}\left(4\sin^2\Omega - 3\sin^2\hat{\beta}_{f(0)}\right) + 4\Delta e_{\perp}\sin\Omega\cos\Omega\right] - 4\,\Delta C^2\sin^2\Omega = 0$$

$$(81) \quad \left(\Delta e_{/\!/}\cos\Omega - \Delta e_{\perp}\sin\Omega\right)\left[\left(\cos^2\Omega - \cos^2\beta\right)\Delta e_{/\!/} - \Delta e_{\perp}\sin\Omega\cos\Omega\right] + \Delta j^2\sin^2\Omega\cos\Omega = 0$$

La courbe représentative des solutions 5a (5b) est donc un arc de la biquadratique d'intersection du cylindre hyperbolique (80) de génératrices parallèles à $\overrightarrow{O\Delta j}$ et du cône du second degré (81) de sommet O (fig. 19). Dans le cas particulier (71) étudié, ces surfaces ont pour équations :

$$(82) \quad \Delta e_{/\!/}\left(0{,}348\ \Delta e_{\perp} - 0{,}0603\ \Delta e_{/\!/}\right) - 0{,}0305\ \Delta C^2 = 0$$

$$(83) \quad \left(0{,}996\,\Delta e_{/\!/} - 0{,}0872\,\Delta e_{\perp}\right)\left(0{,}0218\,\Delta e_{/\!/} - 0{,}087\,\Delta e_{\perp}\right) + 0{,}00757\,\Delta j^2 = 0$$

A cause de (78), seule la partie $\widehat{P_{(0)}S}$:

$$(84) \quad \Delta e_{/\!/} > \left(\Delta e_{/\!/}\right)_6 \qquad (\text{et évidemment } \Delta j \geq 0)$$

convient (fig. 19). Sur cette partie les inégalités (60) sont satisfaites dans le cas particulier choisi ($\Delta a=0$) .

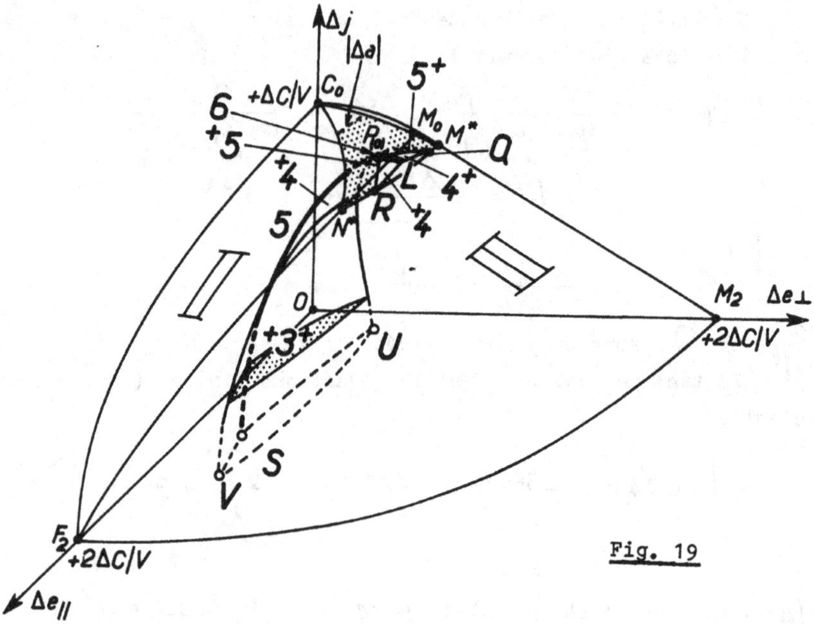

Fig. 19

Selon la remarque du § 3.1.1., à l'approche du plan $\Delta j=0$ la solution 5a (5b) n'est plus définie par cette étude.

4.1.2. - Solutions penta-impulsionnelles de type 5c (5d) -

Dans ce cas :

(85) $f_2 = f_3 \qquad (f_4 = f_5)$

et :

(86) $f_5 < f_4 \qquad (f_2 < f_3)$

d'où l'inéquation :

(87) $$-3\,Y_{(0)}\left(\hat{\beta}_{f(0)}-\frac{\pi}{2}\right)+2\,X_{(0)}<0$$

qui, compte tenu de (34) s'écrit encore :

(88) $$Y_{(0)}<Y_{(0)6} \qquad (\text{Fig. } 12)$$

D'autre part l'inéquation (14) se traduit par :

(89) $$Y_{(0)}\geqslant Y_{(0)lim}=\sin\hat{\beta}_{f(0)} \quad (=\sin 10^\circ=0{,}174)$$

Les équations (56) sont encore valables à condition d'y faire $I''_N = 0\,(I''_0=0)$ et d'y considérer $Y_{(0)}$ comme un paramètre indépendant soumis aux limitations (88) et (89).

Les inégalités (60) sont alors remplacées, pour les solutions 5c, par :

(90) $$-1\leq 1-2\left(I'_0+I''_0\right)=\frac{\Delta a}{2Y_{(0)}\Delta C}\ (=0)\leqslant \operatorname{tg}\Omega\operatorname{cotg}\hat{\beta}_{f(0)}$$

et pour les solutions 5d par :

(91) $$\operatorname{tg}\Omega\operatorname{cotg}\hat{\beta}_{f(0)}\leqslant 1-2\left(I'_0+I''_0\right)=\frac{\Delta a}{2Y_{(0)}\Delta C}\ (=0)\leqslant 1$$

Les inégalités (90) sont évidemment satisfaites alors que les inégalités (91) ne le sont pas. Les solutions penta-impulsionnelles symétriques de type 5d ne peuvent donc pas se présenter dans l'exemple envisagé.

La courbe représentative des solutions 5c est donc la portion $\widehat{P_{(0)}L}$ (fig. 19) de la biquadratique (80) (81) telle que :

(92) $$(\Delta e_{/\!/})_{lim}=2\sin\Omega\,\Delta C\ (=0{,}174\,\Delta C)\leqslant\Delta e_{/\!/}<(\Delta e_{/\!/})_6\ (=0{,}201\,\Delta C)$$

4.2. - Solutions dérivant des solutions penta-impulsionnelles "symétriques"

Nous nous bornerons, comme précédemment, à l'étude des solutions dérivées caractéristiques d'une évolution penta-impulsionnelle.

4.2.1. - Solutions quadri-impulsionnelles avec attente finale 4^+ ou initiale $^+4$

Il est aisé de voir que les points représentatifs de ces solutions sont situés sur les surfaces engendrées par l'arc de biquadratique $\widehat{L P_{(0)} S}$, d'équations (80) et (81), lorsque le point $P_{(0)}$ décrit l'arc $\widehat{P_{(0)} Q}$ $\left[\hat{\beta}_{f(0)} + \Omega \right.$ = constante (= 10° + 5° = 15°) $\left.\right]$ ou l'arc $\widehat{P_{(0)} R}$ $\left[\hat{\beta}_{f(0)} - \Omega \right.$ = constante (= 10° - 5° = 5°) $\left.\right]$. Les surfaces coupent le plan $\Delta j = 0$ selon les arcs $\widehat{SU}$ et $\widehat{SV}$ et le cylindre III selon les arcs $\widehat{LQ}$ et $\widehat{LR}$.

4.2.2. - Solutions tri-impulsionnelles $^+3^+$ avec attente initiale et finale -

Les points représentatifs de ces solutions sont situés dans le volume engendré par l'arc de biquadratique $\widehat{L P_{(0)} S}$ lorsque le point $P_{(0)}$ décrit le triangle curviligne des solutions $^+4^+$. Ce volume est limité, dans le plan $\Delta j = 0$ par le triangle curviligne SUVS et par la portion LQRL du cylindre III.

5 - CONCLUSION

Il a été précédemment démontré [11, 12] que le nombre maximum d'impulsions pouvant intervenir dans un rendez-vous optimal entre orbites képlériennes proches est égal au nombre de paramètres orbitaux finaux imposés, c'est-à-dire égal à 6 dans le cas le plus général.

La présente étude fournit des exemples de tels rendez-vous hexa-impulsionnels.

L'étude partielle des rendez-vous penta-impulsionnels montre dans quelle voie ce travail peut être complété par l'analyse systématique des différents types de solutions dans les cas "symétriques" et "antisymétriques" (6 types d'évolution quadri-impulsionnelle, 4 types d'évolution tri-impulsionnelle, 1 type d'évolution bi-impulsionnelle, sans compter les solutions avec attente qui en dérivent).

Les solutions quasi-planes, voisines du type IVbis, restent

également à étudier à part.

RÉFÉRENCES

1 EDELBAUM, T.N. - "Minimum impulse transfer in the near vicinity of a circular orbit".
The Journal of the Astronautical Sciences, 14, 66-73 (1967).

2 MAREC, J.P. - "Transferts infinitésimaux, impulsionnels, économiques entre orbites quasi-circulaires, non coplanaires".
Note aux Comptes Rendus de l'Académie des Sciences (1966);
aussi : 17ème Congrès International d'Astronautique, Madrid (1966), à paraître dans Astronautica Acta; également : Publication ONERA n° 115 (1966).

3 MAREC, J.P. - "Rendez-vous impulsionnels, optimaux, de longue durée, entre orbites quasi-circulaires, proches, coplanaires ou non".
Colloque de Liège : "Advanced Problems and Methods for Space Flight Optimization" (1966); à paraître chez Pergamon.

4 MAREC, J.P. - "Rendez-vous multi-ipulsionnels, optimaux, de durée moyenne, entre orbites quasi-circulaires, proches, coplanaires".
Communication proposée au 19ème Congrès International d'Astronautique, New-York (1968).

5 PRUSSING, J.E. - "Optimal multi-impulse orbital rendez-vous".
TE. 20 - M.I.T. (1967).

6 PONTRYAGIN, L.S., BOLTYANSKII V.G., GAMKRELIDZE R.V. and MISHCHENKO E.F., The Mathematical Theory of Optimal Processes, Intersciences Publishers, John Wiley and Sons, Inc., New-York (1962).

7 CONTENSOU, P. - "Etude théorique des trajectoires optimales dans un champ de gravitation. Application au cas d'un centre d'attraction unique".
Astronautica Acta 8, 134-150 (1962).

8 MAREC, J.P. - "Transferts optimaux entre orbites elliptiques proches".
Publication ONERA n° 121 (1967).

9 LAWDEN, D.F. - Optimal trajectories for Space Navigation - Butterworths, London (1963).

10 NGUYEN, V.M. - "Rendez-vous en longitude optimal d'un satellite géostationnaire". A paraître dans "La Recherche Aérospatiale".

11 NEUSTADT, L.W. - "Optimization, a moment problem and nonlinear programming". SIAM J. on Control 2, 33-53 (1964).

12 POTTER, J.E., STERN, R.E. - "Optimization of midcourse velocity corrections". IFAC Symposium on Automatic Control in the peaceful uses of space, Stavanger, Norway (1965).

ETUDE DU SENS OPTIMAL DES COMMUTATIONS DANS LA THEORIE DE CONTENSOU-PONTRYAGIN

(Application au cas des transferts de durée indifférente entre orbites elliptiques)

Christian MARCHAL

RESUME

L'utilisation des théories d'optimisation peut se faire en laissant de côté les notions de "commande" et de "domaine de commande" et en ne conservant que celles de "domaine de manoeuvrabilité" et de "vecteur adjoint"; dans ces conditions, il est aisé de démontrer que les commutations, c'est-à-dire les discontinuités des "trajectoires optimales" ou de leurs dérivées premières, ont généralement lieu dans un sens bien déterminé. Toutefois, il y a des cas limites qui conduisent à des solutions singulières telles que les "arcs à poussée intermédiaire" de LAWDEN. Ces solutions singulières ne sont, le plus souvent, pas elles-memes optimales.

L'application de ces résultats aux transferts optimaux de durée indifférente entre orbites elliptiques conduit à une règle de commutation très simple: la commutation a toujours lieu dans le sens qui abaisse la dérivée de l'énergie mécanique (par unité de masse) par rapport à la vitesse caractéristique.

Il y a deux cas limites: le"cas limite de LAWDEN" et le "cas limite de CONTENSOU"; les arcs singuliers auxquels ils conduisent ne sont jamais optimaux.

Un nouveau cas particulier simple est étudié en détail: le transfert entre orbites symétriques par rapport à un plan passant par le centre attractif (les sens de rotation étant symétriques). Les solutions optimales sont de quatre types: 1° "à une impulsion horizontale", 2° "bi-parabolique", 3° "à trois impulsions symétriques", 4° "à deux impulsions dissymétriques"; ce dernier type est rare et ne se présente que si l'exentricité des deux orbites symétriques reliées est supérieure à $\sqrt{\frac{2}{3}}$ (= 0,8165).

STUDY OF THE OPTIMAL DIRECTION OF COMMUTATIONS IN THE CONTENSOU-PONTRYAGIN THEORY

APPLICATION TO THE CASE OF TIME-FREE TRANSFERS BETWEEN ELLIPTIC ORBITS

ABSTRACT

Using optimization theories can be done without the notions of "control" and of "domain of control" but only with those of "manoeuvera-bility domain" and of "adjoint vector". In these conditions it is easy to show that the commutations, that is to say the discontinuities of the optimal trajectories or of their first derivatives are generally in a well determined direction. However there are some limiting cases which lead to singular solutions such as the "intermediate thrust arcs" of LAWDEN. Usually these singular solutions are not optimal.

The application of these results to the optimal time-free transfers between elliptic orbits leads to a very simple rule of commutation : the commutation always occurs in the direction which reduces the derivative of the mechanical energy (per unit of mass) with respect to the characteristic velocity.

There are two limiting cases : the "limit case of LAWDEN" and the "limit case of CONTENSOU". They lead to singular arcs which are never optimal.

A new simple particular case is studied in detail : the transfer between symmetrical orbits with respect to a plane containing the center of mass (the rotations along the orbits being symmetrical). The optimal solutions are of 4 types : 1 - "with one horizontal impulse", 2 - "bi-parabolic", 3 - "with 3 symmetrical impulses", 4 - "with two unsymmetrical impulses" ; this last type is rare and only occurs when the eccentricity of the related symmetrical orbits is greater than $\sqrt{\frac{2}{3}}$ (= 0.8165).

INTRODUCTION

L'étude des commutations est l'un des points les plus délicats des théories d'optimisation : 1° parce qu'elles sont souvent malcommodes à calculer surtout s'il y a un grand nombre de paramètres d'état, 2° parce qu'elles introduisent des discontinuités dans les intégrations, 3° parce que le sens optimal de commutation est malaisé à déterminer.

C'est ce dernier point qui est étudié ici; en général, le sens de la commutation est imposé, il y a toutefois des cas particuliers limites qui conduisent aux "arcs singuliers".

L'application au cas des transferts de durée indifférente entre orbites elliptiques est particulièrement intéressante car elle conduit à une règle de sens de commutation très simple et très commode qui permet d'étudier aisément un nouveau cas particulier: le transfert entre orbites symétriques par rapport à un plan.

1 - EXPOSE SUCCINCT DES THEORIES D'OPTIMISATION

Nous appelerons toujours "prix" ou "coût" la fonction à minimiser (ou à maximiser), ce sera toujours une fonction linéaire des paramètres d'état.

1.1 - Théorie de CONTENSOU [1]

Soit un système évolutif quelconque dont l'état peut être défini par un nombre fini de paramètres:

Nous poserons $\vec{X}$ = vecteur état = $(x_1, x_2, \ldots .\ x_n)$ et $\vec{V} = \frac{d\vec{X}}{dt}$ = "vecteur vitesse" (t étant le paramètre de description).

Si $\vec{V}$ est une fonction univoque de $\vec{X}$ et t il n'y a pas de manoeuvrabilité, chaque trajectoire est déterminée par la donnée de $\vec{X}_0$ et t_0 (exception faite des cas d'intégrales singulières).

Il y a manoeuvrabilité si $\vec{V}$ peut être choisi (par l'intermédiaire de "commandes" qu'il est inutile de préciser) à l'intérieur d'un certain domaine D (dit domaine de manoeuvrabilité) de l'espace des vecteurs $\vec{V}$.

La manoeuvrabilité est canonique si:

1° - D ne dépend que de $\vec{X}$ et t.
2° - D est une fonction continue de $\vec{X}$ et de t.
3° - Le choix de $\vec{V}$ est entièrement libre dans D (en particulier $\vec{V}$ n'est astreint à aucune condition de continuité).

En général, la manoeuvrabilité peut être rendue canonique si l'on inclut dans $\vec{X}$ un nombre suffisant de paramètres d'état.

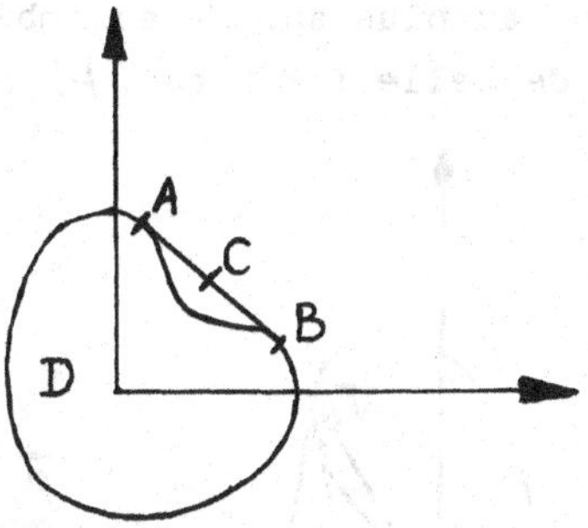

Fig. 1 - Domaine de manoeuvrabilité $\mathcal{D}$ (manoeuverability domain $\mathcal{D}$).

Le domaine de manoeuvrabilité $\mathcal{D}$ est obligatoirement convexe, en effet, si l'analyse des possibilités d'un système conduit à un domaine tel que celui de la figure 1, l'utilisation des points A puis B pendant les intervalles de temps dt_1 et dt_2 est équivalente à celle, pendant le temps $dt_1 + dt_2$, d'un point C sur le segment AB (ainsi un voilier louvoie pour remonter le vent).

Le point essentiel de la théorie du domaine de manoeuvrabilité est le suivant : si la manoeuvrabilité est canonique et si l'on veut qu'à l'instant final une certaine fonction linéaire des paramètres d'état, la fonction de prix, soit maximale ou minimale, il est nécessaire (mais non suffisant) qu'à chaque instant -à l'exception au plus d'un ensemble de mesure nulle- le vecteur $\vec{V}$ soit choisi sur la frontière extérieure du domaine convexe $\mathcal{D}$ cela que l'état final soit entièrement libre ou qu'il soit assujetti à un certain nombre de conditions.

1.2 - Théorie de PONTRYAGIN [2]

Elle s'expose aisément avec le vocabulaire du paragraphe précédent.

Il faut là aussi que la manoeuvrabilité soit canonique.

PONTRYAGIN utilise un vecteur adjoint $\vec{P} = (p_1, p_2, \ldots\ldots p_n)$ qui dépend des conditions limites et est déterminé à posteriori et forme l'Hamiltonien H :

$$H = \vec{P}.\vec{V} = p_1 \frac{dx_1}{dt} + \cdots\cdots + p_n \frac{dx_n}{dt}$$

Nous laisserons de côté les notions de commande et de domaine de commande qui ne sont nullement indispensables.

Les principaux résultats obtenus par PONTRYAGIN sont les suivants :

Si l'on veut qu'à l'instant t_f final (fixé ou non) la fonction linéaire de prix: $f = a_1x_1 + \ldots + a_nx_n$ soit minimale, il faut:

1° - Choisir à chaque instant - sauf au plus sur un ensemble de mesure nulle - $\vec{V}$ dans le domaine $\mathcal{D}$ de telle façon que H soit maximal (il faut donc bien choisir $\vec{V}$, figure 2, sur la frontière de $\mathcal{D}$) $\mathcal{D}$ étant seulement une fonction de $\vec{X}$ et de t, H devient une fonction de $\vec{P}$, $\vec{X}$ et t.

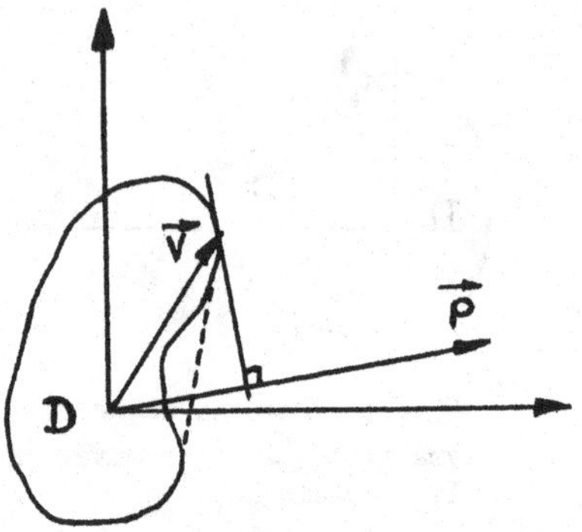

Fig.2 - Domaine $\mathcal{D}$ et vecteur $\vec{P}$
(Domain $\mathcal{D}$ and vector $\vec{P}$)

2° - Faire vérifier au vecteur $\vec{P}$ les équations:

A)

$$(1) \quad \frac{dp_i}{dt} = - \frac{\partial H(\vec{P},\vec{X},t)}{\partial x_i}$$

B) au temps final:

$p_i = -a_i$ si $x_i\ (t_f)$ n'est pas fixé
p_i quelconque si $x_i\ (t_f)$ est fixé
$H(t_f) = 0$ si t_f n'est pas fixé.

Nous n'examinerons pas les cas où les $x_i(t_f)$ sont liés par certaines relations linéaires car on peut toujours se ramener au cas ci-dessus par un changement linéaire des coordonnées d'état.

Tout comme la théorie de CONTENSOU, celle de PONTRYAGIN ne donne que des conditions nécessaires et non suffisantes d'optimalité.

2 - ANALYSE DU SENS OPTIMAL DES COMMUTATIONS

Considérons un cas de commutation (figure 3): les points A et B du domaine $\mathcal{D}$ donnent à $H(=\vec{P}\,\vec{V})$ sa valeur maximale (il en est aussi de même bien entendu de tous les points du segment AB).

Appelons t_o l'instant de la commutation et donnons l'indice A aux résultats obtenus au voisinage de t_o si l'on utilise seulement la portion strictement convexe de $\mathcal{D}$ voisine de A: on obtient ainsi H_A, $\vec{V_A}$, $\vec{P_A}$ et $\vec{X_A}$ et de même symétriquement H_B, $\vec{V_B}$, $\vec{P_B}$ et $\vec{X_B}$.

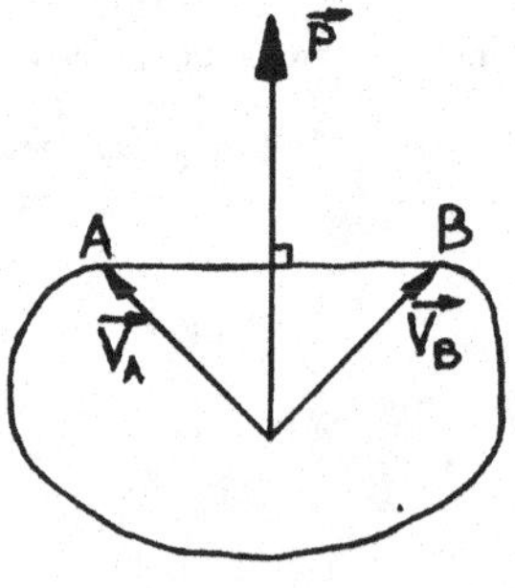

Fig.3 - Cas de commutation
(Commutation case)

Appelons enfin $\vec{V}_B'$ la valeur de $\vec{V}$ correspondant au point de la portion strictement convexe de D voisine de B qui maximise $\vec{P}_A.\vec{V}$, de même $\vec{V}_A'$ correspond au point de la portion strictement convexe voisine de A qui maximise $\vec{P}_B.\vec{V}$ (fig.4)

La commutation ne peut avoir lieu de A vers B que si pour t = to :

1°/ $\frac{d}{dt}\vec{P}_A(\vec{V}_A - \vec{V}_B') \leq 0$

(condition nécessaire pour t < to)

2°/ $\frac{d}{dt}\vec{P}_B(\vec{V}_B - \vec{V}_A') \geq 0$

(condition nécessaire pour t > to).

De même la commutation ne peut avoir lieu de B vers A que si à l'instant t = to :

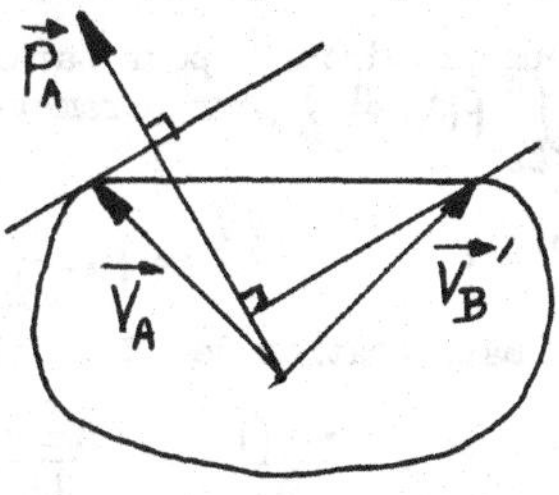

Fig.4 - Voisinage de la commutation (Vicinity of the commutation)

1°/ $\frac{d}{dt}\vec{P}_B(\vec{V}_B - \vec{V}_A') \leq 0$; 2°/ $\frac{d}{dt}\vec{P}_A(\vec{V}_A - \vec{V}_B') \geq 0$

Or, en remplaçant les équations (1) du chapitre précédent par l'expression symbolique :

$$\frac{d\vec{P}}{dt} = / - \frac{\partial H(\vec{P},\vec{X},t)}{\partial \vec{X}}$$

on obtient :

$$\frac{d}{dt}\vec{P}_A(\vec{V}_A - \vec{V}_B') = \frac{d}{dt}\left[H_A(\vec{P}_A,\vec{X}_A,t) - H_B(\vec{P}_A,\vec{X}_A,t)\right]$$

soit, pour t = to :

$$\frac{d}{dt}\vec{P}_A(\vec{V}_A - \vec{V}_B') = \frac{\partial H_A}{\partial t} + \frac{\partial H_A}{\partial \vec{X}}\vec{V}_{B_0} - \frac{\partial H_B}{\partial \vec{X}}\vec{V}_{A_0} - \frac{\partial H_B}{\partial t}$$

et de même :

$$\frac{d}{dt}\vec{P}_B(\vec{V}_B - \vec{V}_A') = \frac{\partial H_B}{\partial t} + \frac{\partial H_B}{\partial \vec{X}}\vec{V}_{A_0} - \frac{\partial H_A}{\partial \vec{X}}\vec{V}_{B_0} - \frac{\partial H_A}{\partial t}$$

Les deux quantités cherchées sont toujours opposées, si elles ne sont pas nulles la commutation examinée ne peut avoir lieu que dans un seul sens (déterminé par leur signe), si elles sont nulles on est dans un cas singulier dont nous allons étudier quelques exemples.

Remarque : il est aisé de vérifier que l'utilisation éventuelle de points intermédiaires du segment AB ne modifie pas les résultats.

3 - EXEMPLES DE CAS SINGULIERS

Soit le problème suivant :

Minimiser C_f :

$$C_f = \int_{t_0}^{t_f} f(x)\, dt$$

f étant une fonction dérivable donnée de x.
avec : $x(t_0) = x_0$

$x(t_f) = x_f$

et la manoeuvrabilité : $$-1 \leqslant \frac{dx}{dt} \leqslant +1$$

Nous prendrons t pour paramètre de description et x et C (avec $C = \int_{t_0}^{t} f(x)\,dt$) pour paramètres d'état; C_f est bien la quantité à minimiser.

D'où : $$H = p_x \cdot \frac{dx}{dt} + p_c \cdot f$$

La maximisation de H conduit à :

Si $p_x > 0$: $\frac{dx}{dt} = +1$

Si $p_x = 0$: $\frac{dx}{dt}$ est quelconque

Si $p_x < 0$: $\frac{dx}{dt} = -1$

(les commutations ne surviennent que pour $p_x = 0$)

Dans chaque cas : $H = |p_x| + p_c \cdot f$

Dérivons p_x et p_c :

$$\frac{dp_c}{dt} = -\frac{\partial H}{\partial C} = 0 \quad \text{donc} \quad p_c = p_c \text{ final} = -1$$

donc $$\frac{dp_x}{dt} = -\frac{\partial H}{\partial x} = -p_c \frac{df}{dx} = \frac{df}{dx} \quad \text{et} \quad H = |p_x| - f$$ est une constante

Examinons enfin $S = \frac{\partial H_A}{\partial t} + \frac{\partial H_A}{\partial \vec{X}} \vec{V}_{Bc} - \frac{\partial H_B}{\partial \vec{X}} \vec{V}_{Ac} - \frac{\partial H_B}{\partial t}$

A désignant par exemple le côté $\frac{dx}{dt} = +1$ et B celui $\frac{dx}{dt} = -1$

On obtient : $S = 2\frac{df}{dx}$

Les "arcs singuliers" ne s'obtiennent donc que pour $p_x = 0$ et $\frac{df}{dx} = 0$

Laissons de côté le cas où $f(x)$ est constant sur certains segments en x, cas qui conduisent évidemment à des "arcs optimaux indifférents" de formes quelconques (respectant $\left|\frac{dx}{dt}\right| \leqslant 1$) dans les segments en x considérés (fig.5).

Ce cas mis à part les arcs optimaux se composent d'"arcs normaux" (vérifiant $\frac{dx}{dt} = \pm 1$) et "d'arcs singuliers" pour lesquels p_x et $\frac{df}{dx}$ sont nuls et donc x est constant, ces derniers arcs ne sont évidemment réellement optimaux que si f correspond à un minimum local.

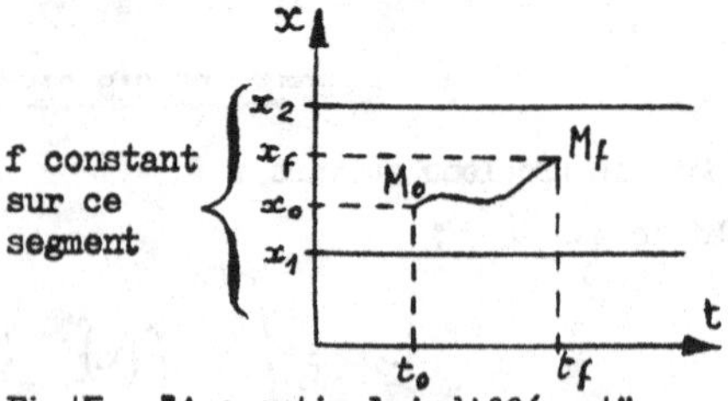

Fig.5 - "Arc optimal indifférent" (Indifferent optimal arc)

Exemple I : $f = x^2$

Les solutions sont de 3 types (figure 6) dont deux utilisent des "arcs singuliers" le long de l'axe $x = 0$; l'étude du problème par la théorie de CONTENSOU-PONTRYAGIN est suffisante.

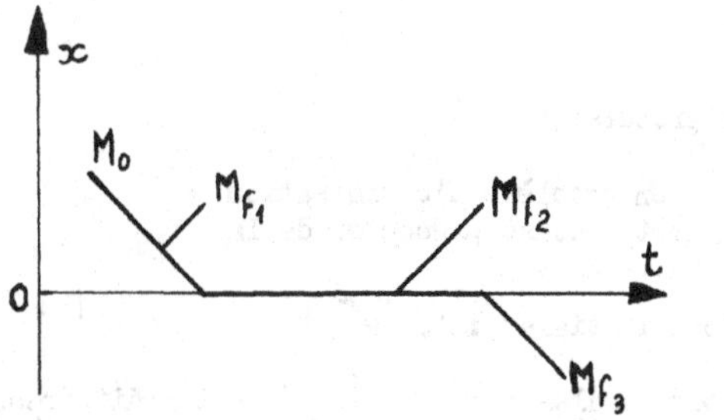

Fig. 6 - Solutions dans le cas $f = x^2$
(Solutions in the case $f = x^2$)

Exemple II : $f = -x^2$

La théorie de CONTENSOU-PONTRYAGIN conduit à des solutions "en dents de scie égales" (figure 7) traversées à mi-hauteur par l'axe $x = 0$; l'arc singulier ($x \equiv 0$) en est un cas limite non optimal. En fait les solutions réellement optimales comportent au plus une seule commutation (solution II, fig.7).

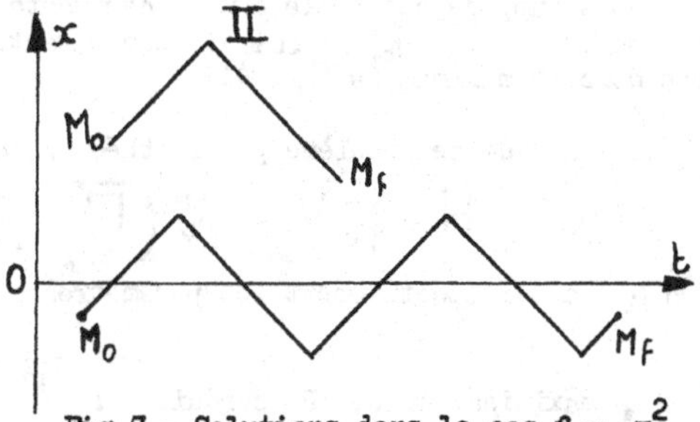

Fig.7 - Solutions dans le cas $f = -x^2$
(Solutions in the case $f = -x^2$)

Si l'on généralise cette étude au cas où f est une fonction de plusieurs paramètres on s'aperçoit que tous les extrémums de f peuvent fournir des arcs singuliers mais seuls les extrémums correspondant à des minimums locaux en donnent qui soient optimaux, ces arcs singuliers jouent alors un rôle très important dans les solutions optimales. D'autre part, dans certaines conditions très particulières, apparaissent des "surfaces singulières" ou des "volumes singuliers" lieux d'arcs singuliers et l'étude générale de ces phénomènes est très complexe.

4 - APPLICATION AUX TRANSFERTS DE DUREE INDIFFERENTE ENTRE ORBITES ELLIPTIQUES

Les transferts entre orbites sont fréquents dans les missions spatiales. Leur optimisation, c'est-à-dire la recherche de la dépense minimale de propulsif, se fait souvent en supposant indifférente la durée du transfert; cela correspond à la réalité dans les cas de courte durée, d'autre part les calculs sont alors très simplifiés.

Dans ces conditions les transferts optimaux sont indépendants du propulseur [3], ce sont les transferts "de vitesse caractéristique minimale" (la vitesse caractéristique étant la somme arithmétique de tous les changements artificiels de vitesse); ils sont toujours composés uniquement d'impulsions [4]; ces impulsions sont bien entendu décomposables en plusieurs arcs de poussée effectués à un tour d'intervalle si les propulseurs l'exigent.

Pour les étudier utilisons les paramètres d'état $\vec{r}$ et $\vec{V}$ (figure 8), respectivement rayon vecteur et vecteur vitesse, et supposons égale à $\vec{\Gamma}$ l'accélération due aux propulseurs.

Un problème d'optimisation de transfert peut se présenter de la manière suivante :

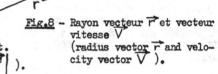

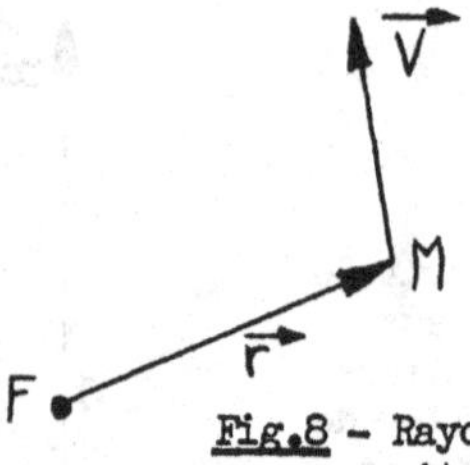

Fig.8 - Rayon vecteur $\vec{r}$ et vecteur vitesse $\vec{V}$ (radius vector $\vec{r}$ and velocity vector $\vec{V}$).

Conditions initiales : $\vec{r}_o, \vec{V}_o, t_o$

Conditions finales : $\vec{r}_f, \vec{V}_f, t_f$; t_f est indifférent.

On veut que $\int_{t_o}^{t_f} \Gamma dt$ soit minimal (avec $\Gamma = |\vec{\Gamma}|$).

Nous supposerons égale à μ la constante gravitationnelle du centre attractif et nous poserons $C = \int_{t_o}^{t} \Gamma\, dt$ = vitesse caractéristique à l'instant t (nous recherchons donc le minimum de C_f).

L'étude de ce problème par la théorie de PONTRYAGIN conduit à l'hamiltonien :

$$H = \vec{p_r}\,\vec{V} + \vec{p_v}\left(\vec{\Gamma} - \frac{\mu \vec{r}}{r^3}\right) + p_c\,\Gamma$$

$\vec{p_r}$, $\vec{p_v}$ et p_c étant respectivement les paramètres adjoints aux paramètres d'état $\vec{r}$, $\vec{V}$ et C.

La maximisation de H conduit à : $\vec{\Gamma}$ parallèle à $\vec{p_v}$ (et de même sens) et à :

1°/ $p_v + p_c > 0$: $\Gamma = \Gamma$ maximum
2°/ $p_v + p_c = 0$: Γ est quelconque
3°/ $p_v + p_c < 0$: $\Gamma = 0$

avec $p_v = |\vec{p_v}|$

Si Γ maximum est constant : $\frac{dp_c}{dt} = -\frac{\partial H}{\partial C} = 0$ donc $p_c \equiv p_{cf} = -1$

Si l'on fait tendre Γ maximum vers l'infini, pour englober tous les cas et en particulier le cas impulsionnel, on obtient :

1°/ On ne peut avoir $p_v > 1$

2°/ Si $p_v < 1$: $\Gamma = 0$, on est donc sur un arc balistique.

3°/ Si $p_v = 1$: Γ est quelconque, on peut donc être sur un arc de poussée ou même en train d'effectuer une impulsion.

Bien entendu la condition $p_v \leqslant 1$ entraîne que pour $p_v = 1$ on doit avoir $\frac{d}{dt}(p_v) = 0$ or $\frac{d}{dt}(\vec{p_v}) = -\frac{\partial H}{\partial \vec{V}} = -\vec{p_r}$ donc pour une impulsion ou un arc de poussée : $p_v = 1$ et $\vec{p_v} \cdot \vec{p_r} = 0$

On obtient de même :

$$\frac{d\vec{p_r}}{dt} = -\frac{\partial H}{\partial \vec{r}} = \frac{\mu}{r^3}\left[\vec{p_v} - 3\vec{r}\left(\frac{\vec{r}\,\vec{p_v}}{r^2}\right)\right].$$

Ces équations permettent d'intégrer le problème, on obtient en particulier les intégrales premières [5] :

1°/ H = constante (donc $H \equiv H_f = 0$ puisque t_f n'est pas fixé)
2°/ $\vec{B} = \vec{r} \wedge \vec{pr} + \vec{V} \wedge \vec{pv}$ = vecteur constant
3°/ $C + 2\vec{r} \cdot \vec{pr} - \vec{V}\,\vec{pv}$ = constante .

Les orbites intermédiares du transfert ne sont jamais des hyperboles car le transfert optimal entre orbites paraboliques est d'une vitesse caractéristique négligeable [3], comme d'autre part la durée du transfert est indifférente on peut toujours, sans sortir de l'optimalité, stopper les propulseurs à un moment quelconque au cours d'une impulsion ou d'un arc de poussée, effectuer balistiquement un nombre entier quelconque de tours sur l'orbite osculatrice puis reprendre le transfert.

On en déduit donc:

1°/ On doit avoir $pv \leq 1$ en tout point de chaque orbite osculatrice.
2°/ Les commutations (discontinuités du point d'application optimal de la poussée) s'obtiennent lorsque pv est égal à 1 en deux points distincts d'une orbite osculatrice (points correspondants de la commutation).

Il est donc très intéressant de connaître l'évolution de pv le long d'une orbite balistique osculatrice.

Nous utiliserons les axes mobiles $\overrightarrow{FXYZ}$ ci-contre (fig.9) avec:
$\vec{FX}$: parallèle à $\vec{r}$ et de même sens,
$\vec{FZ}$ parallèle à $\vec{r} \wedge \vec{V}$ et de même sens,
$\vec{FY}$ parallèle à $(\vec{r} \wedge \vec{V}) \wedge \vec{r}$ et de même sens.

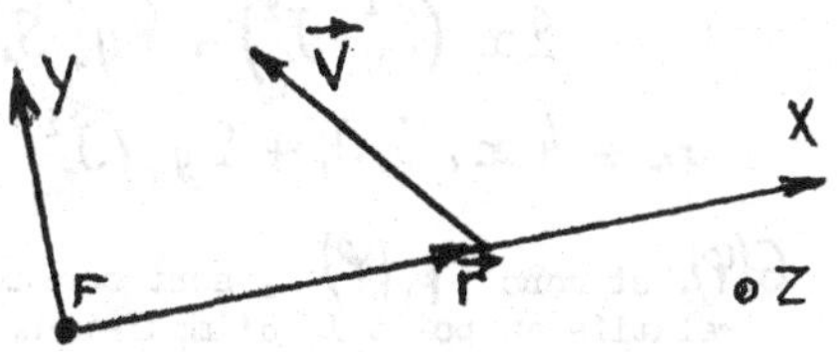

Fig.9 - Axes mobiles $\overrightarrow{FXYZ}$
(Moving axes $\overrightarrow{FXYZ}$)

Le trièdre $\overrightarrow{FXYZ}$ est droit et orthonormé.
Dans ces axes les vecteurs $\vec{r}, \vec{V}, \vec{pr}, \vec{pv}$ ont pour composantes:
$\vec{r} = (r,0,0)$; $\vec{V} = (X,Y,0)$; $\vec{pr} = (\alpha,\beta,\gamma)$; $\vec{pv} = (S,T,W)$.
Remarquons que $r \geq 0$ et $Y \geq 0$.

Nous poserons

1°/ φ = angle $(\vec{FI_1}, \vec{FM})$ = angle, le long de l'orbite osculatrice O, du rayon vecteur $\vec{FI_1}$ du point I_1 d'impulsion et du rayon vecteur $\vec{FM}$ du point courant M (fig.10).

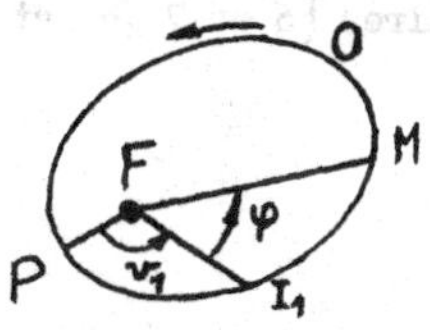

Fig.10 - Notations

2°/ v = anomalie vraie le long de l'orbite 0 (donc $v = v_1 + \varphi$).

3°/ $\left.\begin{array}{l} \frac{1}{\mu} r Y^2 - 1 = x = e \cos v \\ \frac{1}{\mu} r X Y = y = e \sin v \end{array}\right\}$ e étant l'éxcentricité de l'orbite 0.

4°/ $J = \frac{dS}{d\varphi} = T - \frac{\alpha r}{Y}$ $\Big\}$ le vecteur (S, J, 0) est constant le long de l'orbite 0 par rapport à des axes fixes.

H étant nul : $\frac{\mu S}{r^2} = \alpha X + \beta Y$ soit $\frac{SY}{r} = \alpha y + \beta(x+1)$

Enfin nous donnerons l'indice 1 aux éléments concernant le point I_1; donc, puisque I_1 est un point d'impulsion : $p_{v_1}^2 = 1$ soit $S_1^2 + T_1^2 + W_1^2 = 1$ et $\overrightarrow{p_{v_1}} \cdot \overrightarrow{p_{r_1}} = 0$ soit $\alpha_1 S_1 + \beta_1 T_1 + \gamma_1 W_1 = 0$

Des équations du mouvement et de $\frac{d\overrightarrow{p_v}}{dt}$ et $\frac{d\overrightarrow{p_r}}{dt}$ on déduit par intégration $\overrightarrow{p_v}(t)$, on obtient ainsi le long de l'orbite balistique 0 l'expression classique du cas où l'hamiltonien H est nul :

$$\overrightarrow{p_v} = \lambda_1 \overrightarrow{V} + \overrightarrow{\lambda_2} \wedge \overrightarrow{r} + (\overrightarrow{r} \wedge \overrightarrow{V}) \wedge \overrightarrow{\lambda_3} + \overrightarrow{r} \wedge (\overrightarrow{V} \wedge \overrightarrow{\lambda_3})$$; λ_1 étant une constante et $\overrightarrow{\lambda_2}$ et $\overrightarrow{\lambda_3}$ étant deux vecteurs constants.

D'où pv :
$$\left(1 - p_v^2\right) = \frac{1 - \cos\varphi}{(1+x)^2} \cdot G(\varphi)$$

avec :
$$G(\varphi) = G_0 + G_1 \sin\varphi + G_2(1 - \cos\varphi) + G_3 \sin^2\varphi + G_4 \sin\varphi(1 - \cos\varphi)$$

et :
$$G_0 = 2(1+x_1)\left(1 - 3S_1^2 - \frac{r_1^3 p_{r_1}^2}{\mu}\right)$$
$$G_1 = 4 S_1 (1+x_1)(T_1 - 2J_1) + 2 y_1 (3S_1^2 - 1)$$
$$G_2 = (1+x_1)\left(4S_1^2 + 2T_1 J_1 + \frac{r_1^3 \alpha_1^2 + r_1^3 \gamma_1^2}{\mu} - 2\right) + 1 - 4J_1^2 - 2S_1 y_1 \frac{r_1 \alpha_1}{Y_1} + y_1^2 \frac{\alpha_1^2 r_1^2}{Y_1^2}$$
$$G_3 = 2x_1(S_1^2 - J_1^2) + 4y_1 S_1 J_1$$
$$G_4 = 4x_1 S_1 J_1 + 2y_1(J_1^2 - S_1^2)$$

$G(\varphi)$, et donc $p_v(\varphi)$, sont connus en fonction des vecteurs $\overrightarrow{r_1}$, $\overrightarrow{V_1}$, $\overrightarrow{p_{r_1}}$ et $\overrightarrow{p_{v_1}}$ relatifs au point I_1 d'impulsion.

Si l'orbite osculatrice 0 est quelconque $G(\varphi)$ est positif quelque soit φ.

Si l'orbite osculatrice 0 est de commutation il y a une valeur de φ (point correspondant de commutation) pour laquelle $G(\varphi)$ est nul.

Cette expression $G(\varphi)$ permet de résoudre exactement maints problèmes, par exemple elle permet le calcul de l'excentricité limite à partir de laquelle apparaissent les "arcs obligatoirement balistiques" dans les transferts entre orbites coplanaires [6 et 7] (c'est aussi à partir de cette excentricité qu'apparaissent les premiers transferts coplanaires infinitésimaux tri-impulsionnels).

Pour cette excentricité limite le "domaine de manoeuvrabilité" [7] comporte 4 triangles infiniment plats dont le sommet double correspond à un point rencontré au cours de la description de la "spirale de LAWDEN" [8], on doit donc avoir pour ce point :

$$G(\varphi) = k(1-\cos\varphi)\left[1-\cos(\varphi-\varphi_2)\right]$$

Cela impose en particulier $G_0 = 0$ et $G_1 = 0$ et cela permet d'écrire l'équation en z $(= S_1^2)$:

$$8z^3 = 9(1-2z)(81-864z+3225z^2-5646z^3+4760z^4-1568z^5)$$

d'où $z = 0,192.686.0$; on en déduit tous les autres éléments, et en particulier :

$$\left.\begin{aligned} e\sin v_1 &= \pm 0,781.756 \\ e\cos v_1 &= -0,494.439 \end{aligned}\right) \text{ donc } e_{\text{limite}} = 0,924.993$$

Les précédents calculs (graphiques) de cette valeur limite de e se limitaient à la précision de 10^{-3} ou 2.10^{-4} [6 et 7]

4.1 - DETERMINATION DU SENS DES COMMUTATIONS OPTIMALES

L'expression $G(\varphi)$ ne suffit pas bien entendu pour déterminer le sens des commutations, il faut connaître aussi son évolution lors d'une poussée infinitésimale $\Gamma dt = dC$.

Compte tenu de la mobilité des axes $\overrightarrow{FXYZ}$ on obtient (en prenant Γ infiniment grand et dt nul) :

$$dr_1 = 0 \quad ; \quad dX_1 = S_1\, dC \quad ; \quad dY_1 = T_1\, dC$$

$$dS_1 = 0 \quad ; \quad dT_1 = \frac{W_1^2}{Y_1}\, dC \quad ; \quad dW_1 = -\frac{W_1 T_1}{Y_1}\, dC$$

$$d\alpha_1 = 0 \quad ; \quad d\beta_1 = \frac{\gamma_1 W_1}{Y_1}\, dC \quad ; \quad d\gamma_1 = -\frac{\beta_1 W_1}{Y_1}\, dC$$

d'où, en posant $p = \dfrac{r_1^2\, Y_1^2}{\mu}$ = paramètre de l'orbite 0 :

$$\frac{d}{dC}\left(\frac{G(\varphi)}{p}\right) = \frac{2}{p\,Y_1} M(\varphi)$$

avec $M(\varphi) = M_1 \sin\varphi + M_2(1-\cos\varphi) + M_3 \sin^2\varphi + M_4 \sin\varphi\,(1-\cos\varphi)$

et

$$M_1 = T_1 y_1 (1-3S_1^2) + S_1(1+x_1)(5S_1^2 - 2T_1^2 - 3 + 4T_1J_1)$$

$$M_2 = x_1(T_1 - T_1S_1^2 + J_1W_1^2 - J_1T_1^2) + 8T_1J_1^2 + J_1(3S_1^2 - 3 - 2T_1^2)$$

$$M_3 = 2S_1y_1(1-S_1^2-2T_1J_1) + 2J_1x_1(2S_1^2+T_1J_1-1) + 2S_1^2(T_1+J_1) - 2T_1J_1^2$$

$$M_4 = y_1\left(2J_1 - 2J_1S_1^2 + T_1S_1^2 - 3T_1J_1^2\right) + x_1 S_1\left(2 - 3S_1^2 - 2T_1J_1 + J_1^2\right) + S_1\left(J_1^2 + 4J_1T_1 - S_1^2\right)$$

$M(\varphi)$, et donc le signe de $\frac{d\,G(\varphi)}{dC}$ lorsque $G(\varphi) \simeq 0$, sont connus en fonction des vecteurs $\vec{r_1}$, $\vec{V_1}$, $\vec{p_{r_1}}$ et $\vec{p_{v_1}}$.

Imaginons donc un cas de commutation, soit I_2 (obtenu pour $\varphi = \varphi_2$) le point correspondant de commutation, on doit donc avoir :

$$G(\varphi) = \left[1 - \cos(\varphi - \varphi_2)\right] \cdot \left(D_0 + D_1 \cos\varphi + D_2 \sin\varphi\right)$$

avec $D_0 \geqslant \sqrt{D_1^2 + D_2^2} \left[= \sqrt{G_3^2 + G_4^2} = 2e\left(S_1^2 + J_1^2\right)\right]$

Le sens de la commutation est déterminé par le signe de $M(\varphi_2)$: pour $M(\varphi_2) < 0$, p_{v_2} est croissant et la commutation a lieu dans le sens I_1 vers I_2.

Il est bien évident que les expressions de $G(\varphi)$ et $M(\varphi)$ sont suffisament complexes pour qu'il soit difficile de se faire une idée générale, c'est pourquoi il est plus commode d'utiliser des expressions symétriques par rapport à I_1 et I_2.

Utilisons les 9 paramètres S_1, T_1, W_1, J_1, S_2, T_2, W_2, J_2 et φ_2 et calculons les autres paramètres à partir de ceux là. Les commutations ne dépendant que de 4 paramètres (aux questions d'orientation et d'échelle près) les 9 paramètres choisis doivent être liés par 5 relations; quatre d'entre elles sont simples :

$$S_1^2 + T_1^2 + W_1^2 = 1 \quad ; \quad S_2^2 + T_2^2 + W_2^2 = 1 \quad ;$$

$$S_2 = S_1 \cos\varphi_2 + J_1 \sin\varphi_2 \quad ; \quad J_2 = J_1 \cos\varphi_2 - S_1 \sin\varphi_2$$

La cinquième se déduit des trois équations linéaires suivantes permettant de calculer x_1 et x_2 et qui doivent être compatibles :

$$\left(T_1 - T_2 - 2J_1 + 2J_2\right) + x_1\left(T_1 - J_1\right) + x_2\left(J_2 - T_2\right) = 0$$

$$W_1^2 \cos\varphi_2 + S_1 \sin\varphi_2\left(2T_1 - J_1\right) - W_2W_1 + x_1\left(T_1 - J_1\right)\left(S_1 \sin\varphi_2 - T_1 \cos\varphi_2\right) + \\ + x_2\left(1 - S_1^2 - T_1J_1 - W_2W_1\right) = 0$$

$$W_2^2 \cos\varphi_2 + S_2 \sin\varphi_2\left(J_2 - 2T_2\right) - W_2W_1 + x_1\left(1 - S_2^2 - T_2J_2 - W_2W_1\right) + \\ + x_2\left(J_2 - T_2\right)\left(S_2 \sin\varphi_2 + T_2 \cos\varphi_2\right) = 0$$

Connaissant x_1 et x_2 on en déduit y_1 et y_2 par :

$$y_1 = \frac{x_1 \cos\varphi_2 - x_2}{\sin\varphi_2} \quad \text{et } y_2 = \frac{x_1 - x_2 \cos\varphi_2}{\sin\varphi_2}$$

$x_1 = e \cos v_1$
$y_1 = e \sin v_1$

on connaît donc e et v_1; il faut bien entendu $e < 1$ pour le cas elliptique.

Rappelons que $v_2 = v_1 + \varphi_2$.

D'où r_1, X_1, Y_1 et r_2, X_2, Y_2 à un facteur d'échelle près grâce à :

$r_1 = \dfrac{k^2}{\mu(1+x_1)}$; $X_1 = \dfrac{\mu y_1}{k}$; $Y_1 = \dfrac{\mu(1+x_1)}{k}$; $r_2 =$ etc.... avec : $r_1 Y_1 = r_2 Y_2 = k =$ moment cinétique;

Enfin les composantes $\alpha_1, \beta_1, \gamma_1$ et $\alpha_2, \beta_2, \gamma_2$ se déduisent des relations :

$$\alpha_1 = \frac{Y_1}{r_1}(T_1 - J_1) \quad ; \quad \alpha_2 = \frac{Y_2}{r_2}(T_2 - J_2)$$

$$\beta_1 = \frac{\mu S_1}{r_1^2 Y_1} + \frac{X_1}{r_1}(J_1 - T_1) \quad ; \quad \beta_2 = \frac{\mu S_2}{r_2^2 Y_2} + \frac{X_2}{r_2}(J_2 - T_2)$$

$$\gamma_1 = -\frac{S_1\alpha_1 + T_1\beta_1}{W_1} = \frac{W_1(Y_1\cos\varphi_2 - X_1\sin\varphi_2) - Y_2 W_2}{r_1 \sin\varphi_2}$$

$$\gamma_2 = -\frac{S_2\alpha_2 + T_2\beta_2}{W_2} = \frac{Y_1 W_1 - W_2(Y_2\cos\varphi_2 + X_2\sin\varphi_2)}{r_2\sin\varphi_2}$$

Il reste alors à vérifier que l'on a bien :

$$D_0 \geqslant \sqrt{D_1^2 + D_2^2}\ \left[= 2e\left(S_1^2 + J_1^2\right)\right]$$ ce qui est assez long

et exige toujours : $S_1^2 + J_1^2 \left(= S_2^2 + J_2^2\right) \leqslant \dfrac{1}{4}$

Le seul point délicat est d'obtenir la compatibilité des 3 équations donnant x_1 et x_2; d'autre part si $\sin\varphi_2 = 0$ les expressions ci-dessus sont insuffisantes à partir du calcul de y_1 et y_2; heureusement ce cas particulier n'est pas compliqué.

Ces équations et l'analyse faite dans le chapitre 2 permettent de donner une expression simple de $M(\varphi_2)$:

$$\frac{M(\varphi_2)}{1+x_2}\left[T_1(1+x_1) - T_2(1+x_2)\right] = N\left(J_2 - J_1\right)$$

avec : $$N = (1+x_1)\left[2T_1(1-S_2^2) - T_2(1-2S_1^2 - S_1S_2) + 2T_1T_2(J_1 - J_2)\right] + \\ +(1+x_2)\left[T_1(1-2S_2^2 - S_1S_2) - 2T_2(1-S_1^2) + 2T_1T_2(J_1 - J_2)\right]$$

Compte tenu de $T_i^2 + S_i^2 \leqslant 1$ et de $S_i^2 + J_i^2 \leqslant \frac{1}{4}$ il est visible que si $T_1T_2 \leqslant 0$; N est du signe de $(T_1 - T_2)$ et donc $M(\varphi_2)$ est du signe de $J_2 - J_1$.

Ce résultat est général et même les 4 quantités $[-M(\varphi_2)]$; $\left[T_1(1+x_1) - T_2(1+x_2)\right]$; $(J_1 - J_2)$ et N sont toujours du même signe, elles sont positives si la commutation a lieu dans le sens $I_1 \longrightarrow I_2$ et négatives dans le cas contraire.

Cela permet d'énoncer les règles de commutations suivantes :

La commutation a lieu de I_1 vers I_2 si et seulement si :

$$T_1\left(1 + x_1\right) > T_2\left(1 + x_2\right)$$

ce qui revient à : $\dfrac{T_1}{r_1} > \dfrac{T_2}{r_2}$

Une autre règle est : $J_1 > J_2$

Etant donné que : $\dfrac{r_1 Y_1}{2\mu}\left(\vec{V_1}\,\vec{p_{v_1}} - \vec{V_2}\,\vec{p_{v_2}}\right) = J_1 - J_2$

on peut dire que la commutation a lieu de I_1 vers I_2 si et seulement si $\vec{V_1}\,\vec{p_{v_1}} > \vec{V_2}\,\vec{p_{v_2}}$. Comme $\vec{V}\,\vec{p_v}$ est précisément égal à la dérivée de l'énergie mécanique de la fusée (par unité de masse) par rapport à la vitesse caractéristique on peut encore dire que la commutation a lieu dans le sens qui abaisse cette dérivée.

Cas limites

Il y a deux cas limites pour lesquels $M(\varphi_2)$ et donc $(J_1 - J_2)$ et $\left(\dfrac{T_1}{r_1} - \dfrac{T_2}{r_2}\right)$ sont nuls.

Le premier est obtenu pour $\varphi_2 = 0$: I_1 et I_2 sont confondus et ce cas engendre les "arcs singuliers" découverts par LAWDEN [8] ; ces arcs ne sont jamais optimaux [4] (du moins pour les transferts de durée indifférente entre orbites elliptiques).

Nous appelerons ce premier cas limite le "cas limite de LAWDEN".

Le second cas limite est un cas symétrique, on obtient alors:

$$\left.\begin{aligned} x_1 &= x_2 \\ y_1 &= -y_2 \end{aligned}\right\} \text{donc} \left\{\begin{aligned} r_1 &= r_2 \quad ; \quad Y_1 = Y_2 \\ v_1 &= -v_2 \quad ; \quad X_1 = -X_2 \end{aligned}\right.$$

$$S_1 = -S_2 \quad ; \quad \alpha_1 = \alpha_2$$

$$T_1 = T_2 \quad ; \quad \beta_1 = -\beta_2$$

$$W_1 = -W_2 \quad ; \quad \gamma_1 = \gamma_2$$

$$J_1 = J_2 = S_1 \cot v_1$$

D'où, grâce aux équations générales :

$$2\cos v_1\,(1 - T_1^2) - 4S_1T_1\sin v_1 + e\left[1 + W_1^2 + (S_1^2 - T_1^2)\cos 2v_1 - 2S_1T_1\sin 2v_1\right] = 0$$

Ce cas s'étudie aisément en remarquant qu'il correspond à un vecteur $\vec{p_v}\left[= \lambda_1\vec{V} + \vec{\lambda_2}\wedge\vec{r} + (\vec{r}\wedge\vec{V})\wedge\vec{\lambda_3} + \vec{r}\wedge(\vec{V}\wedge\vec{\lambda_3})\right]$ pour lequel le vecteur $\vec{\lambda_3}$ est nul et le vecteur $\vec{\lambda_2}$ est dans le plan perpendiculaire au plan d'orbite le long du grand axe.

Ce second cas limite conduit aux "arcs singuliers alternatifs" imaginés par CONTENSOU, nous l'appellerons donc "cas limite de CONTENSOU", ces arcs singuliers sont composés d'impulsions infiniments petites situées alternativement aux deux positions symétriques optimales, ils ne sont jamais optimaux.

Les "arcs singuliers alternatifs" sont évidemment de durée infinie c'est pourquoi ils ne peuvent se rencontrer que pour des transferts de durée indifférente entre orbites elliptiques. On peut montrer qu'il y a un second type d'arc singulier alternatif obtenu pour une poussée toujours tangentielle et une orbite osculatrice d'excentricité toujours infiniment petite, les impulsions se succèdent à des intervalles de temps réguliers en rapport quelconque avec la période et l'on obtient une sorte de "montée en spirale" (plane). Ce second type est lui aussi non optimal.

Ainsi les arcs singuliers ordinaires de LAWDEN et les "arcs singuliers alternatifs" de CONTENSOU ne sont jamais optimaux, les transferts optimaux de durée indifférente entre orbites elliptiques ont donc toujours un nombre fini d'impulsions; ceci constitue un pas important dans la démonstration de la limitation à 3 du nombre des impulsions dans de tels transferts (à l'exception des transferts "bi-paraboliques" qui comportent 2 impulsions finies et 2 infiniment petites).

Au voisinage des cas limites on peut utiliser l'expression approchée valable dans un cas comme dans l'autre :

$$\frac{M(\varphi_2)}{1+x_2} \simeq (J_2 - J_1).\underbrace{\left(3 - 2S_1^2 - 2S_2^2 - S_1 S_2\right)}_{\text{(facteur compris entre 3 et 7/4)}}$$

5 - ETUDE D'UN CAS PARTICULIER : TRANSFERTS OPTIMAUX ENTRE ORBITES ELLIPTIQUES SYMETRIQUES PAR RAPPORT A UN PLAN PASSANT PAR LE CENTRE ATTRACTIF (LES SENS DE ROTATION ETANT SYMETRIQUES).

Utilisons les notations habituelles.

L'orbite initiale O_o est définie par son demi-grand axe a_o et son excentricité e_o, son plan sert de plan de référence (orienté dans le sens du mouvement) la direction du périgée sert de direction de référence dans ce plan. L'orbite finale O_f est définie par a_f, e_f, i_f, Ω_f, ϖ_f

(fig.11) avec i_f = inclinaison ($0° \leq i_f \leq 180°$); Ω_f = longitude du noeud ascendant; $\varpi_f = \Omega_f + \omega_f$ = longitude du périgée.

Nous utiliserons aussi le moyen mouvement angulaire n égal à $\sqrt{\frac{\mu}{a^3}}$.

Fig. 11 - Notations _ $\varpi_f = \Omega_f + \omega_f$

Le cas étudié : orbites symétriques par rapport à un plan et sens de rotation symétrique, correspond à $a_f = a_o$; $e_f = e_o$; $\varpi_f = 0$.

5.1. - PROBLEME AUXILIAIRE

Recherchons les orientations de plan P d'orbite que l'on peut atteindre à partir de l'orbite O_o pour une vitesse caractéristique C_p donnée.

Ces orientations sont évidemment reliées aux pramètres i_p et Ω_p; nous dessinerons le domaine accessible sur une sphère unitaire S (fig.12) avec:

$$x = \sin i_p \cos \Omega_p$$
$$y = \sin i_p \sin \Omega_p$$
$$z = \cos i_p \ .$$

Le point x,y,z correspond au pôle du plan P d'orbite atteint. Etant donné les symétries du problème le domaine accessible est toujours symétrique par rapport aux plans Oxz et Oyz.

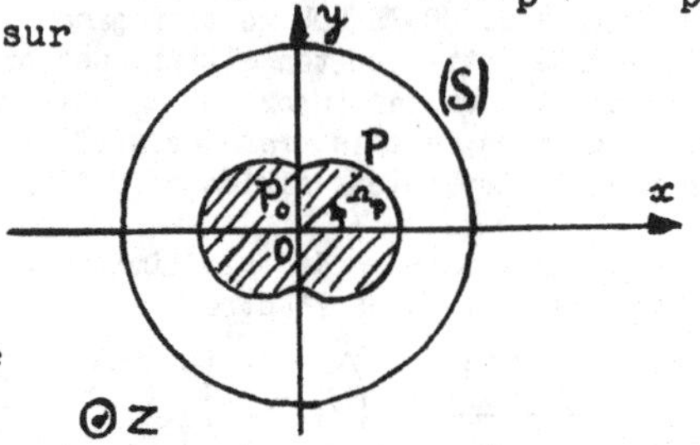

Fig.12 - Sphère S et domaine accessible. (Sphère S and accessible domain)

$$x = \sin i_p \cos \Omega_p$$
$$y = \sin i_p \sin \Omega_p$$
$$z = \cos i_p$$
$$\widehat{P_oP} = i_p$$

Il convient de distinguer deux cas:

1°/ $Cp \geq Cp_{oo} = \dfrac{n_o a_o(\sqrt{2}-\sqrt{1+e_o})}{\sqrt{1-e_o}}$ = impulsion nécessaire pour atteindre, par une accélération tangentielle au périgée, une orbite parabolique.

Dans ce cas, les transferts entre orbites paraboliques étant d'une vitesse caractéristique négligeable, on peut atteindre toute orientation de plan pour la vitesse caractéristique Cp et la sphère unitaire S tout entière fait partie du domaine accessible.

2°/ $Cp < Cp_{oo}$.

Le domaine accessible ne s'étend jamais alors à plus de 30°,0925 du point Po initial (0;0;1); les points limites du domaine sont atteints soit par des transferts mono-impulsionnels soit par des transferts bi-impulsionnels (fig.13).

Les transferts monoimpulsionnels (tel O A) utilisent des impulsions évidement normales au plan d'orbite P atteint (fig.14), il faut bien sûr placer l'impulsion de transfert à l'un

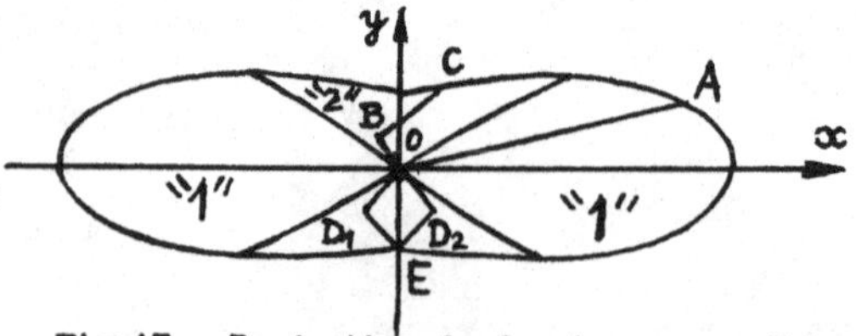

Fig.13 - Projection du domaine accessible sur O x y
(Projection of the accessible domain on O x y).

des deux points d'intersection M_1 et M_2 de l'orbite O_o et du plan P; celui des deux qui est le plus éloigné de F est toujours le plus favorable et l'on obtient :

$$C_P = \frac{n_o a_o}{\sqrt{1-e_o^2}} \sin i_P \left(1 - e_o \left|\cos \Omega_P\right|\right)$$

En conséquence le lieu de A sur la figure 13 fait partie de deux ellipses symétriques de foyer O de grand axe Ox et d'excentricité e_o.

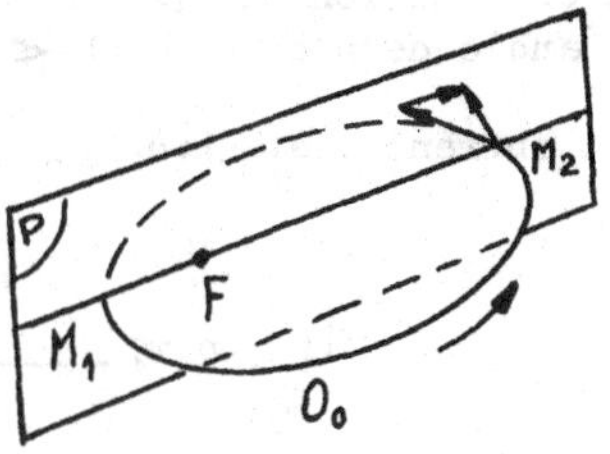

Fig. 14 - Transfert mono-impulsionnel $O_o \to P$ (one-impulse transfer $O_o \to P$)

Les transferts bi-impulsionnels (tel OBC, fig. 13) se terminent par une impulsion analogue à celle du cas mono-impulsionnel mais le point B correspond à une ellipse O_1 de commutation, analysons donc la fonction G (φ) le long de O_1.

L'orbite O_1 est intermédiaire entre les impulsions $\vec{I_1}$ et $\vec{I_2}$ (fig. 15); l'impulsion $\vec{I_2}$ est perpendiculaire au plan P final, elle est donc horizontale ($S_2 = 0$) et T_2 = - sinus de l'angle de plan O_1, P donc $W_2 = \pm \cos$ (angle de plan O_1 P) d'autre part $\vec{p_{r_2}}$ est nul (car $\vec{p_{r_2}}$ doit être normal à P et $\vec{p_{r_2}} \cdot \vec{p_{v_2}} = 0$).

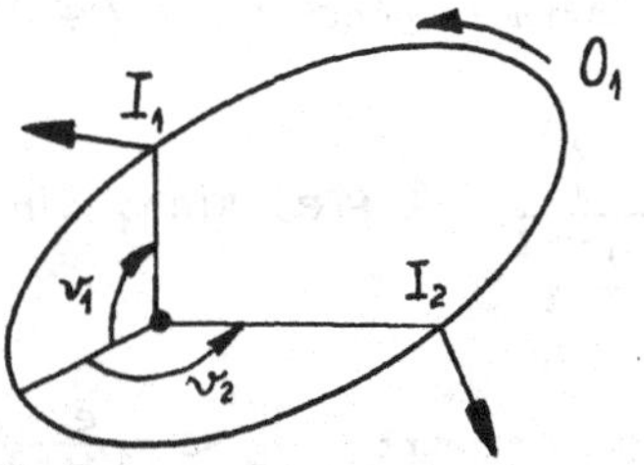

Fig. 15 - Analyse de l'orbite O_1 (Analysis of the orbit O_1)

On en déduit donc, en calculant G (φ) à partir du point I_2 (c'est-à-dire en posant $\varphi = v - v_2$) :

$$G_o = 2\left(1 + e_1 \cos v_2\right)$$: avec e_1 = excentricité de O_1.

$$G_1 = -2 e_1 \sin v_2$$

$$G_2 = -1 - 2T_2^2 - 2e_1 W_2^2 \cos v_2$$

$$G_3 = -2 T_2^2 e_1 \cos v_2$$

$$G_4 = 2 T_2^2 e_1 \sin v_2$$

D'où : $$G(\varphi) = 1 + \cos\varphi + 2e_1 \cos v - 2T_2^2 (1 - \cos\varphi)(1 + e_1 \cos v)$$

$G(\varphi)$ doit être positif pour $v \neq v_1$ et nul pour $v = v_1$, posons donc $v_1 - v_2 = \varphi(I_1) = \varphi_1$ et exprimons tous les éléments de la commutation en fonction de φ_1, T_2 et W_2 [avec $W_2^2 + T_2^2 = 1$ et $T_2 = -\sin$ (angle de plan O_1, P) < 0].

On obtient : $e_1 \sin v_1 = - \dfrac{\sin \varphi_1}{2\left[W_2^2 + T_2^2 \cos \varphi_1\right]^2}$

$$1 + e_1 \cos v_1 = \frac{1 - \cos \varphi_1}{2\left[W_2^2 + T_2^2 \cos \varphi_1\right]}$$

d'où e_1 et v_1 et donc $v_2 = v_1 - \varphi_1$

D'autre part, $S_1 = T_2 \sin \varphi_1$

S_1, T_1, W_1 comme S_2, T_2, W_2 se rapportent bien entendu à l'orbite O_1.

$$\begin{cases} T_1 = -T_2 \left[\cos \varphi_1 + 2W_2^2(1 - \cos \varphi_1)\right] \\ W_1 = W_2 \dfrac{\cos \varphi_1 + e_1 \cos v_1}{1 + e_1 \cos v_1} \end{cases}$$

Enfin la condition $G(\varphi) > 0$ pour $v \neq v_1$ se ramène à :

$$\frac{1}{2T_2^2} - 1 + e_1\left(\sin v_1 \sin \varphi_1 - 1\right) > 0$$

ce qui exige toujours $T_2^2 < \frac{e_1}{2(1+e_1)}$ (donc $0 > T_2 > -\sqrt{\frac{e_1}{2(1+e_1)}} > -\frac{1}{2}$) et $\cos v_2 < -e_1$ (le point I_2 doit donc être dans la moitié supérieure de O_1).

La commutation de l'orbite O_1 s'exprime donc très simplement à partir de φ_1, T_2 et W_2 et cela permet de construire aisément, à partir de O_1, des exemples de transferts bi-impulsionnels optimaux; mais bien entendu le problème ordinaire : trouver O_1 et le transfert à partir de O_o et P n'est résolu qu'implicitement.

<u>Remarque I</u> - Le point E (fig. 13) point pour lequel $\cos \Omega_P = 0$ est accessible par deux transferts optimaux équivalents : $O D_1 E$ et $O D_2 E$ cela donne d'ailleurs la dernière limite pour l'optimalité des transferts bi-impulsionnels : on doit avoir: $W_2 \cdot \cos \Omega_P \leq 0$.

<u>Remarque II</u> - Si l'on effectue le calcul de $G(\varphi)$ directement pour l'orbite O_o, c'est-à-dire si l'on remplace :

e_1 par e_o

T_2 par $T_2' = -\sin i_p$

v_2 par $v_2' = \Omega p$ ou $\Omega p + \pi$ (avec $\cos v_2' \leq 0$)
avec bien entendu $\varphi' = v - v_2'$; on obtient alors:

$G(\varphi') \geq 0$ quel que soit φ': le transfert obtimal est mono-impulsionnel

$G(\varphi') < 0$ pour certaines valeurs de φ': le transfert optimal est bi-impulsionnel

à condition que: $Cp < Cp_{oo}$

En conséquence si $|\cos \Omega p| \leq e_o$ le transfert optimal n'est jamais mono-impulsionnel.

Il est dès lors aisé de construire sur la sphère S les domaines accessibles relatifs aux diverses axcentricités e_o.

Pour $e_o = 0$ (fig.16) il n'y a pas de zone mono-impulsionnelle, la zone "2" (bi-impulsionnelle) est circulaire et s'étend jusqu'à $30^o,0925$ du point P_o, les courbes isovitesse caractéristique y sont des cercles de pole P_o. Au de là de $30^o,0925$ s'étend la zone "∞" utilisant des orbites intermédiaires paraboliques et de vitesse caractéristique Cp_{oo}.

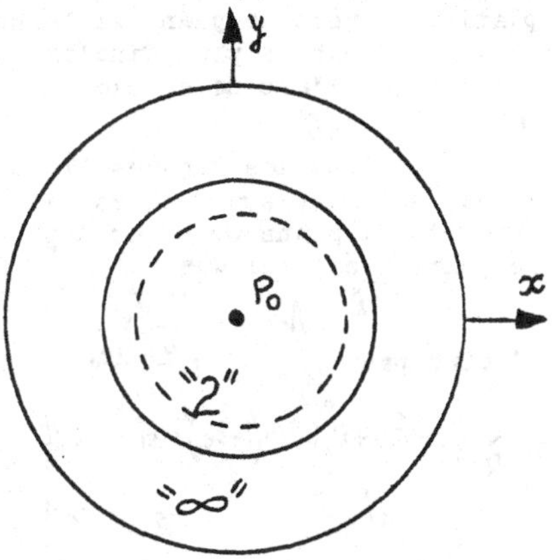

Fig. 16- Domaine accessible pour $e_o = 0$
(Accessible domain for $e_o = 0$)

Pour $0 < e_o < 0,611$ (fig.17) le domaine accessible contient deux zones "1" symétriques (mono-impulsionnelles) entourées d'une zone "2" (bi-impulsionnelle) elle-même entourée de la zone "∞".

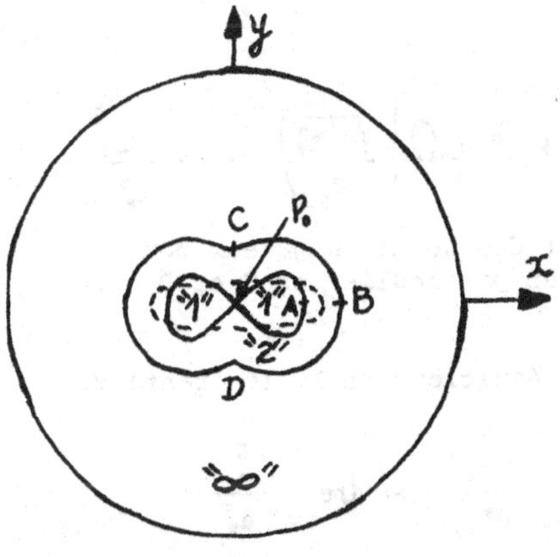

Fig.17 - Domaine accessible pour $0 < e_o < 0, 611$
(Accessible domain for $0 < e_o < 0,611$).

Remarque: pour $e_o > 0,37609...$ (racine de $2x^3+x^2+2x = 1$) certaines courbes iso-vitesse caractéristique (fig.17) coupent en 8 points la limitante des zones "1".

Pour $e_o = 0,611$ les zones "1" atteignent la zone "∞" en 4 points symétriques P_1, P_2, P_3 et P_4 (fig. 18) pour lesquels $i_p = 22°,6$ et $\Omega_p = \pm 31°,3$ ou $\pm 148°,7$.

Pour $0,611 < e_o < 1$ la zone "2" est coupée en 4 morceaux (fig. 19), les zones "1" et "2" s'aplatissent vers le plan Oxz lorsque $e_o \to 1$ pour ne plus finalement occuper que l'arc de cercle $i_p \leq 30°$, $\Omega_p = 0°$ ou $180°$

Dans les figures 17, 18 et 19, qui sont symétriques par rapport aux plans Oxz et Oyz le point A est tel que :

$$\widehat{P_o A} = \text{Arc sin} \sqrt{\frac{e_o}{2+2e_o}}$$

d'autre part :

$$\widehat{P_o B} \begin{cases} \simeq 30° + 0°,0925 (1-8e_o)^2 & \text{si } e_o \leq \frac{1}{8} \\ = 30° & \text{si } e_o \geq \frac{1}{8} \end{cases}$$

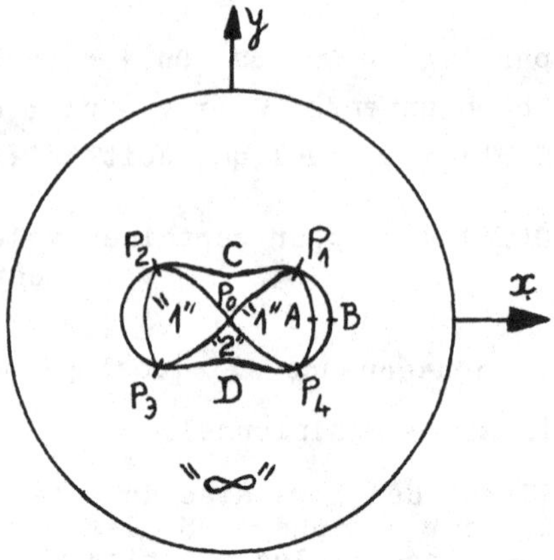

Fig. 18 - Domaine accessible pour $e_o = 0,611$ (Accessible domain for $e_o = 0,611$).

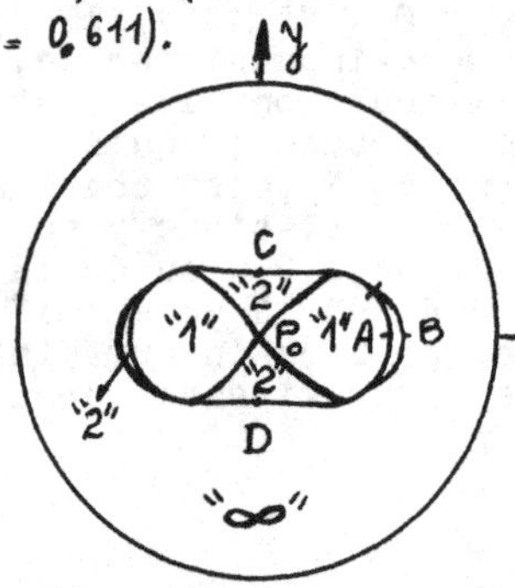

Fig. 19 - Domaine accessible pour $0,611 < e_o < 1$ (Accessible domain for $0.611 < e_o < 1$)

Nous savons déjà que les courbes iso-vitesse caractéristique des zones "1" se projettent sur le plan Oxy selon des ellipses de foyer O d'excentricité e_o et de grand axe Ox; celles des zones "2" ont, au voisinage de P_o l'équation :

$$y = \frac{C_P}{n_o a_o} + \frac{C_P^3}{4 n_o^3 a_o^3} \left(1 - \cot^2 \Omega_P \frac{1-e_o^2}{e_o^2}\right)\left(1 + 3 |\cot \Omega_P| \frac{\sqrt{1-e_o^2}}{e_o}\right) + \text{ordre } \frac{C_P^5}{n_o^5 a_o^5}$$

Cela indique que ces courbes ont des points anguleux pour $\cot \Omega_P = 0$ c'est-à-dire pour $x = 0$ et en particulier en C et D (fig. 17 à 19).

Enfin, au voisinage de P_o, les équations de la limitante des zones "1" et "2" sont :

$$x = \pm \frac{e_o C_P}{n_o a_o \sqrt{1-e_o^2}} \left(1 + \frac{2 C_P^2}{n_o^2 a_o^2 e_o^2}\right) + \text{ordre } \frac{C_P^5}{n_o^5 a_o^5}$$

$$y = \pm \frac{C_P}{n_o a_o} + \text{ordre } \frac{C_P^5}{n_o^5 a_o^5} .$$

5.2. - TRANSFERTS OPTIMAUX ENTRE ORBITES ELLIPTIQUES SYMETRIQUES PAR RAPPORT A UN PLAN (ET DE SENS DE ROTATION SYMETRIQUES).

Ce cas de symétrie correspond à $a_f = a_o$; $e_f = e_o$ et $\varpi_f = 0$.

Imaginons un transfert du chapitre précédent : il conduit optimalement de l'orbite initiale O_o à une orbite du plan P; poursuivons le symétriquement par rapport à P : il aboutit, pour une vitesse caractéristique $C_f = 2 C_P$, à une orbite O_f symétrique de O_o par rapport à P et de sens de rotation symétrique : le transfert de O_o à O_f obtenu est-t-il optimal ?

Examinons les domaines sphériques accessibles pour le coût C_P à partir des orbites O_o et O_f , ils sont disposés symétriquement par rapport au point P qui correspond au plan P (fig. 20); si ces domaines n'ont que le point P en commun, le transfert symétrique de O_o à O_f par P est évidemment optimal, mais ces deux domaines n'étant pas convexes, il se peut qu'ils soient sécants et il faut étudier ce qu'il advient dans ce cas.

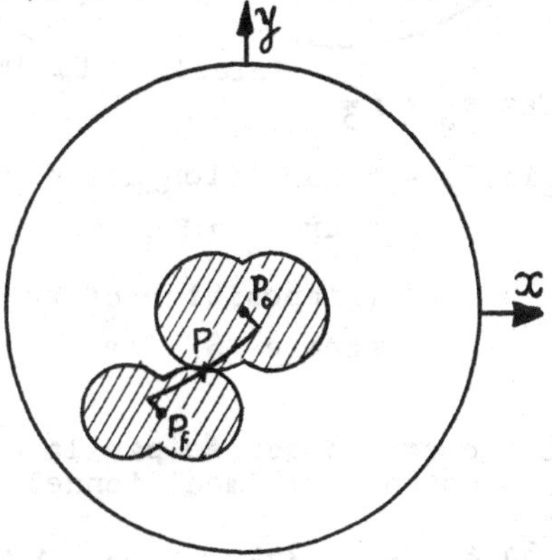

Fig. 20 - Domaines accessibles pour le coût C_P à partir des orbites O_o et O_f
(Accessible domaines from the orbits O_o and O_f for the cost C_P).

On obtient ainsi que le transfert optimal est soit celui que nous venons de définir, soit un transfert bi-impulsionnel (il y a toujours alors deux solutions optimales équivalentes symétriques l'une de l'autre par rapport au plan P).

Il y a deux cas principaux :

1°/ $e_o \leq \sqrt{\frac{2}{3}}$ $(= 0,8165..)$

Les transferts bi-impulsionnels n'apparaissent pas et les solutions optimales sont symétriques par rapport au plan P, elles sont soient mono-impulsionnelles (une impulsion horizontale au point d'intersection le plus élevé des 2 orbites symétriques), soient"tri-impulsionnelles symétriques" (cas correspondant aux zones "2") soient bi-paraboliques (cas de coût $C_{f\infty} = 2 C_{P\infty}$ et correspondant à la zone "∞").

Le domaine accessible s'obtient bien entendu en effectuant sur la sphère S, à partir des domaines des figures 16 à 19 relatives au problème auxiliaire, une homothétie sphérique de centre P_o et de rapport 2 ($i_f = 2 i_P$; $\Omega_f = \Omega_P$) avec $C_f(i_f, \Omega_f) = 2 C_P(i_P, \Omega_P)$.

2°/ $e_o > \sqrt{\frac{2}{3}}$.

Deux zones bi-impulsionnelles apparaissent au voisinage du plan Oyz , il y a deux cas de figure selon que e_o est proche de $\sqrt{\frac{2}{3}}$ ou de 1 (fig. 21).

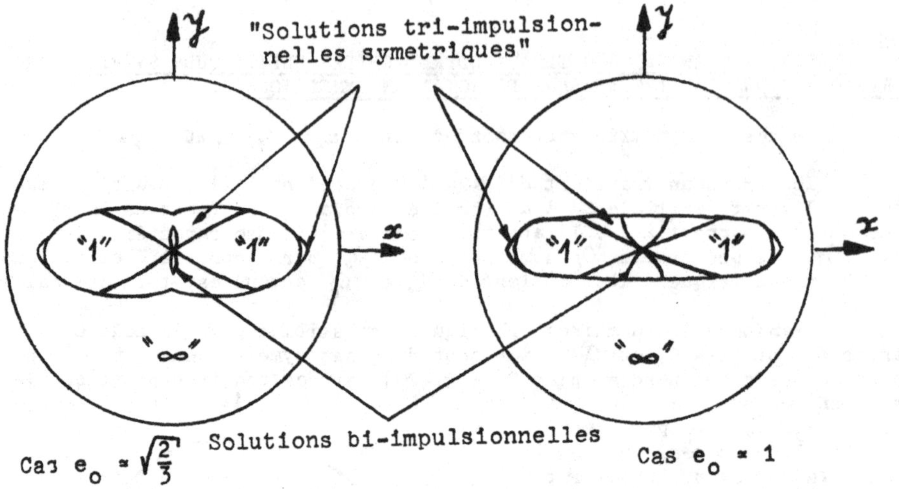

Fig.21 - Répartition des divers types de solutions sur la sphère S pour $e_o > \sqrt{\frac{2}{3}}$.

(Disposition of the different types of solutions on the sphere S for $e_o > \sqrt{\frac{2}{3}}$).

L'économie procurée par la solution bi-impulsionnelle par rapport à la solution "tri-impulsionnelle symétrique" est de:

$$\frac{3n_o a_o}{16}\, y_f^3\left(1-\frac{1-e_o^2}{e_o^2}\cot^2\Omega_f\right)\left(\frac{3e_o^2-2}{4-3e_o^2}-\frac{|\cot\Omega_f|}{e_o}\sqrt{1-e_o^2}\right)+\text{ ordre } n_o a_o y_f^5$$

avec bien entendu $y_f = \sin i_f \sin \Omega_f$.

Remarque I - Dans les zones bi et tri-impulsionnelles le facteur $\left(1-\frac{1-e_o^2}{e_o^2}\cot\Omega_f^2\right)$ est positif ou tout au moins supérieur à (- ordre y_f^2), c'est donc le dernier facteur qui donne le signe de l'économie; son annulation fournit la valeur de $|\cot \Omega_f|$ correspondant aux tangentes en P_o aux limitantes des zones bi et tri-impulsionnelles (fig.21).

Remarque II - Le problème auxiliaire du transfert optimal de O_o au plan P étant résolu (de manière implicite dans le cas bi-impulsionnel et complètement dans les autres cas) il en est de même, pour $e_o \leq \sqrt{\frac{2}{3}}$, du problème du transfert entre orbites symétriques par rapport à un plan (et de sens de rotation symétrique). Par contre pour $e_o > \sqrt{\frac{2}{3}}$ les solutions bi-impulsionnelles, dont l'existence est montrée ci-dessus, ne sont connues que de manière approchée et le problème n'est donc pas entièrement achevé.

Remarque III - Du point de vue pratique on peut noter que les solutions mono-impulsionnelles optimales (une impulsion horizontale au point d'intersection le plus élevé des deux orbites) exigent toujours :

$$i_f \leqslant \operatorname{Arc\,cos} \frac{1}{1+e_0}$$

et :

$$|\cos \Omega_f| > e_0$$

Le point où a lieu l'impulsion doit donc être dans la moitié supérieure des deux orbites O_0 et O_f.

- CONCLUSION -

Les commutations jouent un grand rôle dans les études des problèmes d'optimisation. Elles ont, sauf dans les cas "d'arcs singuliers" un sens optimal bien déterminé dont la recherche analytique est ordinairement assez malaisée.

Les arcs singuliers sont le plus souvent non optimaux mais quand ils sont optimaux ils jouent un rôle particulier important parmi les trajectoires optimales.

Dans le cas des transferts optimaux de durée indifférente entre orbites elliptiques, on aboutit à une règle de sens de commutation très simple exprimée par : $\frac{T_1}{r_1} > \frac{T_2}{r_2}$ ou par $\vec{V_1}\cdot\vec{p_{V_1}} > \vec{V_2}\cdot\vec{p_{V_2}}$; T_1 et T_2 étant les cosinus directeurs circonférentiels (positifs vers l'avant) de la direction de poussée des propulseurs (direction dont le vecteur unitaire est $\vec{p_V}$) et les indices 1 et 2 étant respectivement relatifs aux éléments de la première et de la deuxième impulsion. Il y a deux cas limites : le "cas limite de LAWDEN" et le "cas limite de CONTENSOU" qui aboutissent tous deux à des arcs singuliers non optimaux.

Cette étude permet de résoudre presque complètement le problème du transfert entre orbites elliptiques symétriques par rapport à un plan (et de sens de rotation symétrique).

Il serait très intéressant d'étendre à des cas plus généraux les déterminations de sens optimal des commutations, d'autre part d'autre cas de transferts entre orbites symétriques doivent pouvoir être résolus analytiquement, en particulier celui des transferts entre orbites symétriques par rapport à un plan et de sens de rotation inverse.

REFERENCES

[1] P. CONTENSOU - Etude théorique des trajectoires optimales dans un champ de gravitation. Application au cas d'un centre d'attraction unique. Astronautica Acta - VIII - fasc. 2-3 (1962).

[2] L.S. PONTRYAGIN, V.G. BOLTYANSKII, R.V. GAMKRELIDZE, E.F. MISCHENKO - The mathematical theory of optimal processes. Interscience Publishers, John Wiley and Sons, Inc. New-York (1962).

[3] C. MARCHAL - Transferts optimaux entre orbites elliptiques (durée indifférente). Thèse de Doctorat, A 0 1609, Faculté des Sciences de Paris (1967).

[4] C. MARCHAL - Généralisation tridimensionnelle et étude de l'optimalité des arcs à poussée intermédiaire de LAWDEN. La Recherche Aérospatiale n° 123 (1968).

[5] S. PINES - Constants of the motion for optimum thrust trajectories in a central force field. A.I.A.A. J.2, 2010-2014 (1964).

[6] H.G. MOYER - Necessary conditions for optimal single impulse transfer. A.I.A.A. J., 4, 1405-1410 (1966).

[7] C. MARCHAL - Transferts optimaux entre orbites elliptiques coplanaires. Astronautica Acta, 11, 432-445 (1965).

[8] D.F. LAWDEN - Optimal intermediate thrust arcs in a gravitationnal field Astronautica Acta 8, 106-123 (1962).

CONJUGATE POINTS ON EXTREMAL ROCKET PATHS

H. Gardner Moyer
and
Henry J. Kelley

Summary

Of the methods in use for optimizing rocket trajectories, many do not test whether or not the Jacobi necessary condition is satisifed or otherwise guarantee the minimizing character of the extremal obtained. Thus the possibility of the occurrence of Jacobi conjugate points on rocket extremals motivates the present study of the simplified case of planar vacuum flight in an inverse square law gravity field under constant thrust. The results indicate that conjugate points do indeed occur and shed some light upon the circumstances incident to their appearance.

The first order necessary conditions and the conjugate point condition are reviewed for the rocket trajectory problem. The conjugate point test for Mayer problems is reduced to the examination of a single minor of the usual test matrix plus a Lagrange multiplier.

Initially rocket paths are studied with the central angle ignored. The three state variables are the radius and the radial and circumferential velocity components. Vertical launch trajectories are shown to be extremals. The state variation defined by an extremal adjacent to this trajectory obeys a single second order differential equation which may exhibit an oscillatory solution whose zeros specify conjugate points. When the rocket is launched nonvertically with zero initial velocity, an analytic proof shows that the sign of the circumferential velocity component must change before a conjugate point occurs. It is conjectured that this is also true for trajectories with arbitrary initial velocity. A plausibility argument and numerical results are offered in substantiation.

Rocket paths are also examined with the central angle included with the three state variables mentioned above. Conjugate points are found numerically but no qualitative rules governing them have been deduced.

This research was partially sponsored by the Air Force Office of Scientific Research, Office of Aerospace Research, United States Air Force under AFOSR Contracts Nos. AF 49(638)-1207 and AF 49(638)-1512 and Contract NAS 12-114 with NASA Electronics Research Center, Cambridge, Massachusetts. The complete paper may be found in the Proceedings of the 19th Congress of the International Astronautical Federation published by Pergamon Press.

A MAXIMUM-MINIMUM PRINCIPLE FOR BANG-BANG SYSTEMS

B. Fraeijs de Veubeke

Summary.

Consider the problem of a rocket in vertical flight in a uniform gravitational field, the aerodynamic drag being neglected.
The thrust is provided by several chemical rocket engines working in parallel that can be separated and dropped according to some optimal sequence in order to provide a maximum payload for a given total thrust at departure and a prescribed velocity gain.

The mathematical formulation provides the possibility of a continuous reduction in thrust, that is for the limiting case of an infinity of infinitesimal propulsion units. In this case it is known that, if the velocity performance is set high enough, the optimal sequence consists of a constant thrust arc during which no engines are dropped, followed by a continuous reduction in thrust that keeps the acceleration constant. There is however another type of extremal representing the separation from a finite amount of thrust. The real technical problem involves only this type of extremal and the constant thrust extremal. The optimization problem is then of the bang-bang type, the continuous acceleration type of extremal representing a "chattering" of the control.

It is remarkable that optimal bang-bang solutions, each corresponding to a prescribed number of engine separations, are found by applying a minimum principle for the Hamiltonian (instead of the usual maximum principle) during a portion of the trajectory. More precisely the optimal bang-bang trajectories imply the use of the maximum principle up to the first reversal in the sign of the switching function, then of the minimum principle with a finite number of sign reversals, then of the maximum principle again to the end. Eventually the first or last part (or both) are missing.

The optimality of such bang-bang solutions is established by the analysis of the second variation.

I. THE OPTIMUM STAGING OF ROCKETS IN PARALLEL

I.I. Basic differential equations.

Consider a cluster of chemical rockets, whose instantaneous mass can be conceptually subdivided as follows :

$$M = M_u + \sigma M_1 + M_p + \frac{F}{Kg} \tag{I.I}$$

M_u is the payload mass or useful mass

σM_1 is the structural mass considered to represent a given fraction of the total mass M_1 at departure.

These two are fixed quantities, the other ones are variables :

M_p the instantaneous mass of propellants,

$M_e = \frac{F}{Kg}$ the mass of propulsion equipment, based on the assumption that the thrust F it delivers is proportional (factor K) to its weight gM_e .

If c denotes a fixed effective exhaust velocity of burnt gasses, the thrust is also given by

$$F = - c \frac{dM_p}{dt} \tag{I.2}$$

By elimination of the mass of propellant between (I.I) and (I.2) follows one of the basic differential equations :

$$\frac{dM}{dt} = - \frac{F}{c} + \frac{1}{Kg} \frac{dF}{dt} \tag{I.3}$$

It assumes that the thrust can be continuously reduced by separation of infinitesimal propulsion units. As will appear later, this idealized formulation does not only furnish a method for assessing the optimal performance ceiling that can be reached by the principle of parallel staging of propulsion equipment but also provides a scientific approach to the real problem of discrete staging. A control variable α is now introduced to govern the programming of engine separation by expressing that the thrust can only decrease :

$$\frac{dF}{dt} = - \alpha^2 \qquad (I.4)$$

Finally a performance equation is needed to pose a maningful problem of optimal staging. The simplest one that offers a complete analytical solution is the equation of motion for vertical flight in a uniform gravity field, the aerodynamic drag being neglected :

$$M \frac{dV}{dt} = F - Mg \qquad (I.5)$$

The basic differential system consists of the equations (I.3), (I.4) and (I.5).

<u>I.2. Dimensionless form of the basic system.</u>

Introduce the dimensionless variables

$\omega = V/c$ for the velocity

$\tau = tg/c$ for the time

$\phi = \ell n \frac{M_1}{M}$ for the instantaneous mass

$\beta = \frac{F}{gM}$ an instantaneous acceleration factor.

It is important to note that the acceleration factor has from equation (I.I) an upper limit K when the propellants are used up ($M_p = o$) and M_u and σ approach zero :

$$o < \beta < K \qquad (I.6)$$

In the new variables the basic differential system takes the form

$$\frac{d\phi}{d\tau} = \beta + \gamma^2$$

$$\frac{d\beta}{d\tau} = \beta^2 + \gamma^2 \ (\beta - K)$$

$$\frac{d\omega}{d\tau} = \beta - 1$$

where the control variable α has been changed to γ by

$$\gamma^2 = \frac{c}{K\, g^2 \quad M_1} \quad e^{\phi}\, \alpha^2$$

A further simplification can be introduced, provided no constraints be introduced on the duration of the flight. The time τ is then an ignorable variable and ϕ, which is strictly monotonically increasing, can serve as independant variable. Hence, dividing the last two equations of the basic system by the first, and changing once more the control variable to

$$v = \frac{\beta}{\beta + \gamma^2} \qquad\qquad v \in \left(0,1 \right)$$

we obtain

$$\frac{d\beta}{d\phi} = \beta - K + K\, v \tag{I.7}$$

$$\frac{d\omega}{d\phi} = \left(1 - \frac{1}{\beta} \right) v \tag{I.8}$$

I.3. The optimization problem.

We set up the following optimization problem : the initial velocity being zero and a prescribed terminal velocity having to be reached at burnout $(M_p = o)$, maximize the payload gM_u for a given thrust F_1 available at departure. This is equivalent to minimize the functional

$$J = - \frac{gM_u}{F_1}$$

or, taking M_u from equation (I.I) at burnout

$$M_u = M_2 - \sigma\, M_1 - \frac{F_2}{Kg}$$

and substituting

$$J = \frac{1}{\beta_1} \left(\sigma + \left(\frac{\beta_2}{K} - 1\right) e^{-\phi_2} \right) \quad \text{minimum} \tag{I.9}$$

This is an example of a functional depending on the initial and final values of the independant and state variables.

The Hamiltonian of the problem is, from equations (I.7) and (I.8),

$$H = \lambda_\beta \ (\beta - K) + v\ S \tag{I.10}$$

where the switching function S, which decides on the choice of the control variable v, is

$$S = K\ \lambda_\beta + (1 - \frac{1}{\beta})\lambda_\omega \tag{I.11}$$

The adjoint differential system is

$$\frac{d\lambda_\beta}{d\phi} = -\frac{\partial H}{\partial \beta} = -\lambda_\beta - \frac{v}{\beta^2}\lambda_\omega$$

$$\frac{d\lambda_\omega}{d\phi} = -\frac{\partial H}{\partial \omega} = 0$$

and has an immediate first integral

$$\lambda_\omega = \text{constant} \tag{I.12}$$

A second first integral is provided by the equation

$$\frac{dH}{d\phi} = \frac{\partial H}{\partial \phi} = 0 \qquad \text{whence}$$

$$H = \text{constant}$$

This avoids the necessity of integrating the first equation of the adjoint system. In fact, if λ_β is eliminated between equations (I.10) and (I.11), the switching function can be expressed entirely in terms of the state variable β, the control variable v and the constants H and λ_ω

$$S = \frac{K\ H + (\beta - K)(1-\beta^{-1})\lambda_\omega}{\beta - K + K\ v} \tag{I.13}$$

The discussion of the maximum principle will be further simplified if we introduce the new constant ε defined by

$$\frac{H K}{\lambda_\omega} = 1 + K - 2 \varepsilon \sqrt{K} \tag{I.14}$$

and put the switching function in the form

$$\frac{S}{\lambda_\omega} = \frac{(\beta-\theta_1)(\beta-\theta_2)}{\beta(\beta-K+Kv)} \tag{I.15}$$

where

$$\begin{aligned} \theta_1 &= \sqrt{K} \, (\varepsilon - \sqrt{\varepsilon^2 - 1}) \\ \theta_2 &= \sqrt{K} \, (\varepsilon + \sqrt{\varepsilon^2 - 1}) \end{aligned} \tag{I.16}$$

In addition to the prescribed end values

$$\phi_1 = o \qquad \omega_1 = o \qquad \omega_2 = \bar{\omega}_2$$

we shall need the transversality conditions based on the fact that ϕ_2 , β_1 and β_2 are not prescribed. They are

$$H = H_2 = \frac{\partial J}{\partial \phi_2} = \frac{1}{\beta_1} \left(1 - \frac{\beta_2}{K} \right) e^{-\phi_2} \tag{I.17}$$

$$(\lambda_\beta)_1 = \frac{\partial J}{\partial \beta_1} = - \frac{1}{\beta_1^2} \left(\sigma + \left(\frac{\beta_2}{K} - 1\right) e^{-\phi_2} \right) \tag{I.18}$$

$$(\lambda_\beta)_2 = - \frac{\partial J}{\partial \beta_2} = - \frac{1}{K\beta_1} e^{-\phi_2} . \tag{I.19}$$

A first important conclusion stemming from (I.17) and (I.6) is that the constant of the Hamiltonian is positive

$$H > o \tag{I.20}$$

A second is obtained from equation (I.10) at the end of the trajectory

$$H_2 = (\lambda_\beta)_2 \; (\beta_2 - K) + v_2 \, S_2$$

when we substitute (I.I7) and (I.I9); there comes

$$v_2 S_2 = o \qquad (I.2I)$$

Hence at the end of the trajectory we must have that either the control variable or the switching function vanish.

I.4. Nature of the extremals.

The problem is regular in the sense that the manoeuvrability domain (hodograph space) is convex. It is the straight line segment of figure I.

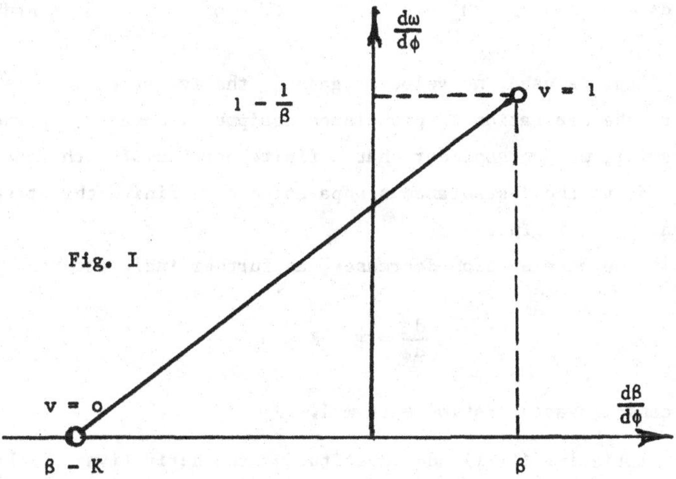

Fig. I

The extremals will be characterized by $v = 1$, $v = o$ or possibly some intermediate value corresponding to the persistence of the switching function to vanish.

a. The constant thrust extremal.

It corresponds to $v = 1$ and, in accordance with the maximum principle, to positive values of the switching function. The old control variables γ and α are zero so that, from (I.4), F is indeed constant. There is no engine separation and ϕ increases only through consumption of propellant. From (I.7)

$$\frac{d\beta}{d\phi} = \beta > o$$

and the acceleration is increasing along this type of arc.

b. The constant time extremal.

It corresponds to $v = o$ and negative values of the switching function. Since the old control variable α tends to infinity, we find, by returning to the original differential equations, that

$$dt = o \qquad dM_p = o \qquad dV = o \qquad dF = KgdM$$

No propellant is used, no velocity gained, the reduction in mass corresponds solely to the separation of propulsion equipment. However, as the time also stands still, we can consider that a finite portion of such an extremal arc corresponds to the instantaneous separation of a finite thrust unit, which is technically meaningful.

Naturally the acceleration decreases, as further indicated by

$$\frac{d\beta}{d\phi} = \beta - K < o$$

c. The constant acceleration extremal.

Differentiating (I.II) and substituting the derivatives of the state variable β and the multipliers, we find in general that

$$\frac{dS}{d\phi} = - S + \lambda_\omega \left(1 - \frac{K}{\beta^2} \right) \qquad \text{(I.22)}$$

Hence if S remains zero for some finite interval of ϕ, we must have either $\lambda_\omega = o$ or $\beta = \sqrt{K}$. The first possibility is ruled out by the consequence that, from (I.II), λ_β should also have to be zero and both multipliers would then vanish along the whole trajectory, together with the Hamiltonian. The second possibility, the constant acceleration one

$$\beta = \sqrt{K} \tag{I.23}$$

gives, when substituted into (I.7), the control value

$$v = 1 - \frac{1}{\sqrt{K}} \tag{I.24}$$

which lies in the possible range. Furthermore, there follows from (I.II) and $S = o$, that

$$\lambda_\beta = - \frac{1}{K} \left(1 - \frac{1}{\sqrt{K}} \right) \lambda_\omega \tag{I.25}$$

and, from (I.IO), that

$$H = \left(1 - \frac{1}{\sqrt{K}} \right)^2 \lambda_\omega \tag{I.26}$$

Equation (I.I4) then gives

$$\varepsilon = 1 \tag{I.27}$$

so that the roots θ_1 and θ_2 of (I.I6) are confluent

$$\theta_1 = \theta_2 = \sqrt{K} \tag{I.28}$$

The velocity gain is given by

$$\frac{d\omega}{d\phi} = \left(1 - \frac{1}{\sqrt{K}} \right)^2 \tag{I.29}$$

Technically speaking, this arc is a limiting case. It implies a continuous separation of infinitesimal thrust units as the rocket is gaining velocity so that the acceleration can be kept constant, despite the reduction in mass due to propellant consumption.

I.5. Synthesis of optimal trajectories.

The general composition of optimal trajectories for any type of functional and boundary conditions is easily obtained from a (β, S) graph based on equation (I.15). This graph depends only on the sign of the constant λ_ω and on the nature and position of the roots (θ_1, θ_2). Here we shall restrict ourselves to the particular problem at hand.

We can first observe that the end condition (I.21) really reduces to

$$S_2 = o \qquad \text{(I.30)}$$

Indeed, since a velocity gain is imposed, the trajectory must contain at least one arc of the constant thrust or constant acceleration type. If the trajectory ends on a constant thrust arc $(v = 1)$, condition (I.30) is needed to implement (I.21). If it ends on a constant acceleration arc, (I.30) is actually satisfied. In both cases we can switch to $v = o$ and, provided the switching function becomes negative, add a final constant time arc. However this terminal arc does not change the terminal velocity nor the value of the functional; in fact, as shown by equation (I.7) for $v = o$

$$(\beta/K - 1)\, e^{-\phi} \qquad \text{remains constant.}$$

Hence the only thing that can be achieved by such an extension of the trajectory is a separation of the payload and structure from a part or from the total of the remaining propulsion units. This new solution is not essentially different and there is no loss in generality in ending the trajectory on the constant thrust or constant acceleration arc.

Then from (I.30) follows

$$\lambda_\omega = - \frac{\beta_2}{\beta_2 - 1} K (\lambda_\beta)_2$$

or, taking (I.19) into account

$$\lambda_\omega = \frac{\beta_2}{\beta_2 - 1} \; \frac{e^{-\phi_2}}{\beta_1} \qquad \text{(I.31)}$$

It is also obvious that we must have

$$\beta_2 > o \tag{I.32}$$

For, if this were not true, the final arc would be a constant thrust one with continuous reduction in velocity and continuous increase in the value of the functional; the required terminal velocity would already have been reached earlier with a smaller value of the functional. From (I.3I) and (I.32) it follows that

$$\lambda_\omega > o \tag{I.33}$$

The nature of the initial arc can be fixed by considering a physical limitation : the optimal payload for given F_1 must still be positive or, stated otherwise, the optimal value of the functional must be negative. Comparing (I.9) and (I.I8) this holds only if

$$(\lambda_\beta)_1 > o \tag{I.34}$$

This condition in turn is compatible with (I.20) and $\beta_1 < K$ only when $v_1 S_1 > o$. Hence, from the maximum principle,

$$v_1 = 1 \tag{I.35}$$

and the initial arc is of constant thrust type. This conclusion also permits to write

$$H = (\lambda_\beta)_1 (\beta_1 - K) + S_1 = (\lambda_\beta)_1 (\beta_1 - K) + (1 - \frac{1}{\beta_1}) \lambda_\omega$$

When in this relation we substitute H from (I.I7), $(\lambda_\beta)_1$ from (I.I8) and λ_ω from (I.3I), we find

$$\sigma \frac{\beta_1}{\beta_1 - 1} = e^{-\phi_2} \frac{\beta_2}{\beta_2 - 1} \tag{I.36}$$

This relation completes with $\omega_1 = o$, $\omega_2 = \overline{\omega}_2$, $\phi_1 = o$, the set of boundary conditions needed for the two basic differential equations (I.7) and (I.8). It also shows in conjunction with (I.32) that

$$\beta_1 > 1 \tag{I.37}$$

This inequality was otherwise necessary to obtain a lift-of capability.
The correct (β,S) graph can now be constructed with this information. The sign of λ_ω is known from (I.33).
Further we have

$$1 < \varepsilon < \frac{K+1}{2\sqrt{K}} \tag{I.38}$$

The lower bound is justified by the fact that the trajectory ends with $S_2 = o$ and from (I.I5) this can only occur for $\beta = \theta_1$ or θ_2. Hence the roots cannot be complex conjugate nor negative. The upper bound is justified from the definition (I.I4) of ε and the fact that, according to (I.20) the Hamiltonian is, like λ_ω, positive.
In the range (I.38) of ε values we have

$$1 < \theta_1 < \sqrt{K} < \theta_2 < K \tag{I.39}$$

and the (β,S) graph based on (I.I5) is as depicted on figure 2.

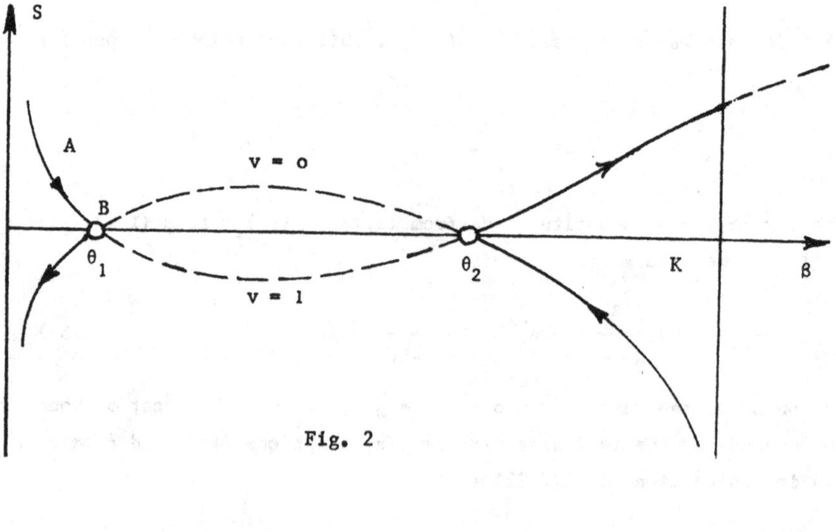

Fig. 2

The part of the branches $v = o$ and $v = 1$ which violate the maximum principle are drawn interrupted. From the sense of description of the branches it is clear that an optimal trajectory can only consist of a single constant thrust arc, like the one represented by the segment AB. The technical variant consisting in adding a constant time arc is possible.
For such a solution it is found that

$$\beta_2 = \beta_1 e^{\phi_2} = \theta_1 < \sqrt{K} \qquad \beta_1 = 1 + \sigma(\beta_2 - 1)$$

$$\omega_2 = \phi_2 + \frac{1}{\beta_2} - \frac{1}{\beta_1}$$

$$J = \frac{\sigma}{\beta_1} + \frac{1}{K} - \frac{1}{\beta_2}$$

The functional is negative (payload positive) if

$$\sigma < \frac{K - \beta_2}{\beta_2(\beta_2 - 1) + K}$$

This solution is optimal until β_2 reaches $\sqrt{K}$ and

$$\omega_2 = \ln \frac{\sqrt{K}}{1+\sigma(\sqrt{K}-1)} + \frac{1}{\sqrt{K}} - \frac{1}{1+\sigma\sqrt{K}}$$

For higher values of ω_2 we must consider the limiting case $\varepsilon = 1$, for which $\theta_1 = \theta_2 = \sqrt{K}$. The corresponding (β, S) graph is shown on figure 3, where it can be seen that the optimal trajectory consists of an initial constant thrust arc until the tangency point where $S = o$ is reached. By switching then to $v = 1 - \frac{1}{\sqrt{K}}$ we stay at the tangency point which represents a constant acceleration arc, until the required velocity is reached.
Again a constant time arc can be added afterwards to separate the payload. In this case we find

$$\frac{1}{\beta_1} = 1 - \sigma(1 - \frac{1}{\sqrt{K}})e^{\phi_2} \qquad \beta_2 = \sqrt{K}$$

$$\omega_2 = \frac{1}{\sqrt{K}}(2 - \frac{1}{\sqrt{K}}) \ln(\frac{\sqrt{K}}{\beta_1}) + \frac{1}{\sqrt{K}} - \frac{1}{\beta_1} + \phi_2(1 - \frac{1}{\sqrt{K}})^2$$

The solution becomes meaningless when the velocity required is so large that the functional J becomes positive.

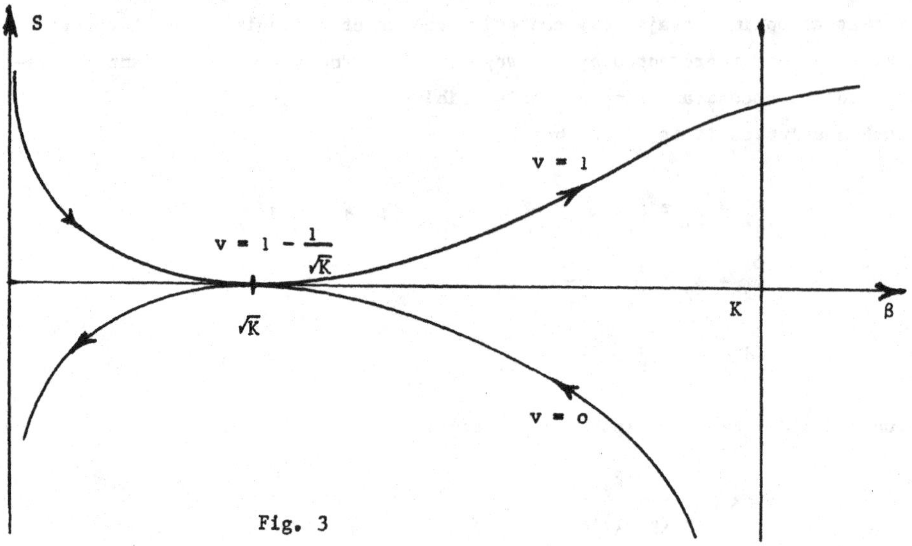

Fig. 3

Figure 4 gives an example of optimal values as functions of the terminal velocity for $\sigma = 0.I$ and $K = 25$.

2. THE MAXIMUM-MINIMUM PRINCIPLE

2.I. Chattering-free solutions of the staging problem.

Any optimal solution involving chattering is of theoretical interest only. In practice it can perhaps be approximated closely if the physical implication of chattering is a high frequency commutation of an electronic relay switch.
In the present case, a too large number of small propulsion units would bring about weight increases and loss of reliability which could only be introduced in the formulation at the price of considerable mathematical complications.
On the other hand, if the number of units into which the propulsion equipment is subdivided has been specified beforehand, physical intuition suggests that an optimal programming for the sequence of separation and the size of separating units must still exist. Such an optimal solution would consist only on

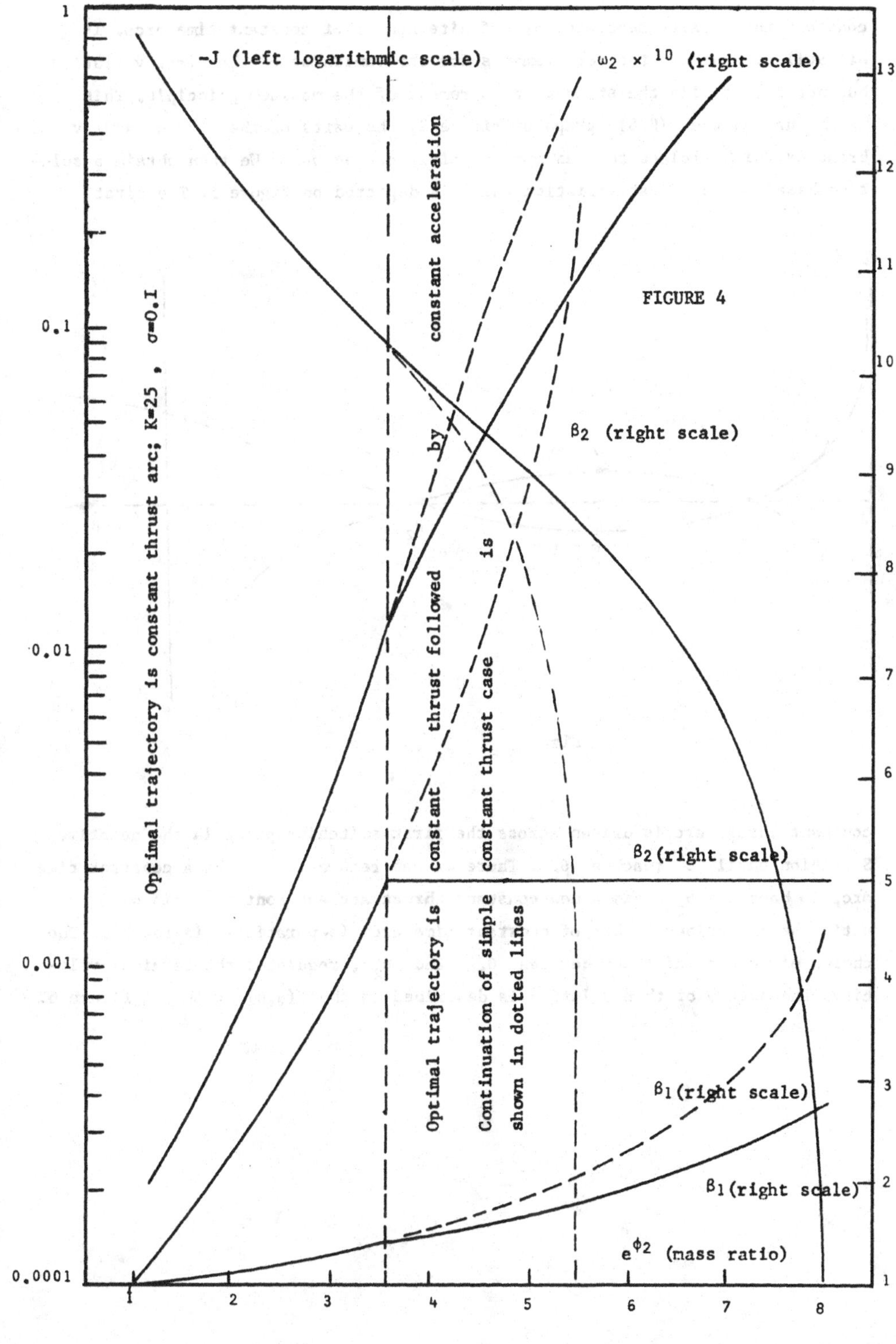
-J (left logarithmic scale)
ω2 × 10 (right scale)
FIGURE 4
β2 (right scale)
β2 (right scale)
β1 (right scale)
β1 (right scale)
eφ2 (mass ratio)
Optimal trajectory is constant thrust arc; K=25 , σ=0.1
Optimal trajectory is constant thrust followed by constant acceleration
Continuation of simple constant thrust case is shown in dotted lines
1
0.1
0.01
0.001
0.0001
1
2
3
4
5
6
7
8
9
10
11
12
13

constant thrust arcs separated by a finite number of constant time arcs. It has still to satisfy the requirements for the vanishing of the first variation but not necessarily the stronger requirement of the maximum principle. This means that on the (β, S) graph of figure 2, the parts of the $v = o$ and $v = 1$ branches which violate the maximum principle can be used. We then obtain a solution based on the first variation which is depicted on figure 5. The first

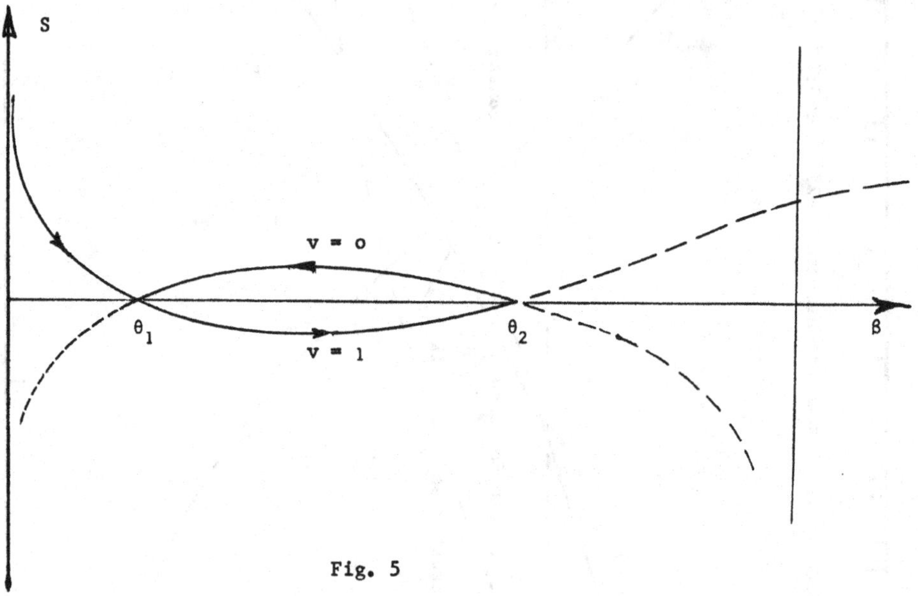

Fig. 5

constant thrust arc is driven across the first switching point in the negative S region until β reaches θ_2. There we can return to θ_1 by a constant time arc, go back to θ_2 via a new constant thrust arc and continue this game until the prescribed number of constant time arcs (separations) is reached. The choice of ε, or distance between θ_1 and θ_2, regulates the terminal velocity. The nature of this solution is described on the (ϕ, β) graph of figure 6.

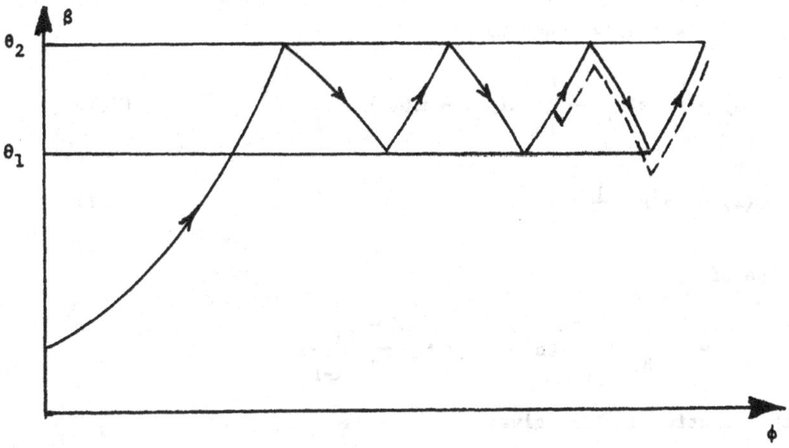

Fig. 6

It will be observed that, up to the first switching point, the choice of controls follows the maximum principle, after that a minimum principle (for the Hamiltonian). The reason for this is also clear when one remembers that the maximum principle follows from the strong variation criterion.
A strong variation applied before the first switching point actually increases the value of the functional. Applied after, as shown in dotted lines on figure 6, it reduces the value of the functional and so theoretically provides a better trajectory. However the number of constant time arcs has been thereby increased by one unit and does no more correspond to the prescribed number.
It is physically obvious that the use of this maximum-minimum principle provides the optimal solution in the case of a prescribed number of constant time arcs.
A mathematical proof is given in the next section.

2.2. The second variation test for the discrete staging problem.
Because all differential equations of this problem have elementary integrals, the optimization of the discrete staging can be reduced to an algebraic problem of constrained minimum. The prescribed number of constant time arcs is denoted by n . The $n+1$ constant thrust arcs have, as yet unknown, initial and final β values denoted by the sequence (β_1,γ_o) , (α_1,γ_1) ... (α_n,γ_n) with $\gamma_n \equiv \beta_2$.

The terminal velocity is easily found to be

$$\omega_2 = \sum_{o}^{n} m(\gamma_i) - \sum_{1}^{n} m(\alpha_i) - m(\beta_1) \tag{2.I}$$

where $$m(x) = \ln x + \frac{1}{x} \tag{2.2}$$

The terminal value of ϕ

$$\phi_2 = \ln \frac{\gamma_o}{\beta_1} + \sum_{1}^{n-1} \ln \frac{\gamma_i}{\alpha_i} + \sum_{1}^{n} \ln \frac{K - \alpha_i}{K - \gamma_{i-1}}$$

inserted into the functional (I.9) gives

$$J = \frac{\sigma}{\beta_1} - \frac{1}{K} L(\alpha, \gamma) \tag{2.3}$$

where $$\ln L = \sum_{1}^{n} p(\alpha_i) - \sum_{o}^{n} p(\gamma_i) \tag{2.4}$$

$$p(x) = \ln \frac{x}{K - x} \tag{2.5}$$

Equating to zero the partial derivatives of the augmented function $f = J + \mu\omega_2$ with respect to the unknowns α_i and γ_i, we find that each of them satisfies the same algebraic equation

$$- \lambda x^2 + x (\lambda + \lambda K + K) - K \lambda = o \tag{2.6}$$

where $$\lambda = \frac{K \mu}{L}$$

Denoting by θ_1 and θ_2 the roots of (2.6) we have

$$K = \theta_1 \theta_2$$

$$\frac{K}{\lambda} = \frac{L}{\mu} = - (\theta_1 - 1)(\theta_2 - 1) \tag{2.7}$$

Since $\gamma_i > \alpha_i$, we must chose

$\alpha_i = \theta_1$ the smallest root

$\gamma_i = \theta_2 = \frac{K}{\theta_1}$ the largest root

and we can compute

$$L = \frac{(\theta_1 - 1)^{n+1}}{(\theta_2 - 1)^n} > o \qquad (2.8)$$

and
$$\mu = - \frac{L}{(\theta_1 - 1)(\theta_2 - 1)} = - \frac{(\theta_1 - 1)^n}{(\theta_2 - 1)^{n+1}} < o \qquad (2.9)$$

The partial derivative of the augmented function with respect to β_1, which plays a special role, gives

$$\frac{\partial f}{\partial \beta_1} = - \frac{\sigma}{\beta_1^2} - \mu\, m'(\beta_1) = o \qquad \text{or}$$

$$\sigma + \mu(\beta_1 - 1) = o \qquad (2.10)$$

and finally, in view of (2.9)

$$\beta_1 = 1 + \sigma \frac{(\theta_2 - 1)^{n+1}}{(\theta_1 - 1)^n} \qquad (2.11)$$

and
$$J = \frac{\sigma}{\beta_1} - \frac{1}{K} \frac{(\theta_1 - 1)^{n+1}}{(\theta_2 - 1)^n} \qquad (2.12)$$

This solution coincides with the one obtained by the maximum-minimum principle. To prove its optimality we apply the second variation test.
The constraint $\delta\omega_2 = o$ on the first variations is

$$\sum_o^n m'(\gamma_i)\delta\gamma_i - \sum_1^n m'(\alpha_i)\delta\alpha_i - m'(\beta_1)\delta\beta_1 = o$$

Because of the first variation conditions $\partial f/\partial x = o$, which require

$$L\, p'(x) + K\, \mu\, m'(x) = o \qquad x = \alpha_i\, ,\ \gamma_i$$

it can be placed in the more convenient form

$$\frac{L}{K} \sum_o^n p'(\gamma_j)\delta\gamma_j - \frac{L}{K} \sum_1^n p'(\alpha_i)\delta\alpha_i - \frac{\sigma}{\beta_1^2} \delta\beta_1 = o \qquad (2.13)$$

The second variation of the augmented function is

$$\delta^2 f = \frac{1}{2}\ \frac{\partial^2 f}{\partial \beta_1^2}\ (\delta\beta_1)^2 - \frac{1}{2} \sum_1^n \left(\frac{L}{K}\ p''(\alpha_i) + \mu\ m''(\alpha_i)\right)\ (\delta\alpha_i)^2$$

$$+ \frac{1}{2} \sum_o^n \left(\frac{L}{K}\ p''(\gamma_i) + \mu\ m''(\gamma_i)\right)\ (\delta\gamma_i)^2$$

$$- \frac{1}{2}\ \frac{L}{K}\ \left(\sum_o^n\ p'(\gamma_j)\delta\gamma_j - \sum_1^n\ p'(\alpha_i)\delta\gamma_i\right)^2$$

Its last term can be simplified in view of (2.I3) and the resulting form of the second variation contains only squares of variations :

$$\delta^2 f = \frac{1}{2}\ B(\delta\beta_1)^2 + \frac{1}{2}\ A \sum_1^n\ (\delta\alpha_i)^2 + \frac{1}{2}\ C \sum_o^n\ (\delta\gamma_i)^2 \qquad (2.I4)$$

where

$$A = -\frac{L}{K}\ p''(\alpha_i) - \mu\ m''(\alpha_i) = -\frac{L}{K}\ p''(\theta_1) - \mu\ m''(\theta_1)$$

which, after computation is seen to be

$$A = \frac{L\ \theta_2(\theta_2 - \theta_1)}{K\ \theta_1^2(\theta_1-1)(\theta_2-1)^2} > o$$

$$C = \frac{L}{K}\ p''(\gamma_i) + \mu\ m''(\gamma_i) = \frac{L}{K}\ p''(\theta_2) + \mu\ m''(\theta_2)$$

or

$$C = \frac{L\ \theta_1(\theta_2 - \theta_1)}{K\ \theta_2^2(\theta_1-1)^2(\theta_2-1)} > o$$

$$B = \frac{\partial^2 f}{\partial\beta_1^2} - \frac{K\ \sigma^2}{L\ \beta_1^4} = \frac{1}{\beta_1^3}\ \left(2\ \sigma - 2\ \mu + \mu\ \beta_1\ \ - \frac{K\ \sigma^2}{L\ \beta_1}\right)$$

To show that this coefficient is also positive we use (2.3) and (2.I0) to find

$$B = \beta_1^{-3}\ \left(-\mu - \frac{\sigma K}{L}\ J\right)$$

But from (2.9) μ is negative, from (2.8) L is positive and, to have physical significance, J must be negative (If it is positive the velocity performance was set too high for a solution with positive payload to exist).
Hence B is also positive and it is clear without further calculations that the constraint (2.13) cannot prevent the second variation (2.14) from being positive definite.
Although it is unnecessary in this case, the constraint can be applied to transform the second variation test in an eigenvalue problem [3,4]. Substracting the eigenvalue parameter ζ to the (diagonal) matrix of second derivatives of f and bordering by the coefficients of the constraint, we obtain the eigenvalues from the determinant

$$\left|\begin{array}{c|ccc|ccc|c}
B-\zeta & & & & & & & m'(\beta_1) \\
\hline
 & A-\zeta & & & & & & m'(\theta_1) \\
 & & \ddots & & & & & \vdots \\
 & & & A-\zeta & & & & m'(\theta_1) \\
\hline
 & & & & C-\zeta & & & -m'(\theta_2) \\
 & & & & & \ddots & & \vdots \\
 & & & & & & C-\zeta & -m'(\theta_2) \\
\hline
m'(\beta_1) & m'(\theta_1) & \cdots & m'(\theta_1) & -m'(\theta_2) & \cdots & -m'(\theta_2) & 0
\end{array}\right| = 0$$

This is equivalent to the equation

$$\frac{(m'(\beta_1))^2}{B-\zeta} + \frac{n\,(m'(\theta_1))^2}{A-\zeta} + \frac{(n+1)\,(m'(\theta_2))^2}{C-\zeta} = 0$$

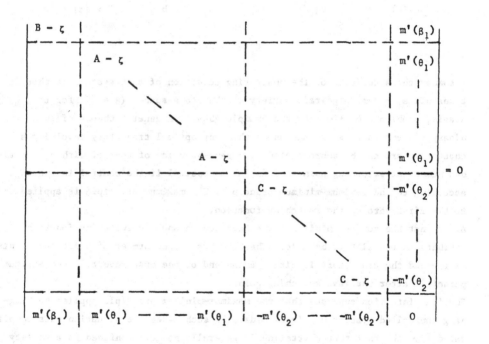

which shows that there are exactly two eigenvalues, one in each of the intervals defined by the numbers (A, B, C) . They are certainly positive if A, B and C are positive, hence the second variation is positive definite.

2.3. The maximum-minimum principle in general.

There are many other binary systems, with a bang-bang control, which, for certain functionals and boundary conditions, exhibit optimal trajectories involving chattering. One of the simplest examples is

$$\dot{q}_1 = u \qquad \dot{q}_2 = q_1^2 \qquad u = \pm 1$$

$$q_1(o) = a \qquad q_2(o) = o \qquad q_1(T) = b \qquad T > |a| + |b|$$

$$q\ (T) \text{ minimum}$$

A common characteristic of the chattering condition of such systems is that it takes place in some algebraic variety of the state space ($\beta = \sqrt{K}$ for the staging problem, $q_1 = o$ for the example above). A general theory of such binary systems is possible and shows that any optimal trajectory involving a chattering arc can be approximated by a bang-bang law of control with a prescribed number of switching points. This law is optimal if the control is chosen according to the maximum-minimum principle. The maximum principle is applied up to the first zero of the switching function.

After that the control minimizes the Hamiltonian and the switching function exhibits an oscillatory behavior. When the prescribed number of switching points is reached the trajectory is either at an end or one must revert to the maximum principle after the last switching point.

Physical intuition suggests that the maximum-minimum principle applies to bang-bang controlled systems of higher order. In such cases the situation gets complicated from the fact that chattering is generally no more confined to a variety of the state space. For instance, if we complicate the rocket staging problem by imposing a terminal total energy (kinetic plus potential) per unit mass instead of a terminal velocity, the differential equation

$$\frac{de}{d\phi} = \omega\ V$$

governing the total energy e must be added to the system.
The condition of persistance of a zero value of the switching function does no more result in some holonomic constraint between state variables but gives directly the control as a function of the state variables :

$$v = f(\beta,\omega)$$

Chattering can in principle take place so long as this function has a value between zero and one.

References :

I. B. Fraeijs de Veubeke, Chapter I2, Rocket Propulsion, Elsevier, I960.
2. ———————— L'étagement optimum des groupes de fusées en fonctionnement parallèle, Astronautica Acta, VII-5-6, I96I.
3. ———————— Une condition suffisante de minimum relatif dans le problème des extrémés liés, Acad. R. de Belgique, Bull. Cl. des Sciences, 5I2-I7, I966-4.
4. ———————— The second variation test with algebraic and differential constraints, Proceedings of the first international colloquium on Optimization Methods, June I967, Liège, Pergamon Press I968, p. I9I-2I9.

WEIGHT LIMITATION INFLUENCE ON OPTIMUM MOTION PARAMETERS OF A BODY OF VARIABLE MASS

G.L. Grodzovsky
and
B.N. Kiforenko

The weight limitation influence on optimum motion parameters of a multistage body of variable mass with the limited exhaust velocity is considered. As well as in the case of the onestage system presented previously, the general variational problem for the multistage system is divided into two aspects: dynamics and weight, along the arcs of a trajectory. The algorithm of solving the problem of optimum motion of a multistage body of variable mass with regard for weight limitations is given. The example of similarity solution on maximum energy storage at the vertical climb in a constant gravitational field are given.

I. The Problem of Motion Optimization of a Multistage Body of Variable Mass with Limited Exhaust Velocity

The variational problem considered is one of the fundamental problems of space flight mechanics the object of which is the combined choice of optimum weight parameters, optimum propulsion system control and all optimum trajectories definition. These aspects were most developed in the mechanics of low-thrust space flight with a sustained propulsion system of limited power the weight of which is comparable to the vehicle take-off weight (see ref. 1 and oth.). In Refs. 2 and 3 dealing with the onestage body of variable mass the sufficient influence of the propulsion system weight (with an account for the design factor) was shown also on the optimum motion parameters of a body of variable mass with propulsion systems of limited jet exhaust velocity and low specific weight. In this paper the results of the investigations are extended to the case of a multistage body of variable mass.

We consider the variational problem of maximum payload delivery G_π for time T between two points of phase space $(\vec{r}_o, \dot{\vec{r}}_o)$ and $(\vec{r}_1, \dot{\vec{r}}_1)$ by an n-stage system with a power limited propulsion system $P_i \leq P_{i\,max}$ and limited exhaust velocity $V_i \leq V_{i\,max}$ ($P_{i\,max}$ and $V_{i\,max}$ are the maximum thrust and exhaust velocity of the i-stage). The stages are characterised by the propulsion system specific weight $\gamma_i = G_{xi}/P_{i\,max}$ and design factor $\beta_i = G_{\beta i}/G_{\mu i}$, where G_{xi} is propulsion system weight, $G_{\beta i}$ is construction weight and $G_{\mu i}$ is fuel initial weight of the i-stage. The system initial weight is given by

$$G_o = G_\pi + \sum_{i=1}^{n} [\gamma_i P_{i\,max} + (1+\beta_i) G_{\mu i}] \qquad (1.1)$$

We write the general equations of the variational problem considered with given specific weights γ_i and β_i as follows:

$$\dot{G} = g\frac{P(t)}{V(t)} \qquad G(0) = G_o \qquad G_\pi = G(T) - \gamma_n P_{n\,max} - \beta_n G_{\mu n} = \max$$

$$\dot{\vec{r}} = \vec{v} \qquad \vec{r}(0) = \vec{r}_o \qquad \vec{r}(T) = \vec{r}_1 \qquad (1.2)$$

$$\dot{\vec{v}} = g\frac{P(t)}{G(t)}\vec{e} + \vec{R} \quad \vec{v}(0) = \vec{v}_o \qquad \vec{v}(T) = \vec{v}_1$$

$$(0 \leq P_i(t) \leq P_{i\,max} \qquad 0 \leq V_i(t) \leq V_{i\,max} \qquad |\vec{e}| \equiv 1)$$

where $\vec{e}$ is a unit vector of thrust direction, $\vec{R}$ is an acceleration vector due to gravitational forces. When $\gamma_i = \beta_i = 0$ we have a known problem of rocket dynamics, that is definition of optimum motion trajectory with an ideal power-limited propulsion system (without taking into account weight factors). For the case of multistage system this problem of rocket dynamics was considered in detail in ref. 4 . To solve the variational problem considered (1.2) conditions obtained in ref. 4 should be completed by the relations for defining optimum values of governing parameters $P_{i\ max}$ and $G_{\mu i}$ which can be obtained by using formal relations in the equations of the problem

$$\dot{P}_{i\ max} = 0 \quad , \quad \dot{G}_{\mu i} = 0 \qquad (1.3)$$

For the case when $n = 1$ an algorithm is given in ref 3 ; this algorithm allows to reduce the variational problem considered to the well known problem of maximum G_π with the fixed $P_{i\ max}$, and then with $P_{i\ max}$ being chosen from the relation

$$G_{\pi\ max} = \max_{P_{i\ max}} \left[G_\pi (P_{i\ max})\right] \qquad (1.4)$$

A similar method can be used for solving the present problem, considering subsequent stages as a payload of the preceding stage and varying stage function parameters.

2. As an example we consider the problem of maximum energy storage E_{max} at the vertical climb in a uniform gravitational field g_0 = const in vacuum. Now write the variational problem equations with $V_{max} = V_{i\ max}$ = const in dimensionless form

$$\dot{G} = -\frac{a_1 P(t)}{V(t)} \qquad G(0) = 1 \qquad G(T) = G_\pi + \gamma_n P_{n\ max} + \beta_n G_{\mu n}$$

$$\dot{v} = \frac{a_1 P(t)}{G(t)} - 1 \qquad v(0) = 0 \qquad v(T) = \text{opt} \qquad (2.1)$$

$$\dot{E} = v\,\frac{a_1 P(t)}{G(t)} \qquad E(0) = 0 \qquad E(T) = \max\ , \quad T = \text{opt}$$

In these equations the current value of weight G is related to initial weight G_o ; the velocity of mass center motion v is related to maximum exhaust velocity V_{max} jet exhaust velocity V is related to V_{max} ; total energy E is related to V^2_{max} ; thrust P is related to $P_{1\ max}$ $(0 \leq P \leq 1)$; $a_1 = P_{1\ max}/G_o$; time t is related to V_{max}/G_o .

We can show, that $\text{opt}\ V = V_{max}$. We also succeed to show that the optimum trajectory does not involve coasting arcs and special control arcs, i.e. $\text{opt}\ P \equiv 1$.

Eqs. (2.1) are completely integrated 5 . Energy increment along the thrust arc of the K-stage is obtained as follows:

$$\Delta E_K = \ell_n \frac{G_K}{G_K - G_{\mu K}} \left(v_K + \frac{1}{2}\ell_n \frac{G_K}{G_K - G_{\mu K}} - \frac{G_K}{a_K}\right) + \frac{G_{\mu K}}{a_K}$$

$$G_K = 1 - \sum_{i=1}^{K-1} \beta a_i + (1+\beta_i) G_{\mu i} \qquad (2.2)$$

$$v_K = \sum_{i=1}^{K-1} \left(\ell_n \frac{G_i}{G_i - G_{\mu i}} - \frac{G_{\mu i}}{a_i}\right)$$

where G_K and v_K are the weight and velocity at the moment of the engine start of the K-stage.

Figs. 1 and 2 show an example of calculation for the case when $n = 4$ with $G_\pi = 0{,}02$ and $\beta = 0{,}005$. A fourstage system is seen to be optimum up to $\gamma \leq 0{,}006$ and then a threestage system becomes optimum. This example illustrates the influence of parameter γ on optimum redistribution of the weight and maximum thrust of stages. In fig. 3 the variation of the maximum available motion energy E_{max} is given as a function of parameter γ for the example considered.

I, Гродзовский Г.Л., Иванов Ю.И., Токарев В.В. Механика космического полета с малой тягой. Изд-во "Наука", М., 1966.

2. Гродзовский Г.Л. Некоторые вариационные задачи механики космического полета. Инж.ж., Механика тв.тела № 5; 1966.

3. Гродзовский Г.Л. Вариационные задачи механики космического полета. Доклад на УШ Международном Симпозиуме по космической науке и технике, Токио, май, 1967.

4. Троицкий В.А. Об оптимальных режимах движения многоступенчатых ракет. Космические исследования, т. У, вып. 2, 1967.

5. Миеле А. Механика полета, т. I, Изд.-во "Наука", М., 1965.

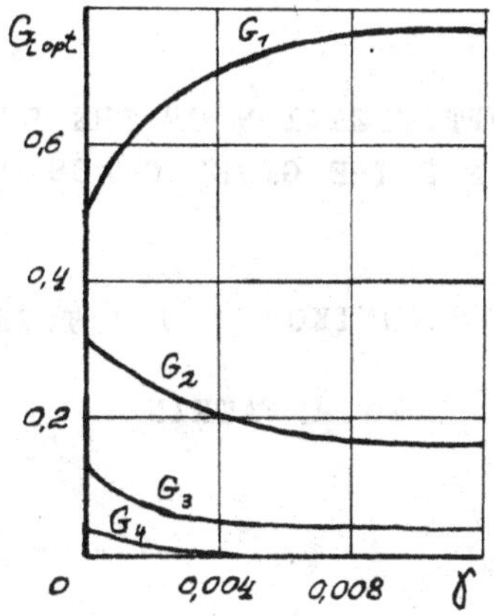

Fig. 1

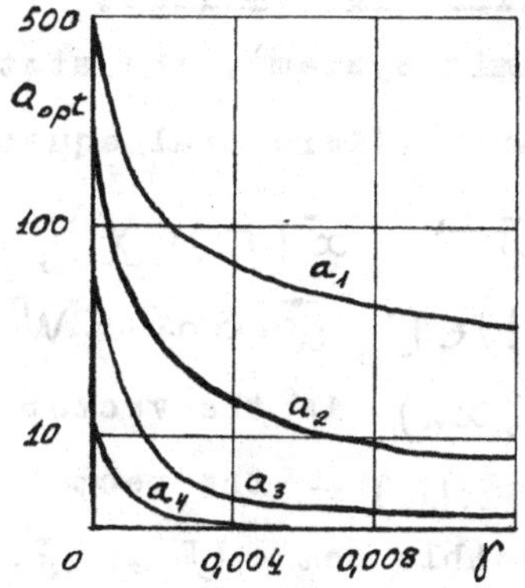

Fig. 2

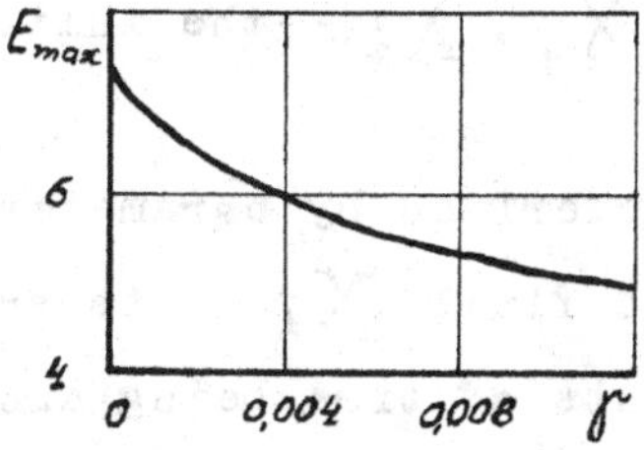

Fig. 3

PARAMETERS OPTIMIZATION OF THE DYNAMIC SYSTEM, UNIVERSAL FOR THE GIVEN CLASS OF MANEUVERS

R.N. OVSYANNIKOV • V.V. TOKAREV

Yu.M. FATKIN

I. Maneuver and system parameters. Let there be a controllable object (a dynamic system), the state of which is described by the ordinary differential equations of the form:

$$\frac{d\vec{x}}{dt} = \vec{f}(\vec{x}, \vec{u}, \vec{w}, t), \quad \vec{x}(t_0) \in X_0, \; \vec{x}(t_1) \in X_1 \qquad (1)$$
$$(\vec{u}(t) \in U, \; \vec{w} = \text{const} \in W),$$

where $\vec{x} = (x_0, x_1, \ldots, x_n)$ — is the vector of the system phase coordinates, $\vec{u} = (u_1, \ldots, u_r)$ — the vector of the control functions from an admissible set U, $\vec{w} = (w_1, \ldots, w_q)$ — the vector of the control parameters from an admissible set W, t — the time and X_0, X_1 — the initial and final sets of system states.

Maneuver is characterized by parameter series defining the initial X_0 and final X_1 sets and by the initial t_0 and final t_1 instants of time being among these series. These qualities in combination are denoted by the vector:

$$\vec{b} = \vec{b}\,(X_0, X_1, t_0, t_1) \qquad (2)$$

Dynamic system is characterized by the series of the constant control parameters values w_ℓ (the limit values of the control functions $U_{k\,min}$, $U_{k\,max}$ may also refer to the latters) and is denoted by the vector

$$\vec{w} = (w_1, w_2, \ldots, w_q) \tag{3}$$

The vectors $\vec{w}$, differing by at least the only component, correspond to physically different dynamic system. Various control functions $\vec{u}(t)$ may be realized in the same system, of course, at invariant values of those limit quantities, which enter the series $\vec{w}$. Adjustment parameters cohditionally, are also involved in the control $\vec{u}$. Those control parameters are meant which can not vary with time in the process of maneuver realization, but can vary before the realization of the next maneuver (system adjustment).

2. The optimal system for a single maneuver. The quality of a single maneuvre realization (the vector $\vec{b}$ is fixed) is characterized by a final phase coordinate value

$$x_0(t_1) = x_{01} = extr \tag{4}$$

(maximum and minimum).

In the routine formulation of the optimization problem for each fixed maneuver $\vec{b}$ it is required to choose such controls $\vec{u}(t) \in U$ and to specify such parameters $\vec{w} \in W$, which would enable to realize conditions (1) and to deliver an extremum to a functional (4). We shall divide this problem into two parts.

The first part deals with dynamics. The parameters of the maneuvre $\vec{b}$ and of the system $\vec{w}$ are specified, and the optimal control $\vec{u}(t)$ is being chosen

$$\underset{\vec{u}(t) \in U}{\operatorname{extr}} x_0(t_1) = x_{01}(\vec{b}, \vec{w}) \qquad \vec{u}(t) = u_{opt}(t, \vec{b}, \vec{w}) \tag{5}$$

The dynamic part of the problem is common for both the routine formulation of the problem and for that of the universal system. After its solution the functional (4) becomes a function of the maneuver and system. The solution (5) will be considered to be established and the function $x_{01}(\vec{b}, \vec{w})$ – to be known in the whole interested range of the parameter values $\vec{b}$ and $\vec{w}$.

The second part deals with choosing of the system parameters. The Maneuver parameters $\vec{b}$ are specified, the solution (5) of the dynamic part of the problem is known, and the optimal system parameters $\vec{w}$ are being defined.

$$\underset{\vec{w} \in W}{\operatorname{extr}} x_{01}(\vec{b}, \vec{w}) = x_{01}(\vec{b}) \qquad \vec{w} = \vec{w}_{opt}(\vec{b}) \tag{6}$$

Here, for every maneuver $\vec{b}$, its own system $\vec{w}_{opt}(\vec{b})$ is being chosen and the functional attaines its highest (lowest) value and depends only on the maneuver parameters ($\vec{b}$). We shall call the solution (6) an ideal one and thereafter it is assumed to be known.

3. Universal system requirement. Let now assume that not one but a series of maneuvres are to be accomplished. The maneuver parameters $\vec{b}$ may take different values with different frequency within a fixed set **B**. A normed frequency $\nu(B')$ of the maneuver frequency $\vec{b}$ within any subset **B'⊂B** is introduced, i.e. the probability of the maneuvres, belonging to **B'** (probability measure):

$$0 \leq \nu(B') = P\left\{\vec{b} \in B' \subset B\right\} \leq 1 \qquad (7)$$

Both continuous and discrete distributions are admitted here. In the first case only sets of full dimensionality (coinciding with the dimensionality of the vector $\vec{b}$) have nonzero frequency, but in the second case there exist sets of dimensionality, smaller than the dimensionality $\vec{b}$ but having nonzero frequency.

If, in such a situation, we proceed to follow the ideal solution (6), then it will be required to create a large number of non-identical systems (from "the number" of different values of the maneuver parameters vector $\vec{b}$). This may appear to be unprofitable from an economical point of view because of the additional design and manufacture expenditures. To reduce these expenditures the system universality requirement is set forward. It is formulated in two versions.

<u>The first version.</u> The Ω - number of non-identical systems, less than the maneuvres number, is specified. It means that the system parameters vector $\vec{w}$, for all values of the maneuver-parameters vector $\vec{b} \in B$ may admit only Ω different values:

$$\vec{w} = \vec{w}^{(\omega)} \qquad (\omega = 1, 2, \ldots, \Omega \ ; \ \vec{w}^{(\omega)} = const) \tag{8}$$

One may dispose of the series $\{\vec{w}^{(1)}, \ldots, \vec{w}^{(\Omega)}\}$ and, for every maneuvre $\vec{b}$, choose the most suitable value $\vec{w}^{(\omega)}$ of this series.

The second version is possible, when the system is assembled of elementary sections. The parameters $\vec{w}$ of such an assembled system are defined by the parameters $\Delta\vec{w}$ of the elementary section and by the sections number n

$$\vec{w} = (n, \Delta\vec{w}) \tag{9}$$

So, for the components w_ℓ of the vector $\vec{w}$, which are the additive quantities (geometrical dimensions, mass and so on) this relation is linear: $w_\ell = n \Delta w_\ell$. The components w_q, which are the physical parameters of the processes (temperature, electrical field strength and so on), do not depend on the sections number: $w_q = \Delta w_q$

Other relations are possible.

The parameters $\Delta \vec{w}$ of the elementary section must be the same for all the maneuvers $\vec{b} \in B$ (the universal modulus). But instead, for every maneuver $\vec{b}$ one may choose his own number of the sections n.

In both versions of the universality problem formulation for every maneuver, its own controls $\vec{u}(t)$ are assumed to be chosen (of course, within the limitations, imposed by specifying the vector $\vec{w}$). This problem is the dynamic part (1.5) of the optimization problem, which is assumed to be solved, as it has already been pointed out. The formulated problems refer to the operations research to the aspect of so-called combined operations(1).

4. Optimality criteria for the universal system. As compared with the ideal solution (6) the universal system, will provide the worse value of the functional (5) for every single maneuvre. The exclusion may be only those maneuvers, for which the universal system parameters will appear to be optimum in the sense of the criterion (6).

For a single maneuvre the loss in the functional when using the universal system is measured by the difference between the best value (6) and the value (5), provided by the universal system

$$\Delta x_{01}(\vec{b},\vec{w}) = x_{01}(\vec{b}) - x_{01}(\vec{b},\vec{w}) \quad \begin{cases} \geq 0 & x_{01}=max \\ \leq 0 & x_{01}=min \end{cases} \qquad (10)$$

Instead of the absolute difference (10) the relative one may be used.

$$\frac{\Delta x_{01}(\vec{b},\vec{w})}{x_{01}(\vec{b})}=\frac{x_{01}(\vec{b})-x_{01}(\vec{b},\vec{w})}{x_{01}(\vec{b})} \tag{11}$$

The total loss in the functional, for the whole set of the maneuvres B, repeating with the frequency (7), must be characterized in different manners in dependence on the physical essence of the problem. However we may remain on common positions, if we use the concept of closeness of two elements $x_{01}(\vec{b})$ and $x_{01}(\vec{b},\vec{w})$ of the functional space. The functions are determined within the set B of the vector argument $\vec{b}$, the maneuver frequency (7) specifies the measure in the arguments space.

The closeness of the elements is defined through the difference norm (10)

$$\|\Delta x_{01}\|=\|x_{01}(\vec{b})-x_{01}(\vec{b},\vec{w})\| \tag{12}$$

or (11)

$$\left\|\frac{\Delta x_{01}}{x_{01}}\right\|=\left\|1-\frac{x_{01}(\vec{b},\vec{w})}{x_{01}(\vec{b})}\right\| \tag{13}$$

Further three types of norms will be used: the norm L_p, the norm C and the norm M.

According to the norm L_p the eloseness to the ideal solution is defined by the Stieltjes integral of the functionals differences (10) and (11) for all the maneuvres from **B** with measure (7)

$$\|\Delta x_{01}\|_{L_p} = \Big[\int_{\vec{b}\in B} | x_{01}(\vec{b}) - x_{01} \ (\vec{b},\vec{w})|^p \nu\,(d\vec{b}) \Big]^{1/p}, \qquad (14)$$

$$\Big\| \frac{\Delta x_{01}}{x_{01}} \Big\|_{L_p} = \Big[\int_{\vec{b}\in B} \Big| 1 - \frac{x_{01}(\vec{b},\vec{w})}{x_{01}(\vec{b})} \Big|^p \nu\,(d\vec{b}) \Big]^{1/p}$$

At $p = 1$ the integrals (14) give mathematical expectation (mean values) of the functional deviation from the ideal solution.

In an attempt to reach closeness in the integral sense (14) at low values p the case may be encounted, when the functional values $x_{01}(\vec{b},\vec{w})$ for the individual maneuvres become too poor. Such a situation seems to be the most probable for the low-frequency maneuvers. In order to provide the reasonable level of the functional values for the every single maneuver an additional condition of the type of the inequality may be introduced into the functional (5)

$$\varphi\,(x_{01}(\vec{b},\vec{w})) \geq 0 \qquad (15)$$

where $\vec{b}$ takes all the values from the domain **B**, excluding zero frequency subdomains (the measure set zero).

According to the norm C closeness to the ideal solution is defined by the exact upper boundary of the modulus of the differences (10) and (11) for all the maneuvers of the admissible domain

$$\|\Delta x_{01}\|_C = \sup_{\vec{b}\in B} |x_{01}(\vec{b}) - x_{01}(\vec{b},\vec{w})| ,$$

$$\left\|\frac{\Delta x_{01}}{x_{01}}\right\|_C = \sup_{\vec{b}\in B} \left|1 - \frac{x_{01}(\vec{b},\vec{w})}{x_{01}(\vec{b})}\right| \tag{16}$$

The distance in the space C does not depend on the maneuver frequency distribution (7) and may be used for upper estimation of the distance in the space L_p .

$$\|\Delta x_{01}\|_{L_p} \leq \|\Delta x_{01}\|_C , \qquad \left\|\frac{\Delta x_{01}}{x_{01}}\right\|_{L_p} \leq \left\|\frac{\Delta x_{01}}{x_{01}}\right\|_C \tag{17}$$

(the parameters $\vec{w}$ in the left and right sides of the inequalities are the same). In addition, the distance in the space C gives the estimation of the functional for every single maneuver.

If there are maximum deviations 16 from the ideal solution for the maneuver of the zero frequency domain (measure set zero), then the estimations (17) may prove to be too rough. They may be improved by discarding of all the measure set zero B $(\nu(B_k)=0)$ from the set B, when seeking the exact upper boundaries (16). This will be the norm M

$$\|\Delta x_{01}\|_M = \inf_{B_k \subset B} \left[\sup_{\vec{b}\in B\setminus B_k} |x_{01}(\vec{b}) - x_{01}(\vec{b},\vec{w})|\right]$$

$$\left\|\frac{\Delta x_{01}}{x_{01}}\right\|_M = \inf_{B_k \subset B} \left[\sup_{\vec{b}\in B\setminus B_k} \left|1 - \frac{x_{01}(\vec{b},\vec{w})}{x_{01}(\vec{b})}\right|\right] \quad (\nu(B_k)=0) \qquad (18)$$

The norms L_p , M and C are related by the following inequality chain, which stems from their determination

$$\|\Delta x_{01}\|_{L_p} \leq \|\Delta x_{01}\|_M \leq \|\Delta x_{01}\|_C ,$$

$$\left\|\frac{\Delta x_{01}}{x_{01}}\right\|_{L_p} \leq \left\|\frac{\Delta x_{01}}{x_{01}}\right\|_M \leq \left\|\frac{\Delta x_{01}}{x_{01}}\right\|_C . \qquad (19)$$

Besides, the norm M is limit for the norms L_p :

$$\|\Delta x_{01}\|_M = \lim_{p\to+\infty} \|\Delta x_{01}\|_{L_p} , \quad \left\|\frac{\Delta x_{01}}{x_{01}}\right\|_M = \lim_{p\to+\infty} \left\|\frac{\Delta x_{01}}{x_{01}}\right\|_{L_p} \qquad (20)$$

The norm M may estimate the functional more accurately than the norm C for every single maneuver from the nonzero measure domain. The norm M is less universal, than the norm C with regard to the maneuver frequency distribution (7), but more universal than L_p . The universality of the norm C with regard to any maneuver frequency distribution allows to use it as an optimization criterion of the universal system in the game technique. In this case the information of the maneuver frequency is not available, and it must be specified in a werse from in a definite sense, with similtaneous choosing the optimal (best) parameters of the dynamic system. The game problem, formulated in such a manner, is solved in reference [2]. The solution of the continuous (infinite) game is obtained in a final form. For the case when such a solution may not be obtained, one may use the iteration method of solving the discrete game, suggested by Neumann [3] and generalized to the infinite games. The examples of the problem solution with a given maneuver frequency distribution are presented in references [4] and [5].

R E F E R E N C E S

1. Germeyer Yu. B. "Methodological and Mathematical Fundamentals of the Operations Research and Theory of Games" MGU Vitsh. Tsentr Acad. Nauk SSSR 1967.

2. Tokarev V.V., Fatkin Yu.M. "A Similarity Approach to the Problem of Choosing the Optimum Parameters for Dynamic System" Inzh. Zh. MTT no 6 1966.

3. Von Neumann, John. "A Numerical Method to Determine Optimum Strategy" Naval Res. Log. Q. 1954 v. 1, p. 109-115.

4. Tokarev V.V. "On Choosing the Parameters for Dynamic System, Universal for a Given Class of Maneuvers" Isv. Acad. Nauk SSSR Mekh. i Mashin. no 5, 1964; Proceedings of XVIth International Astronautics Congress, Athens 1965.

5. Ovsyannikov R.N., Shumilkin V.G. "Some Questions of Universality of the Parameters of the Propulsion System of Limited Power with a Propellant Accumulator" Inzh. Zh. MTT no 4, 1968 (to be published).

AN INVESTIGATION OF OPTIMAL ROCKET FLIGHT IN THE VICINITY OF A PLANET

V.K. Isaev • B.Kh. Davidson
V.V. Sonin

Summary

Some problems of optimal thrust magnitude and thrust direction control for a rocket vehicle maneuver in a vacuum in the vicinity of a spherical planet are considered. The paper consists of three sections.

In Section I the problem of optimum thrust magnitude and thrust direction programming for a rocket travelling in a central gravitational field is investigated. It is shown that for the optimal motion in a rather near vicinity of a planet there exists the control plane, i.e., the optimum thrust vector lies in a plane with constant orientation in the inertial (Galilei) reference system. In other words, the existence of an invariant direction with a zero optimal thrust vector projection on it is proved.

In Section II a family of approximate analytical solutions for the problem of optimal transfer between near circular orbits with low thrust engine and ideal exhaust velocity control is studied. Some results on transfer control synthesis (with full information) are given.

In Section III a numerical analysis of a vehicle optimal landing problem (in the sense of propellant expenditure) at a predicted impact-point on the Lunar surface is made. Planar descent trajectories of a vehicle from low-altitude selenocentric orbits in a central gravitational field are studied. The optimal thrust direction and thrust magnitude programs are found. The influence of some parameters (thrust-weight ratio, the altitude of the initial satellite orbit, range angle of a descent) on mass expenditure is considered. A similar problem is solved for the putting of Lunar vehicle into a Moon satellite orbit.

Sections I and III are written by V.K. Isaev and B. Kh. Davidson, Section II is written by V.K. Isaev and V.V. Sonin.

Introduction

The object of this paper is to study optimal control properties for a variable mass point maneuver in a vacuum in a central gravitational field. These problems are solved by using the Pontryagin maximum principle. In the first section the optimal attitude control structure in a central gravitational field is considered. The existence of an analytical solution for the second problem (transfer of the exhaust velocity-unlimited vehicle between near circular orbits) makes it possible to obtain optimal control synthesis. The comparison with an exact solution shows an applicability range of approximately one. In the third section the numerical solution of the variational problem connected with soft landing of a thrust-limited rocket vehicle from the Moon satellite orbit is presented.

These common problems are characterized by one feature: optimal trajectories lie in some near vicinity of a spherical planet (in the third problem this assumption bounds only the hight of the orbits under consideration).

I. On one feature of the optimal attitude program for the three-dimensional motion of a variable mass point in a central gravitational field

Let us consider the optimal three-dimensional motion of a variable mass point in a central gravitational field in a vacuum. The equations of motion in the Cartesian coordinate system with origin at the centre of attraction are

$$\dot{u} = \frac{P}{m} \cos \vartheta \cos \Phi - \frac{\mu}{R^3} x ,$$

$$\dot{v} = \frac{P}{m} \sin \vartheta \qquad - \frac{\mu}{R^3} y ,$$

$$\dot{w} = \frac{P}{m} \cos \vartheta \sin \Phi - \frac{\mu}{R^3} z , \qquad (1.1)$$

$$\dot{x} = u, \quad \dot{y} = v, \quad \dot{z} = w,$$

$$\dot{m} = - \frac{P}{c}$$

Here

u,v,w - velocity vector components,

x,y,z - point coordinates,

R - $(x^2 + y^2 + z^2)^{1/2}$,

m - point mass, $m(t) = \frac{M(t)}{M(0)}$

μ - planet gravitational constant,

P - engine thrust referred to $M(0)$,

c - jet exhaust velocity,

ϑ, Φ - thrust orientation angles.

Control functions subject to optimization are P, c, ϑ and Φ. The structure of optimal control ϑ and Φ does not depend on the type of P and c programs, optimum or given a priori. Further, for certainty a case $c = const$ is considered, P can be varied from zero to P_{max}.

By using the Pontryagin maximum principle, Refs 1,2, optimal control functions are obtained from condition of minimum of the Hamiltonian $\mathfrak{H}$.

$$P = \begin{cases} P_{max}, & \Theta > 0 , \\ 0 , & \Theta < 0 , \end{cases} \qquad (1.2)$$

$$\left. \begin{aligned} \sin \vartheta &= - \frac{p_v}{\rho} , \quad \cos \vartheta = \frac{r}{\rho} , \\ \sin \Phi &= - \frac{p_w}{r} , \quad \cos \Phi = - \frac{p_u}{r} , \end{aligned} \right\} \qquad (1.3)$$

$$\Theta = \rho + \frac{m p_m}{c} , \quad \rho = (p_u^2 + p_v^2 + p_w^2)^{1/2} , \quad r = (p_u^2 + p_w^2)^{1/2} .$$

From Eq. (1.3) follows that the thrust vector P is colinear to the vector with components $-p_u, -p_v, -p_w$ (Fig. 1), i.e., to the radius-vector of $\mathfrak{p}$-trajectory, Ref. 3. The properties of optimal attitude control program are completely determined by the type of p-trajectory.

Adjoint variables p_i are described by the following system of equations:

$$\dot{p}_u = -p_x, \quad \dot{p}_v = -p_y, \quad \dot{p}_w = -p_z,$$

$$\dot{p}_x = \frac{\mu}{R^3}\left[p_u - \frac{3x}{R^2}(xp_u + yp_v + zp_w)\right],$$

$$\dot{p}_y = \frac{\mu}{R^3}\left[p_v - \frac{3y}{R^2}(xp_u + yp_v + zp_w)\right], \tag{1.4}$$

$$\dot{p}_z = \frac{\mu}{R^3}\left[p_w - \frac{3z}{R^2}(xp_u + yp_v + zp_w)\right],$$

$$\dot{p}_m = -\frac{P}{m^2}\rho .$$

The necessary conditions of control optimality (1.2) and (1.3) should be taken into account in Eqs. (1.1) and (1.4).

The right hands of Eqs. (1.4) depend on state variables explicitly, so (1.4) cannot be integrated separately. Due to this fact it is impossible to investigate the optimal control structure for a general case of motion in a central field. Ref. 4 presents a uniform-spherical model of the gravitational field which describes approximately this field in a certain spherical layer, the depth of which is rather small compared to the mean distance R_m from the centre of attraction.

Gravitational acceleration components in the uniform-spherical field are assumed to be linear functions of the coordinates:

$$g_x = -\nu^2 x, \quad g_y = -\nu^2 y, \quad g_z = -\nu^2 z, \tag{1.5}$$

R_m is the radius of a circular orbit in the neighborhood of which motion takes place, $\nu^2 = \frac{g_m}{R_m}$.

In this case the adjoint system has a simplest form

$$\begin{aligned} \dot{p}_u &= -p_x, & \dot{p}_x &= \nu^2 p_u, \\ \dot{p}_v &= -p_y, & \dot{p}_y &= \nu^2 p_v, \\ \dot{p}_w &= -p_z, & \dot{p}_z &= \nu^2 p_w, \quad \dot{p}_m = -\frac{P}{m^2}\rho \end{aligned} \tag{1.6}$$

The first six equations of this system which completely determine the optimum attitude control properties do not depend on the state variables. As a result of the integration we have

$$p_u = p_u^o \cos \nu t - \frac{p_x^o}{\nu} \sin \nu t ,$$

$$p_v = p_v^o \cos \nu t - \frac{p_y^o}{\nu} \sin \nu t , \qquad (1.7)$$

$$p_w = p_w^o \cos \nu t - \frac{p_z^o}{\nu} \sin \nu t ,$$

Primer-vector components change according to a sine-law with the same period $T_m = \frac{2\pi}{\nu}$. Therefore the $\mathfrak{p}$-trajectory ($\mathfrak{p}$-hodograph) is an ellipse, the plane of which passes through the origin of coordinates and has fixed orientation relative to the reference system, Fig. 2. The orientation angles of the ellipse plane in Oxyz Cartesian frame are determined by the initial values of adjoint variables. The latter ones are found from the solution of a two-point boundary-value problem for Eqs. (1.1), (1.6), where P, θ, Φ are prescribed by the Pontryagin maximum principle, Eqs. (1.2), (1.3).

Thus for the case of optimal variable mass point motion in a rather narrow spherical layer of the central field, both the $\mathfrak{p}$-vector and P (thrust vector) belong to a plane with fixed orientation in the (Oxyz) inertial system. This plane moving progressively together with a variable mass point is the so-called control plane, Fig. 3.

II. The optimal control synthesis for transfer between near circular orbits.

In the analysis of the interplanetary transfer problem the so-called transport system of coordinates was used, Ref. 5 . The principle of this system is as follows: equations of motion and boundary conditions are given in some moving system which is chosen in such a way that a linear approximation to the solution of the optimal transfer problem may be easily found. With a proper choice of the transport coordinate

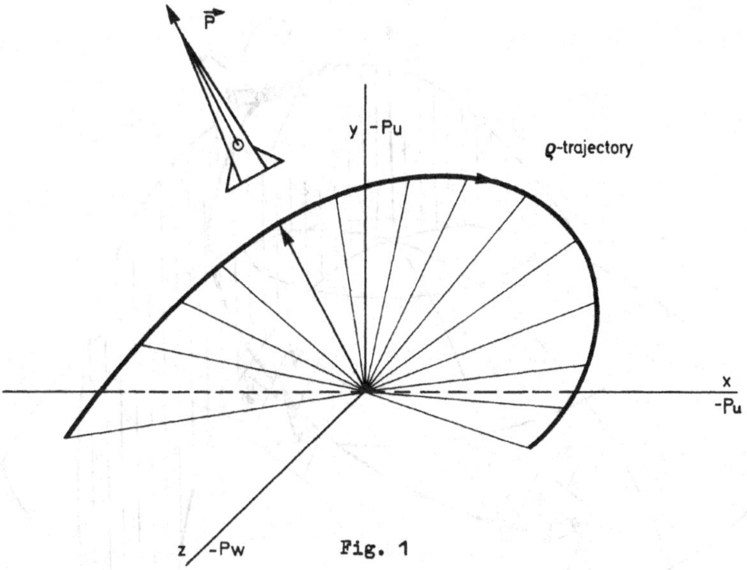
P
y -Pu
ρ-trajectory
x
-Pu
z -Pw

Fig. 1

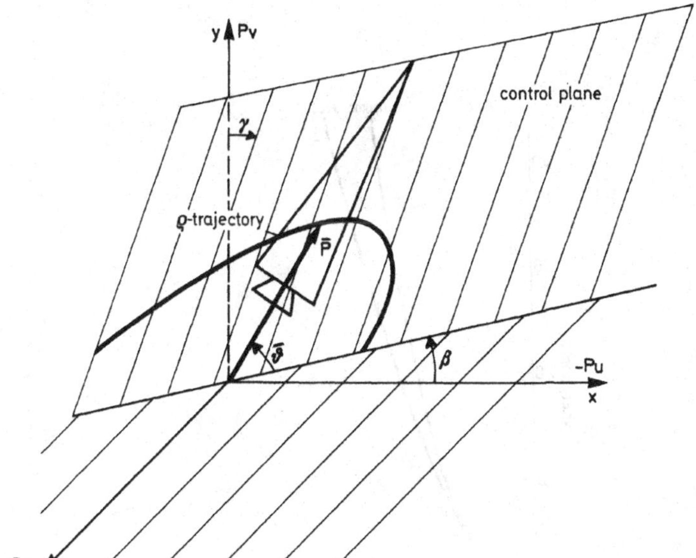
y Pv
control plane
γ
ρ-trajectory
P
ϑ
β
-Pu
x
-Pw z

Fig. 2

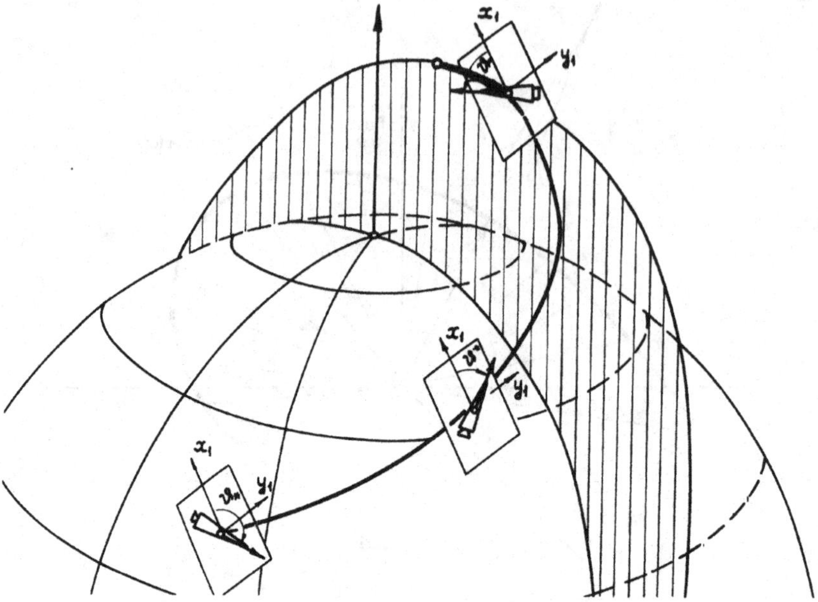

x_1
y_1
x_1
y_1
x_1
y_1

Fig. 3

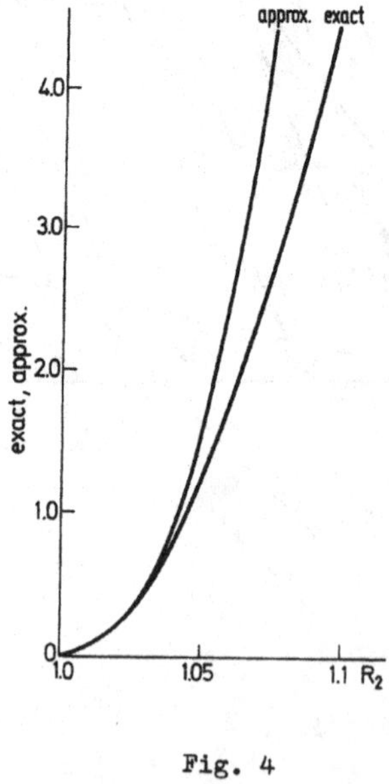

approx.
exact
4.0
3.0
2.0
1.0
0
exact, approx.
1.0
1.05
1.1
R_2

Fig. 4

system the solution of the problem even in linear approximation may be of sufficient accuracy for practical purposes.

In Ref. 5 a trihedron of a Keplerian orbit was chosen as a transport coordinate system for the problem of optimal transfers between Earth and Mars orbits. The origin of the trihedron at the launching time coincided with the Earth and at the end point with the Mars projection into an ecliptic plane.

In this section two-dimensional motion between near circular orbits is studied by introducing the Oxy-transport system, the origin of which moves along a circular orbit. The Ox-axis is directed along the velocity vector and the Oy-axis along the radius-vector. Let the mass $M(0) = M_o$, coordinates and velocities of a vehicle be given at the initial moment of time $t = 0$ in the Oxy-transport system

$$u(0) = u^o , \quad v(0) = v^o , \quad x(0) = x^o , \quad y(0) = y^o \tag{2.1}$$

with correspondence of the vehicle movement along the initial circular orbit. It is necessary to determine the thrust magnitude and thrust direction programs for the minimum mass expenditure transfer with given time T into a position moving along a near circular orbit and having the coordinates

$$u(T) = u^1, \quad v(T) = v^1, \quad x(T) = x^1, \quad y(T) = y^1 \tag{2.2}$$

We assume that during motion the vehicle deviation from the origin of the transport coordinate system is small compared with the radius of the transport orbit R_o . Then neglecting the terms of the order $\frac{x^2 + y^2}{R_o^2}$, the equations of the vehicle motion can be written down as:

$$\begin{aligned} \dot{u} &= \frac{\tilde{N}u_1}{mc} \cos\varphi - 2\omega v , \\ \dot{v} &= \frac{\tilde{N}u_1}{mc} \sin\varphi + 2\omega u + 3\omega^2 y , \\ \dot{x} &= u , \quad \dot{y} = v , \quad \dot{m} = - \frac{\tilde{N}u_1}{c^2} \end{aligned} \tag{2.3}$$

(see Refs. 6, 7), where u, v are velocity projections on the axes of the transport coordinate system, $u_1 = \frac{N}{N_{max}}$ is dimensionless power, c - variable exhaust velocity, φ - thrust inclination angle to the Ox-axis, $\widetilde{N} = \frac{2N_{max}}{M(0)}$ and $m(t) = \frac{M(t)}{M(0)}$ - dimensionless mass of a vehicle, where

$$m(0) = 1 .$$

Let us consider the optimal control of a variable mass point, the motion of which is described by a system of equations (2.3) and must satisfy boundary conditions, Eqs. (2.2) and (2.4), dimensionless final mass $m(T) = m^1$ being a maximized functional, and jet exhaust velocity may vary from 0 to infinity

$$0 < c < \infty .$$

Then the optimal control is defined by the following relations (see Refs. 7,8)

$$u_1 \equiv 1 , \quad \cos\varphi = -\frac{p_u}{\rho} , \quad \sin\varphi = -\frac{p_v}{\rho} , \tag{2.4}$$

$$c = -\frac{2mp_m}{\rho} ,$$

where $\rho = (p_u^2 + p_v^2)^{1/2}$ and $p = (p_u, p_v, p_x, p_y, p_z)$ is adjoint vector with components satisfying the following system of differential equations:

$$\left.\begin{aligned} &\dot{p}_u = - p_x - 2\omega p_v , \\ &\dot{p}_v = - p_y + 2\omega p_u , \\ &\dot{p}_x = 0 , \quad \dot{p}_y = - 3\omega^2 p_v , \quad \dot{p}_m = \frac{\widetilde{N}\rho^2}{2m^3 p_m} \end{aligned}\right\} \tag{2.5}$$

$$p_m(T) = - 1 . \tag{2.6}$$

For the given boundary conditions in Ref. 7 the formulas are obtained which allow to find $c(t)$ and $\varphi(t)$-optimal programs and to solve the boundary-value problem. We present them here.

At first from the system of linear equations:

$$\|A\| C = \widetilde{B} \tag{2.7}$$

we find a fourdimendional vector C . $\|A\|$-matrix and $\widetilde{B}$-vector elements are computed from formulas given in Appendix I.

After determination of C_i , $i = 1,\ldots,4$, from Eqs. (2.7) we find, assuming $t = T$, the value of $I(T)$ from the following formula

$$I(t) = \frac{3(C_1^2 - C_2^2)}{2\omega} \sin \omega t \cos \omega t + \frac{3C_1C_2}{2\omega} \cos 2\omega t + \frac{4}{\omega}(C_1C_3 - \frac{4C_2C_4}{\omega}) \sin \omega t$$
$$+ \frac{4}{\omega}(C_2C_3 + \frac{4C_1C_4}{\omega}) \cos \omega t + \frac{12C_4}{\omega}(C_1 \sin \omega t + C_2 \cos \omega t)t +$$
$$+ [\frac{5}{2}(C_1^2 + C_2^2) + C_3^2 + \frac{4C_4^2}{\omega^2}]t + 3C_4(C_3 + C_4t)t^2 - \frac{3C_1C_2}{2\omega} - \frac{4}{\omega}(C_2C_3 + \frac{4C_1C_4}{\omega}) \quad (2.8)$$

Further we define constants, $C_i^* = \frac{Ci}{\omega}$ (i=1,..., 4) see Ref. 8,

$$m(T) = \frac{1}{1 + \frac{I(T)}{N}} ,$$

$$\nu_1 = - \frac{\tilde{N}}{2m^2(T)} , \qquad \beta = - m^2(T) ,$$

in terms of which the solutions of equations (2.3), (2.5) are explicitly expressed. Corresponding formulas are given in Appendix II.

Using the relations obtained, it is not difficult to find explicit control programs $\varphi(t)$ and $c(t)$ for each particular mission.

This solution was used for computation of transfers between an orbit with a radius equal to 1.0 and near circular orbits. The time of transfer is prescribed to be equal to a half of the orbital period along an original orbit and the transfer angle in intertial coordinates is 180°. The origin of a transport coordinate system is assumed to coincide with the impact point. In fig. 4 the functional is plotted against the radius of the final orbit. For comparison the exact solution is given. It was obtained by solving the boundary problem for an exact set of equations which described the optimum motion in a central field. The correspondence is fairly good up to $R^1 = 1{,}05$.

III. On optimal rocket landing on the Moon and its optimal putting into a lunar satellite orbit

In this section the exact solutions of variational problems of the rocket vehicle landing on the Moon, its takeoff from the Moon surface and putting into a lunar satellite orbit are examined. The following general assumptions were made:

1. The Moon's gravitational field is central and a Newtonian one.
2. Lunar satellite orbits are circular.
3. The thrust magnitude and thrust direction control is inertialess; the thrust magnitude is limited:

$$0 \leq P \leq P_{max}$$

4. The exhaust velocity is constant.
5. Launching and landing trajectories are planar.

Launching and landing maneuvers are shown in Fig. 5. Here Oxy is the inertial Cartesian coordinate system with its origin at the Moon's centre, the Oy-axis passes through a point of the landing (launch). The final mass value m^1 is the functional to be maximized. The range angle ψ_n in the landing arc (ψ_B respectively) is an important parameter.

The optimal motion is governed by the sets of equations (1.1) for u,v,x,y,m coordinates, (1.4) for adjoint variables and the necessary conditions of control optimality (1.2), where $\Phi \triangleq 0$, $p_w = p_z \triangleq 0$.

The boundary conditions for the landing problems are as follows:

$$\begin{aligned}
&\underline{t = 0:}\\
&H = H_o, \quad L = L = R_M \psi_n, \quad V_\tau = V_{kp}(H_o),\\
&V_r = 0, \quad m = 1.\\
&\underline{t = T_n:} \qquad\qquad\qquad\qquad\qquad\qquad (3.1)\\
&H = 0, \quad L = 0, \quad V_\tau = u^1 = 0, \quad V_r = v^1 = 0, \quad p_m^1 = -1.
\end{aligned}$$

Similarly, for the launching

$$\underline{t = 0 :}$$

$$H = 0\ ,\quad L = 0\ ,\quad V_\tau = 0\ ,\quad V_r = 0\ ,\quad m = 1\ . \tag{3.2}$$

$$\underline{t = T_B :}$$

$$H = H^1\ ,\quad L = L^1 = R_M\psi_B\ ,\quad V_\tau = V_{kp}(H^1)\ ,\quad V_r = 0\ ,\quad p_m = -1\ ,$$

where $R = R_M + H$, H are the orbit radius and altitude, R_M is the Moon's radius, L is the selenocentric range (over the Moon surface), V_r and V_τ are radial and transversal velocity components.

While solving the boundary-value problems (Eqs. (1.1) to (1.4)) with the boundary condition (3.1) (or (3.2) respectively) we find the values p_u^o, p_v^o, p_x^o, p_y^o, p_m^o unknown at the initial instant of time $t = 0$. Taking into consideration the integral $\mathfrak{H} = 0$, the free time T_n, T_B is determined and a number of unknown parameters is decreased up to four.

The boundary-value problem was solved by the modified Newton method, Ref. 9, while optimal control, i.e., the thrust magnitude and thrust orientation programs, and corresponding optimal descent (launching) trajectories being automatically determined.

It is convenient to start the presentation of calculation results from description of $\mathfrak{p}$-trajectories, or the curves $\mathfrak{p} = -(\bar{i}p_u(t) + \bar{j}p_v(t))$.

It is seen in Fig. 6 that for the value $\psi_n = 100^o$ the $\mathfrak{p}$-trajectory is similar to a circular arc, for $\psi_n = 60^o$, to an ellipse arc, for $\psi_n = 10^o$, to a portion of a straight line (or an ellipse arc with a large excentricity) and this is in agreement with the theory of the uniform-spherical, Ref. 4, and the constant fields, Ref. 3 (see also Section I).

In the process of the numerical calculations the altitudes of the orbit and the range angles ψ_n, ψ_B were varied within the diapason from 15 km to 200 km and from 10° to 180° respectively. In this region optimal trajectories consist of three arcs: two maximum thrust arcs (each of t_1 and $t_3 - t_2$ duration) devided by a coasting arc (see Fig. 7 for landing and Fig. 8 for launching). This result is in agreement also

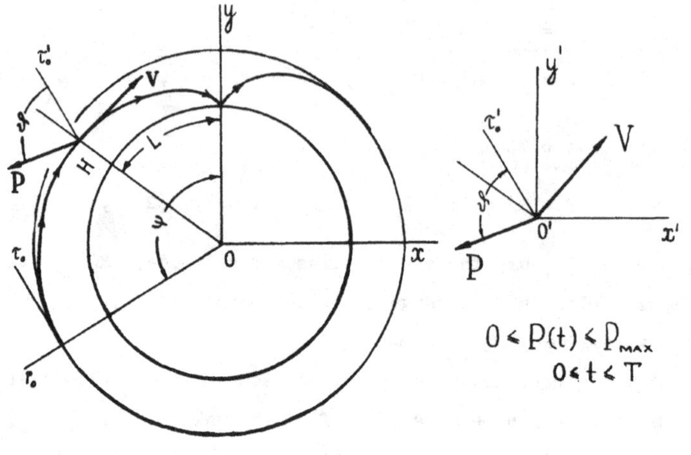

Fig. 5

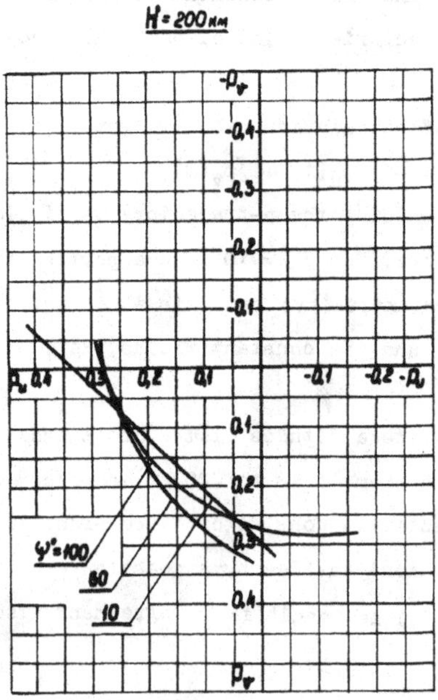

Fig. 6

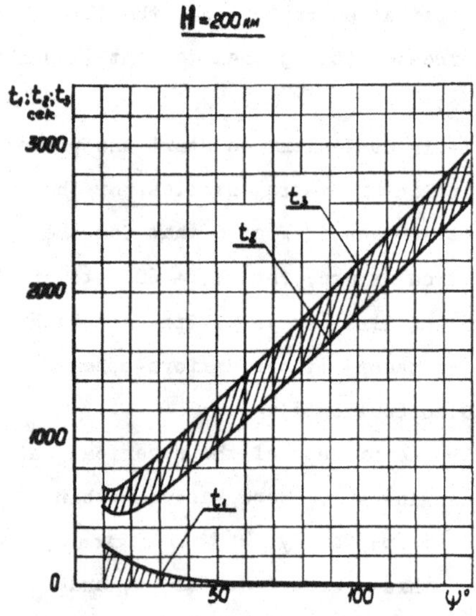

Fig. 7

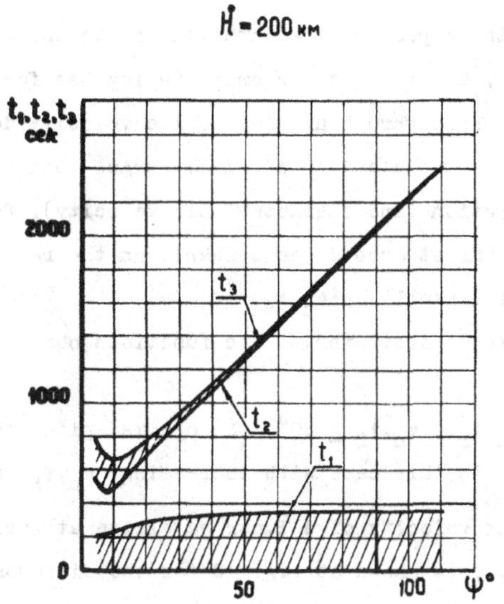
H = 200 км
t1, t2, t3
сек
2000
1000
0
50
100
ψ°
t3
t2
t1

Fig. 8

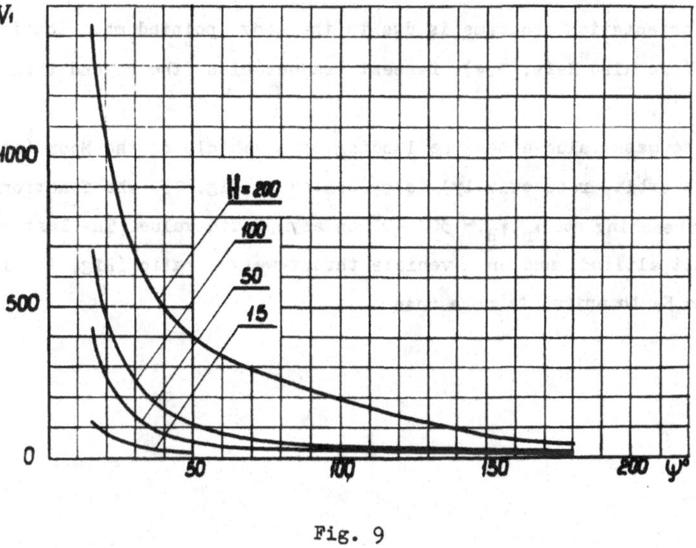
ΔV1
1000
500
0
50
100
150
200
ψ°
H=200
100
50
15

Fig. 9

with Ref. 4. Within the interval of 15°-20° the first thrust arc of the landing trajecotry is small with respect to the duration and to the characteristic velocity of the whole maneuver (Fig. 9). In this arc only the impulse for departure from an initial orbit is applied. The second thrust arc (where the velocity decays by about 98%) is the principal one. Within the diapason of small ranges both thrust arcs are compatible with respect to the duration (and characteristic velocity). For $\psi_n, \psi_B \geq 15^{o}$-$20^{o}$ the total duration of both thrust arcs depends feebly on the range angle, and the maneuver duration depends almost linearly on ψ_n, ψ_B.

At the putting into an orbit thrust arc functions change, the second arc being correcting.

With increasing ψ_n, ψ_B; $\psi_n, \psi_B \geq 20^{o}$, the optimal pitching program tends to the linear function of t. In this case with increasing ψ_n, ψ_B the average magnitude of $\frac{d\theta}{dt}$ approaches an angular velocity of a lunar satellite at zero altitude. The average pitching angular velocity is shown at Fig. 10 for the diapason where such an averaging is worth while. Due to a small value of t_{1n} a thrust direction is approximately constant at the orbit ejection arc. Corresponding relationship is presented in Fig. 11 for the reference acis connected with a horizon at the point of orbit ejection. With decreasing ψ_n this impulse tends to the transversal. If the orbit altitude grows a divergence from the transversal increases.

When the range is small the optimal program $\theta(t)$ is a non-linear one. Such an evolution of orientation programs is due to the above pointed modification of p-trajectory type (see also Refs. 3,4). Descent trajectories (the second thrust arc) are shown in Fig. 12.

The finite mass value after the landing of a vehicle on the Moon (and on the Moon satellite orbit, respectively) is presented in Fig. 13. The functional m^1 quickly grows reaching at $\psi_n, \psi_B \sim 30^{o}$-$40^{o}$ an asymptotic value. The last one depends feebly on orbit altitude and on a vehicle thrust-weight ratio (Fig. 14) if this ratio $A_3 = \frac{P}{g}$ (g in Earth units) is more than 0.5.

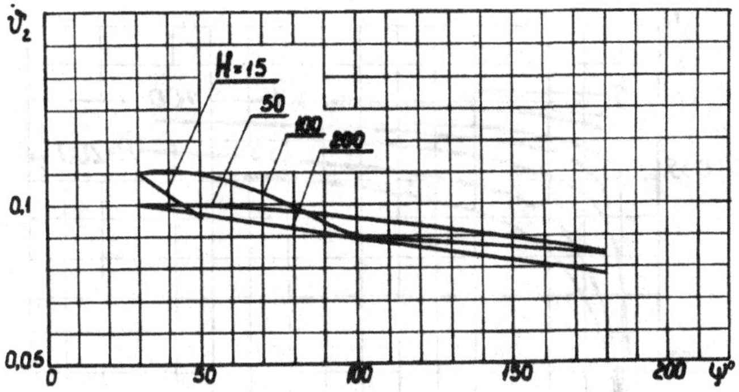

Fig. 10

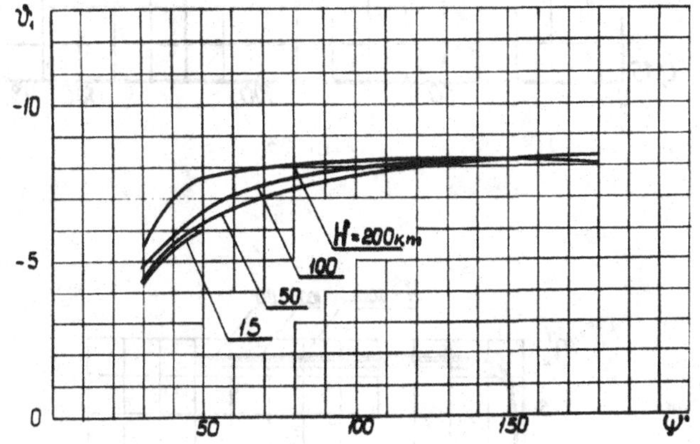

Fig. 11

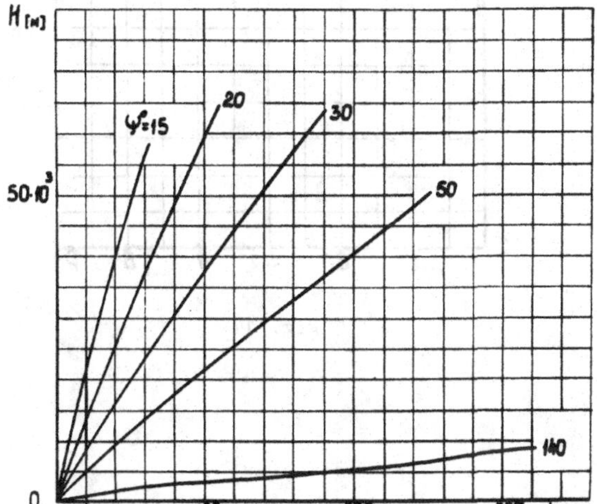

Fig. 12

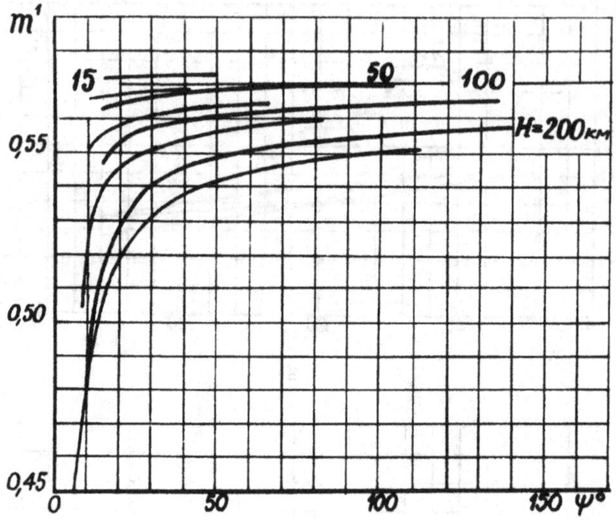
m'
15
50
100
H=200км
0,55
0,50
0,45
0
50
100
150 ψ°

Fig. 13

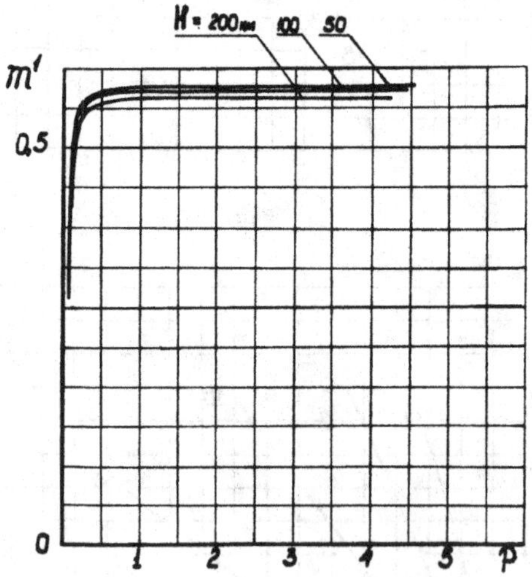
H = 200км
100
50
m'
0,5
0
1
2
3
4
5
p

Fig. 14

Appendix

I. Elements of $\|A\|$-matrix and $\tilde{B}$

$$a_{11} = 5T \cos \omega T - \frac{3}{\omega} \sin \omega T ,$$

$$a_{12} = \frac{6(1-\cos \omega T)}{\omega} - 5T \sin \omega T ,$$

$$a_{13} = \frac{4 \sin \omega T}{\omega} - 3T ,$$

$$a_{14} = \frac{16(1-\cos \omega T)}{\omega^2} - \frac{9T^2}{2} ,$$

$$a_{21} = \frac{5}{2} T \sin \omega T ,$$

$$a_{22} = \frac{5}{2} T \cos \omega T - \frac{3 \sin \omega T}{2\omega} ,$$

$$a_{23} = \frac{2(1-\cos \omega T)}{\omega} ,$$

$$a_{24} = \frac{2}{\omega}(3T - \frac{4}{\omega}\sin \omega T) ,$$

$$a_{31} = \frac{1}{\omega}\left[- \frac{8}{\omega}(\cos \omega T-1) + 5T \sin \omega T\right] ,$$

$$a_{32} = \frac{1}{\omega}(5T \cos \omega T - \frac{11}{\omega} \sin \omega T+6T) ,$$

$$a_{33} = \frac{4}{\omega^2}(1-\cos \omega T) - \frac{3T^2}{2} ,$$

$$a_{34} = - \frac{16}{\omega^2}(\frac{1}{\omega} \sin \omega T-T) - \frac{3T^3}{2} ,$$

$$a_{41} = \frac{5}{2\omega}(\frac{1}{\omega} \sin \omega T - T \cos \omega T) ,$$

$$a_{42} = \frac{1}{\omega}\left[\frac{4}{\omega}(\cos \omega T-1) + \frac{5T}{2} \sin \omega T\right] ,$$

$$a_{43} = \frac{2}{\omega}(T - \frac{1}{\omega} \sin \omega T) ,$$

$$a_{44} = \frac{1}{\omega}\left[3T^2 + \frac{8}{\omega^2}(\cos \omega T-1)\right] .$$

$$\tilde{B}_1 = u^1 - 2(2u^o+3\omega y^o)\cos\omega T + 2v^o \sin\omega T + 3u^o + 6\omega y^o ,$$

$$\tilde{B}_2 = v^1 - v^o \cos\omega T - 2u^o \sin\omega T - 3\omega y^o \sin\omega T ,$$

$$\tilde{B}_3 = x^1 - (\frac{4}{\omega}\sin\omega T - 3T)u^o - \frac{2}{\omega}(\cos\omega T-1)v^o + 6(\omega T - \sin\omega T)y^o - x^o ,$$

$$\tilde{B}_4 = y^1 + \frac{2}{\omega}(\cos\omega T-1)u^o - \frac{v^o}{\omega}\sin\omega T + (3\cos\omega T-4)y^o .$$

II. Solutions of Eqs. (2.3), (2.5)

$$u(t) = (2B_2 - \frac{3C_2}{\omega} + 5C_1 t)\cos\omega t - (2B_1 + \frac{3C_1}{\omega} + 5C_2 t)\sin\omega t -$$

$$- \frac{9}{2}C_4 t^2 - 3C_3 t - \frac{3}{2}\omega B_3 + \frac{4}{\omega^2}C_4 ,$$

$$v(t) = (B_1 + \frac{5}{2}C_2 t)\cos\omega t + (B_2 + \frac{5}{2}C_1 t)\sin\omega t + \frac{6}{\omega}C_4 t + \frac{2}{\omega}C_3 ,$$

$$x(t) = (\frac{2B_1}{\omega} + \frac{8C_1}{\omega^2} + \frac{5C_2}{\omega}t)\cos\omega t + (\frac{2B_2}{\omega} - \frac{8C_2}{\omega^2} + \frac{5C_1}{\omega}t)\sin\omega t -$$

$$- \frac{3}{2}C_4 t^3 - \frac{3}{2}C_3 t^2 + (\frac{4C_4}{\omega^2} - \frac{3\omega B_3}{2})t + B_4 ,$$

$$y(t) = -(\frac{B_2}{\omega} - \frac{5C_2}{2\omega^2} + \frac{5C_1}{2\omega}t)\cos\omega t + \frac{3C_4}{\omega}t^2 + (\frac{B_1}{\omega} + \frac{5C_1}{2\omega^2} + \frac{5C_2}{2\omega}t)\sin\omega t +$$

$$+ \frac{2C_3}{\omega}t + B_3 .$$

$P_u(t) = 2C_1^* \cos\omega t - 2C_2^* \sin\omega t + 3C_4^* t + C_3^*$; $P_v(t) = C_1^* \sin\omega t + C_2^* \cos\omega t - \frac{2}{\omega}C_4^*$; $C_i^* = \frac{Ci}{v}$ $(i=1,\dots,4)$.

$$B_1 = v^o - \frac{2C_3}{\omega} ,$$

$$B_2 = 2u^o + 3\omega y^o - \frac{3C_2}{2\omega} - \frac{8C_4}{\omega^2} ,$$

$$B_3 = \frac{2u^o}{\omega} + 4y^o - \frac{4C_2}{\omega^2} - \frac{8C_4}{\omega^3} ,$$

$$B_4 = x^o - \frac{2v^o}{\omega} - \frac{8C_1}{\omega^2} + \frac{4C_3}{\omega^2} .$$

ЛИТЕРАТУРА

I. Л.С.Понтрягин, В.Г.Болтянский, Р.В.Гамкрелидзе, Е.Ф.Мищенко.

Математическая теория оптимальных процессов. Физматгиз,Москва,I96I г.

2. Л.И.Розоноэр.

Принцип максимума Л.С.Понтрягина в теории оптимальных систем I,П.Автоматика и телемеханика,т.ХХ,№ I0,II,I959 г.

3. В.К.Исаев.

Принцип максимума Л.С.Понтрягина и оптимальное программирование тяги ракет.
Автоматика и телемеханика,т.ХХП,№ 8,I96I г.,т.ХХШ,№ I, I962 г.

4. Г.Е.Кузмак,В.К.Исаев, Б.Х.Давидсон.

Оптимальные режимы движения точки переменной массы в однородном центральном поле. ДАН СССР,т.I49, № I,I963 г.

5. В.В.Белецкий,В.А.Егоров.

Межпланетные полеты с двигателями постоянной мощности "Космические исследования",т.П,вып.3,I964,стр.36I-39I.

6. А.И.Лурье.

Аналитическая механика § II№I3.
Физматгиз,Москва,I96I,стр.6I6-622.

7. В.К.Исаев,Ю.М.Копнин.

Обзор некоторых качественных результатов,полученных в динамике полета с помощью теории оптимальных процессов. Доклад на ХУП конгрессе МАФ,Мадрид,I966 г.

8. В.К.Исаев,В,В. Сонин.

Об одной нелинейной задаче оптимального управления. Автоматика и телемеханика,т.ХХШ,№ 9,стр.III7-II29.

9. В.К.Исаев, В.В.Сонин.

Об одной модификации метода Ньютона численного решения краевых задач.Журн. вычисл.матем. и матем.физики.т.3, № 6,I963 , III4.

The problems of optimalization in dynamic multi-branch industrial models

Yu.P. Ivanilov • A.A. Petrov

We shall consider an open dynamic multi-branch model comprising branches or industries. Let, for brevity, each industry produce only one item, the output being limited by the productive capacity of the industry. In reality there are many ways of increasing or in the general case of altering the productive capacity such as: construction, conversion, conservation and deconservation. In our model (π-model) all these ways are described, but in this paper we restrict the discussion to construction and conversion only.

Let the planning period T be divided into $T-1$ equal subperiods, called stages. In the present paper for the sake of simplicity the lead time or the production cycle is taken equal to one stage in all the industries.

1. THE MODEL EQUATIONS

Let the gross output of the i-th item be denoted $x_i(t)$, its consumption $w_i(t)$, and the investment part of the i-th item output $z_i(t)$. Then the input-output equation for the i-th item at a stage t has the form:

$$z_i(t) = x_i(t) - \sum_{j=1}^{N} a_{ij}(t)x_j(t) - w_i(t) \tag{1.1}$$

where $a_{ij}(t)$ are the input coefficients.

To make the argument shorter and without loss of generality we shall not consider here the cases when an industry produces several items or the production cycles in the industries are unequal or are not equal to one stage.

THE INVENTORY INCREMENT EQUATIONS

We denote the inventory entry of the i-th item at the time t $q_i(t)$, the input of the i-th item for increasing and conversion of production at the time t $k_i(t)$. Then the inventory increment of the i-th item is described by the equation:

$$q_i(t+1) = q_i(t) + z_i(t) - k_i(t) \quad (1.2)$$

THE PRODUCTIVE CAPACITY INCREMENT EQUATIONS

We define the basic productive capacity of the i-th industry at the stage t as the maximum output of this industry completely operative plants by the same t. This capacity we denote with $\xi_i(t)$ and its increment at the stage t with $\zeta_i(t)$. Then the productive capacity increment equation is of the form:

$$\xi_i(t) = \xi_i(t-1) + \zeta_i(t) \quad (1.3)$$

2. THE CONSTRAINTS ON THE MODEL VARIABLES

The productive capacity constraints. By our definition the efficient productive capacity of the i-th industry at the stage t is the maximum gross output of the i-th industry plants which are in the process of construction and conversion. Denote this capacity with $\overline{\zeta_i(t)}$. It is evident that the gross output of the i-th industry at the time t cannot exceed the total capacity of the industry, this capacity consisting of basic and efficient ones at the stage t ; thus

$$x_i(t) \leq \xi_i(t) + \overline{\zeta_i(t)} \quad (2.1)$$

The phase constraints. It is evident that the following constraints are to be met:

$$x_i(t) \geq 0 \quad (2.2)$$

$$q_i(t) \geq 0 \quad (2.3)$$

The inequalities

$$q_i(t+1) - q_i(t) \geq 0 \quad (2.4)$$

$$z_i(t) \geq 0 \quad (2.5)$$

are to be added to (2.3) if the stocks on the inventory lists cannot be utilized as the input or for consumption at the stages to follow. Here (2.4) expresses a simple fact that these stocks cannot decrease. If the rate of consumption is not given the inequality

$$w_i(t) \geq 0 \quad (2.6)$$

is to be satisfied.

The resources constraints. In our model labour is considered to be a primary factor. We introduce a labour coefficient*) for the i-th industry at time t $c_i(t)$ equal to labour expenditure in men-hours per unit item. Denoting the labour available at the stage t $\pi(t)$ as we have the following inequalities:

$$\gamma\pi(t) \leq \sum_{i=1}^{N} c_i(t)x_i(t) \leq \pi(t) \tag{2.7}$$

where $\gamma \leq 1$ is the unemployment coefficient.

The relations (1.1), (1.2), (1.3), (2.1) - (2.7) are the fundamental relations for all the input-output models, the latter varying in specific forms of the parameters $\zeta_i(t)$, $\overline{\zeta_i(t)}$, and $k_i(t)$.

3. EXPRESSIONS FOR PRODUCTIVE CAPACITY INCREMENT AND CONSTRUCTION AND CONVERSION EXPENDITURES (simplified π-model)

As mentioned above only construction and conversion of industries are dwelt upon in this paper for the sake of brevity.

Construction. Let us denote with $\theta_i(t)$ the productive capacity the construction of which was started at time t , and with n_i - construction lead-time in the i-th industry. So, the productive capacity $\theta_i(t-n_i)$ at time t turns to be a basic capacity and increments the basic capacities be the value

$$\zeta_i(t) \sim \theta_i(t-n_i) \tag{3.1a}$$

The lead-time period consists of the construction and putting into production steps. Let construction cover l_i stages. Production output starts after m_i stages from the beginning of construction. The increase of the productive capacity efficiency is described by function $\alpha_i(t,\tau)$ which is plotted on Fig. 1. We replace the continuous curve $\alpha_i(t,\tau)$ by a stepwise one introducing the efficiency coefficient $\alpha_i(t,\tau)$ at the τ-th step of putting the productive capacity into production. The argument t implies that the efficiency factor depends on time. As the result we have the following expression for the efficient productive capacity

---

*) Note that this coefficient can be defined as labour expenditure per unit of the total productive capacity of the industry.

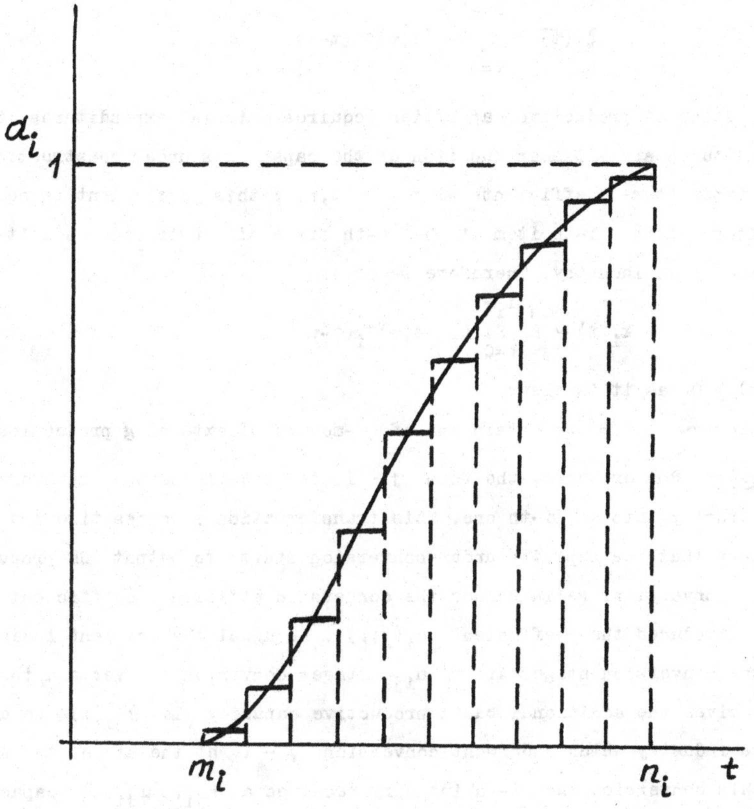

Fig. 1

$$\overline{\zeta_i(t)} \sim \sum_{\tau=m_i}^{n_i-1} \alpha_i(t,\tau)\theta_i(t-\tau) \tag{3.2a}$$

Construction of productive capacities requires material expenditures. Suppose these expenditures are a linear function of the capacities under construction. We denote these expenditure coefficients with $b_{ij}(t,\tau)$; this coefficient is equal to the expenditure of the i-th item at the τ-th stage of constructing a unit productive capacity in a j-th industry. Therefore we obtain

$$k_i(t) \sim \sum_{j=1}^{N} \sum_{\tau=0}^{l_i-1} b_{ij}(t,\tau)\theta_i(t-\tau) \tag{3.3a}$$

where $\theta_i(t) \geq 0$ as it is clear.

Replacing $\sim$ by $=$ we obtain one of π-models of extending production.

<u>Conversion</u>. Conversion of the form $j \to i$ is by definition a transformation of the j-th capacity into an i-th one. This transformation requires time and material. Let us suppose that the capacity under conversion starts to output the product with the beginning of conversion. We introduce the conversion efficiency coefficient $\alpha_{ij}(t,\tau)$ as we have introduced the coefficient $\alpha_i(t,\tau)$. As usual the argument indicates the number of the conversion stage. After n_{ij} stages converion is over and the j-th industry receives the additional basic productive capacity. Let $\mu_{ji}(t)$ be a capacity of the j-th industry which underwent conversion $j \to i$ at the stage t . As the result of this conversion the i-th industry receives a $\varkappa_{ji}(t)\mu_{ji}(t)$ capacity. Here $\varkappa_{ji}(t)$ is a capacity which appeares in the i-th industry as the result of coefficient conversion of the unit capacity of the j-th industry ($j \to i$).

Now we can write an expression for the basic capacity increment resulting from the conversion in the form:

$$\zeta_i(t) \sim \sum_{j=1}^{N} \varkappa_{ji}(t-n_{ji})\mu_{ji}(t-n_{ji}) - \sum_{j=1}^{N} \mu_{ij}(t) \tag{3.1b}$$

and an expression for the effective capacity of the production capacities under conversion:

$$\overline{\zeta_i(t)} \sim \sum_{j=1}^{N} \sum_{\tau=0}^{n_{ji}-1} \alpha_{ji}(t,\tau)\varkappa_{ji}(t-\tau)\mu_{ji}(t-\tau) \tag{3.2b}$$

We assume that the total conversion input linearly depends on the capacity under conversion and the proportionality coefficient $b_{ijk}(t,\tau)$. This coefficient is equal to the i-th product input for the $j \to k$ conversion of a unit capacity. The argument τ

again indicates the stage number of the conversion. Thus we have:

$$k_i(t) \sim \sum_{j=1}^{N} \sum_{k=1}^{N} \sum_{\tau=0}^{n_{jk}-1} b_{ijk}(t,\tau)\mu_{jk}(t,\tau) \tag{3.3b}$$

The following inequalities take place:

$$\mu_{jk}(t) \geq 0 \quad , \quad \xi_j(t) \geq \sum_{k=1}^{N} \mu_{jk}(t) \tag{3.5}$$

Summing up the right parts of the Equ. (3.1a), (3.1b), (3.2a), (3.2b), (3.3a), and (3.3b) and adding the inequalities (3.4) and (3.5) we obtain the expressions for $\zeta_i(t)$, $\overline{\zeta_i(t)}$, and $k_i(t)$:

$$\zeta_i(t) = \theta_i(t-n_i) + \sum_{j=1}^{N} \varkappa_{ji}(t-n_{ji})\mu_{ji}(t-n_{ji}) - \sum_{j=1}^{N} \mu_{ij}(t) \tag{3.1}$$

$$\overline{\zeta_i(t)} = \sum_{\tau=m_i}^{n_i-1} \alpha_i(t,\tau)\theta_i(t-\tau) + \sum_{j=1}^{N} \sum_{\tau=0}^{n_{ji}-1} \alpha_{ji}(t,\tau)\varkappa_{ji}(t-\tau)\mu_{ji}(t-\tau) \tag{3.2}$$

$$k_i(t) = \sum_{j=1}^{N} \sum_{\tau=0}^{l_j-1} b_{ij}(t,\tau)\theta_j(t-\theta) + \sum_{j=1}^{N} \sum_{k=1}^{N} \sum_{\tau=0}^{n_{jk}-1} b_{ijk}(t,\tau)\mu_{jk}(t-\tau) \tag{3.3}$$

4. THE FINAL CONSUMPTION MODEL

The consumption is supposed to be given. It can also be considered to be an independent variable (controlled variable) of the problem. But it does not extend the model to the general one. In fact, consumption can be expressed as the sum of the given minimal norm w_i^o , and the increment of the stock at the stage t :

$$w_i(t) = w_i^o + q_i(t+1) - q_i(t)$$

5. ON SOME FORMULATIONS OF THE PROBLEM

The number of the variables exceeds the number of the model equations. All the free variables can be considered to be controlled ones. It is natural to choose as the controlled variables the following ones: the gross output $x_i(t)$; the initial capacities $\theta_i(t)$, and the capacities under conversion $\mu_{ji}(t)$.

The controlled variables are determined after the optimisation problem is stated.

As the result of the existence of the time delay arguments the problem has no unique solution. To receive the unique solution the controlled variables are to be determined by the end of the plan period. We have not yet found a method of doing it. To find it is an important and a complex problem involving the properties of the "main paths" of the model.

Various forms of functionals are widely discussed in the literature, so we shall not dwell upon this problem here.

6. APPLICATION OF THE MAXIMUM PRINCIPLE TO THE CALCULATION OF DYNAMIC MODELS

The discrete maximum principle was used to solve the optimization problem within the frame of our model. This method allowes us to reduce the dimentionality of the problem, and hence the computer memory.

In the general case we have the following optimization problem: find the maximum of the functional $\Phi(y(t),u(t))$ where the phase $y(t)$ variable vectors $u(t)$ and the controlled variable vectors must satisfy the following equation:

$$y(t+1) = f(t,y(t),u(t))$$

and the condition:

$$g(t,y(t),u(t)) \geq 0$$

We shall not discuss here different forms of the discrete maximum principle, rather, we shall describe the procedure which was used to solve the optimization problems in our Computer Centre.

We introduce a penalty functional:

$$\Psi(y,u) = -e(\Phi_a-\Phi)^2 H(\Phi_a-\Phi) - \sum_{t=0}^{T-1} \sum_{m=1}^{M} e_m^t \left[g_m^t(y,u)\right]^2 H(-g_m^t)$$

where Φ_a is a properly chosen value; $e > 0$ and $e_m^t > 0$ are weight coefficients, and

$$H(\xi) = \begin{cases} 1 & \text{if } \xi \geq 0 \\ 0 & \text{if } \xi < 0 \end{cases}$$

The maximum of a nonpositive functional Ψ is equal to H zero and is reached for such values of y and u which satisfy the inequalities:

$$\Phi(y,u) \geq \Phi_a \qquad g(y,u) \geq 0$$

Let be a set of Φ_a for which the maximum of functional Ψ is equal to zero. One can show that the supremum of this set under some restrictions is equal to the maximum of the functional $\Phi(y,u)$ under the restriction: $g(y,u) \geq 0$.

Now we shall describe the algorithm of finding the maximum of Φ The Pontrjagin function:

$$\Pi(y,u,p) = \Psi(y,u) - \sum_{t=0}^{T-1} p(t+1)f(t,y(t),u(t))$$

and the conjugated system:

$$p(t) = \frac{\partial f}{\partial y(t)} p(t+1) - \frac{\partial \Phi}{\partial y(t)}$$

where p is the impulse vector, is considered.

Let us choose an initial value of Φ_a^o , and an initial approximation for controlled variables u_a^o . Having solved the system of equations in $y(t)$ and $u(t)$, we calculate the functional Ψ^o . The gradient of function Π with y and p constant is known to give the direction of the maximum increase of the functional value. Therefore the following approximation for the controlled variable vector is choosen according to the rule:

$$u^1(t) = u^o(t) + \rho\frac{\partial \Pi}{\partial u(t)}$$

where ρ is the step value. Then we again solve the system in y and p and calculate a new approximation for the functional Ψ^1 . If ρ is small enough then $\Psi^1 > \Psi^o$. We repeat the procedure until $\Psi = 0$. Then we increase the value of Φ_a and repeat the described procedure. At some iteration the increase in Φ_a will result in producing a negative maximum of Ψ . This value of Φ_a and the corresponding value $u(t)$ of the controlled variable vectors give the approximate solution of the initial optimization problem.

The described method was realized as an ALGOL procedure ("максизвие"), which was put into practice on a БЭСМ- 6 computer.

Offsetdruck: Julius Beltz, Weinheim/Bergstr.

Lecture Notes in Mathematics

Bisher erschienen/Already published

Vol. 1: J. Wermer, Seminar über Funktionen-Algebren. IV, 30 Seiten. 1964. DM 3,80 / $ 1.10

Vol. 2: A. Borel, Cohomologie des espaces localement compacts d'après. J. Leray. IV, 93 pages. 1964. DM 9,– / $ 2.60

Vol. 3: J. F. Adams, Stable Homotopy Theory. Third edition. IV, 78 pages. 1969. DM 8,– / $ 2.20

Vol. 4: M. Arkowitz and C. R. Curjel, Groups of Homotopy Classes. 2nd. revised edition. IV, 36 pages. 1967. DM 4,80 / $ 1.40

Vol. 5: J.-P. Serre, Cohomologie Galoisienne. Troisième édition. VIII, 214 pages. 1965. DM 18,– / $ 5.00

Vol. 6: H. Hermes, Eine Termlogik mit Auswahloperator. IV, 42 Seiten. 1965. DM 5,80 / $ 1.60

Vol. 7: Ph. Tondeur, Introduction to Lie Groups and Transformation Groups. Second edition. VIII, 176 pages. 1969. DM 14,– / $ 3.80

Vol. 8: G. Fichera, Linear Elliptic Differential Systems and Eigenvalue Problems. IV, 176 pages. 1965. DM 13,50 / $ 3.80

Vol. 9: P. L. Ivănescu, Pseudo-Boolean Programming and Applications. IV, 50 pages. 1965. DM 4,80 / $ 1.40

Vol. 10: H. Lüneburg, Die Suzukigruppen und ihre Geometrien. VI, 111 Seiten. 1965. DM 8,– / $ 2.20

Vol. 11: J.-P. Serre, Algèbre Locale. Multiplicités. Rédigé par P. Gabriel. Seconde édition. VIII, 192 pages. 1965. DM 12,– / $ 3.30

Vol. 12: A. Dold, Halbexakte Homotopiefunktoren. II, 157 Seiten. 1966. DM 12,– / $ 3.30

Vol. 13: E. Thomas, Seminar on Fiber Spaces. IV, 45 pages. 1966. DM 4,80 / $ 1.40

Vol. 14: H. Werner, Vorlesung über Approximationstheorie. IV, 184 Seiten und 12 Seiten Anhang. 1966. DM 14,– / $ 3.90

Vol. 15: F. Oort, Commutative Group Schemes. VI, 133 pages. 1966. DM 9,80 / $ 2.70

Vol. 16: J. Pfanzagl and W. Pierlo, Compact Systems of Sets. IV, 48 pages. 1966. DM 5,80 / $ 1.60

Vol. 17: C. Müller, Spherical Harmonics. IV, 46 pages. 1966. DM 5,– / $ 1.40

Vol. 18: H.-B. Brinkmann und D. Puppe, Kategorien und Funktoren. XII, 107 Seiten, 1966. DM 8,– / $ 2.20

Vol. 19: G. Stolzenberg, Volumes, Limits and Extensions of Analytic Varieties. IV, 45 pages. 1966. DM 5,40 / $ 1.50

Vol. 20: R. Hartshorne, Residues and Duality. VIII, 423 pages. 1966. DM 20,– / $ 5.50

Vol. 21: Seminar on Complex Multiplication. By A. Borel, S. Chowla, C. S. Herz, K. Iwasawa, J.-P. Serre. IV, 102 pages. 1966. DM 8,– / $ 2.20

Vol. 22: H. Bauer, Harmonische Räume und ihre Potentialtheorie. IV, 175 Seiten. 1966. DM 14,– / $ 3.90

Vol. 23: P. L. Ivănescu and S. Rudeanu, Pseudo-Boolean Methods for Bivalent Programming. 120 pages. 1966. DM 10,– / $ 2.80

Vol. 24: J. Lambek, Completions of Categories. IV, 69 pages. 1966. DM 6,80 / $ 1.90

Vol. 25: R. Narasimhan, Introduction to the Theory of Analytic Spaces. IV, 143 pages. 1966. DM 10,– / $ 2.80

Vol. 26: P.-A. Meyer, Processus de Markov. IV, 190 pages. 1967. DM 15,– / $ 4.20

Vol. 27: H. P. Künzi und S. T. Tan, Lineare Optimierung großer Systeme. VI, 121 Seiten. 1966. DM 12,– / $ 3.30

Vol. 28: P. E. Conner and E. E. Floyd, The Relation of Cobordism to K-Theories. VIII, 112 pages. 1966. DM 9,80 / $ 2.70

Vol. 29: K. Chandrasekharan, Einführung in die Analytische Zahlentheorie. VI, 199 Seiten. 1966. DM 16,80 / $ 4.70

Vol. 30: A. Frölicher and W. Bucher, Calculus in Vector Spaces without Norm. X, 146 pages. 1966. DM 12,– / $ 3.30

Vol. 31: Symposium on Probability Methods in Analysis. Chairman. D. A. Kappos.IV, 329 pages. 1967. DM 20,– / $ 5.50

Vol. 32: M. André, Méthode Simpliciale en Algèbre Homologique et Algèbre Commutative. IV, 122 pages. 1967. DM 12,– / $ 3.30

Vol. 33: G. I. Targonski, Seminar on Functional Operators and Equations. IV, 110 pages. 1967. DM 10,– / $ 2.80

Vol. 34: G. E. Bredon, Equivariant Cohomology Theories. VI, 64 pages. 1967. DM 6,80 / $ 1.90

Vol. 35: N. P. Bhatia and G. P. Szegö, Dynamical Systems. Stability Theory and Applications. VI, 416 pages. 1967. DM 24,– / $ 6.60

Vol. 36: A. Borel, Topics in the Homology Theory of Fibre Bundles. VI, 95 pages. 1967. DM 9,– / $ 2.50

Vol. 37: R. B. Jensen, Modelle der Mengenlehre. X, 176 Seiten. 1967. DM 14,– / $ 3.90

Vol. 38: R. Berger, R. Kiehl, E. Kunz und H.-J. Nastold, Differentialrechnung in der analytischen Geometrie IV, 134 Seiten. 1967 DM 12,– / $ 3.30

Vol. 39: Séminaire de Probabilités I. II, 189 pages. 1967. DM 14,– / $ 3.90

Vol. 40: J. Tits, Tabellen zu den einfachen Lie Gruppen und ihren Darstellungen. VI, 53 Seiten. 1967. DM 6.80 / $ 1.90

Vol. 41: A. Grothendieck, Local Cohomology. VI, 106 pages. 1967. DM 10,– / $ 2.80

Vol. 42: J. F. Berglund and K. H. Hofmann, Compact Semitopological Semigroups and Weakly Almost Periodic Functions. VI, 160 pages. 1967. DM 12,– / $ 3.30

Vol. 43: D. G. Quillen, Homotopical Algebra. VI, 157 pages. 1967. DM 14,– / $ 3.90

Vol. 44: K. Urbanik, Lectures on Prediction Theory. IV, 50 pages. 1967. DM 5,80 / $ 1.60

Vol. 45: A. Wilansky, Topics in Functional Analysis. VI, 102 pages. 1967. DM 9,60 / $ 2.70

Vol. 46: P. E. Conner, Seminar on Periodic Maps. IV, 116 pages. 1967. DM 10,60 / $ 3.00

Vol. 47: Reports of the Midwest Category Seminar I. IV, 181 pages. 1967. DM 14,80 / $ 4.10

Vol. 48: G. de Rham, S. Maumary et M. A. Kervaire, Torsion et Type Simple d'Homotopie. IV, 101 pages. 1967. DM 9,60 / $ 2.70

Vol. 49: C. Faith, Lectures on Injective Modules and Quotient Rings. XVI, 140 pages. 1967. DM 12,80 / $ 3.60

Vol. 50: L. Zalcman, Analytic Capacity and Rational Approximation. VI, 155 pages. 1968. DM 13.20 / $ 3.70

Vol. 51: Séminaire de Probabilités II. IV, 199 pages. 1968. DM 14,– / $ 3.90

Vol. 52: D. J. Simms, Lie Groups and Quantum Mechanics. IV, 90 pages. 1968. DM 8,– / $ 2.20

Vol. 53: J. Cerf, Sur les difféomorphismes de la sphère de dimension trois (Γ_4= O). XII, 133 pages. 1968. DM 12,– / $ 3.30

Vol. 54: G. Shimura, Automorphic Functions and Number Theory. VI, 69 pages. 1968. DM 8,– / $ 2.20

Vol. 55: D. Gromoll, W. Klingenberg und W. Meyer, Riemannsche Geometrie im Großen. VI, 287 Seiten. 1968. DM 20,– / $ 5.50

Vol. 56: K. Floret und J. Wloka, Einführung in die Theorie der lokalkonvexen Räume. VIII, 194 Seiten. 1968. DM 16,– / $ 4.40

Vol. 57: F. Hirzebruch und K. H. Mayer, O (n)-Mannigfaltigkeiten, exotische Sphären und Singularitäten. IV, 132 Seiten. 1968. DM 10,80 / $ 3.00

Vol. 58: Kuramochi Boundaries of Riemann Surfaces. IV, 102 pages. 1968. DM 9,60 / $ 2.70

Vol. 59: K. Jänich, Differenzierbare G-Mannigfaltigkeiten. VI, 89 Seiten. 1968. DM 8,– / $ 2.20

Vol. 60: Seminar on Differential Equations and Dynamical Systems. Edited by G. S. Jones. VI, 106 pages. 1968. DM 9,60 / $ 2.70

Vol. 61: Reports of the Midwest Category Seminar II. IV, 91 pages. 1968. DM 9,60 / $ 2.70

Vol. 62: Harish-Chandra, Automorphic Forms on Semisimple Lie Groups X, 138 pages. 1968. DM 14,– / $ 3.90

Vol. 63: F. Albrecht, Topics in Control Theory. IV, 65 pages. 1968. DM 6,80 / $ 1.90

Vol. 64: H. Berens, Interpolationsmethoden zur Behandlung von Approximationsprozessen auf Banachräumen. VI, 90 Seiten. 1968. DM 8,– / $ 2.20

Vol. 65: D. Kölzow, Differentiation von Maßen. XII, 102 Seiten. 1968. DM 8,– / $ 2.20

Vol. 66: D. Ferus, Totale Absolutkrümmung in Differentialgeometrie und -topologie. VI, 85 Seiten. 1968. DM 8,– / $ 2.20

Vol. 67: F. Kamber and P. Tondeur, Flat Manifolds. IV, 53 pages. 1968. DM 5,80 / $ 1.60

Vol. 68: N. Boboc et P. Mustață, Espaces harmoniques associés aux opérateurs différentiels linéaires du second ordre de type elliptique. VI, 95 pages. 1968. DM 8,60 / $ 2.40

Vol. 69: Seminar über Potentialtheorie. Herausgegeben von H. Bauer. VI, 180 Seiten. 1968. DM 14,80 / $ 4.10

Vol. 70: Proceedings of the Summer School in Logic. Edited by M. H. Löb. IV, 331 pages. 1968. DM 20,– / $ 5.50

Vol. 71: Séminaire Pierre Lelong (Analyse), Année 1967 – 1968. VI, 19 pages. 1968. DM 14,– / $ 3.90

Bitte wenden / Continued

Beschaffenheit der Manuskripte
Die Manuskripte werden photomechanisch vervielfältigt; sie müssen daher in sauberer Schreibmaschinenschrift geschrieben sein. Handschriftliche Formeln bitte nur mit schwarzer Tusche eintragen. Notwendige Korrekturen sind bei dem bereits geschriebenen Text entweder durch Überkleben des alten Textes vorzunehmen oder aber müssen die zu korrigierenden Stellen mit weißem Korrekturlack abgedeckt werden. Falls das Manuskript oder Teile desselben neu geschrieben werden müssen, ist der Verlag bereit, dem Autor bei Erscheinen seines Bandes einen angemessenen Betrag zu zahlen. Die Autoren erhalten 75 Freiexemplare.

Zur Erreichung eines möglichst optimalen Reproduktionsergebnisses ist es erwünscht, daß bei der vorgesehenen Verkleinerung der Manuskripte der Text auf einer Seite in der Breite möglichst 18 cm und in der Höhe 26,5 cm nicht überschreitet. Entsprechende Satzspiegelvordrucke werden vom Verlag gern auf Anforderung zur Verfügung gestellt.

Manuskripte, in englischer, deutscher oder französischer Sprache abgefaßt, nimmt Prof. Dr. A. Dold, Mathematisches Institut der Universität Heidelberg, Tiergartenstraße oder Prof. Dr. B. Eckmann, Eidgenössische Technische Hochschule, Zürich, entgegen.

Cette série a pour but de donner des informations rapides, de niveau élevé, sur des développements récents en mathématiques, aussi bien dans la recherche que dans l'enseignement supérieur. On prévoit de publier

1. des versions préliminaires de travaux originaux et de monographies
2. des cours spéciaux portant sur un domaine nouveau ou sur des aspects nouveaux de domaines classiques
3. des rapports de séminaires
4. des conférences faites à des congrès ou à des colloquiums

En outre il est prévu de publier dans cette série, si la demande le justifie, des rapports de séminaires et des cours multicopiés ailleurs mais déjà épuisés.

Dans l'intérêt d'une diffusion rapide, les contributions auront souvent un caractère provisoire; le cas échéant, les démonstrations ne seront données que dans les grandes lignes. Les travaux présentés pourront également paraître ailleurs. Une réserve suffisante d'exemplaires sera toujours disponible. En permettant aux personnes intéressées d'être informées plus rapidement, les éditeurs Springer espèrent, par cette série de »prépublications«, rendre d'appréciables services aux instituts de mathématiques. Les annonces dans les revues spécialisées, les inscriptions aux catalogues et les copyrights rendront plus facile aux bibliothèques la tâche de réunir une documentation complète.

Présentation des manuscrits
Les manuscrits, étant reproduits par procédé photomécanique, doivent être soigneusement dactylographiés. Il est recommandé d'écrire à l'encre de Chine noire les formules non dactylographiées. Les corrections nécessaires doivent être effectuées soit par collage du nouveau texte sur l'ancien soit en recouvrant les endroits à corriger par du verni correcteur blanc.

S'il s'avère nécessaire d'écrire de nouveau le manuscrit, soit complètement, soit en partie, la maison d'édition se déclare prête à verser à l'auteur, lors de la parution du volume, le montant des frais correspondants. Les auteurs recoivent 75 exemplaires gratuits.

Pour obtenir une reproduction optimale il est désirable que le texte dactylographié sur une page ne dépasse pas 26,5 cm en hauteur et 18 cm en largeur. Sur demande la maison d'édition met à la disposition des auteurs du papier spécialement préparé.

Les manuscrits en anglais, allemand ou francais peuvent être adressés au Prof. Dr. A. Dold, Mathematisches Institut der Universität Heidelberg, Tiergartenstraße ou au Prof. Dr. B. Eckmann, Eidgenössische Technische Hochschule, Zürich.